THE RECLAMATION OF FORMER COAL MINES AND STEELWORKS

Studies in Environmental Science 56

THE RECLAMATION OF FORMER COAL MINES AND STEELWORKS

I.G. Richards
J.P. Palmer
P.A. Barratt
Richards, Moorehead & Laing Ltd.,
55 Well Street, Ruthin,
Clwyd LL15 1AF, Wales, U.K.

ELSEVIER
Amsterdam – London – New York – Tokyo 1993

ELSEVIER SCIENCE PUBLISHERS B.V.
Molenwerf 1
P.O. Box 211, 1000 AE Amsterdam, The Netherlands

ISBN: 0-444-81703-4

This book is printed on acid-free paper.

Printed in the Netherlands

Studies in Environmental Science

Other volumes in this series

1 **Atmospheric Pollution 1978** edited by M.M. Benarie
2 **Air Pollution Reference Measurement Methods and Systems** edited by T. Schneider, H.W. de Koning and L.J. Brasser
3 **Biogeochemical Cycling of Mineral-Forming Elements** edited by P.A. Trudinger and D.J. Swaine
4 **Potential Industrial Carcinogens and Mutagens** by L. Fishbein
5 **Industrial Waste Management** by S.E. Jørgensen
6 **Trade and Environment: A Theoretical Enquiry by** H. Siebert, J. Eichberger, R. Gronych and R. Pethig
7 **Field Worker Exposure during Pesticide Application** edited by W.F. Tordoir and E.A.H. van Heemstra-Lequin
8 **Atmospheric Pollution 1980** edited by M.M. Benarie
9 **Energetics and Technology of Biological Elimination of Wastes** edited by G. Milazzo
10 **Bioengineering, Thermal Physiology and Comfort** edited by K. Cena and J.A. Clark
11 **Atmospheric Chemistry. Fundamental Aspects** by E. Mészáros
12 **Water Supply and Health** edited by H. van Lelyveld and B.C.J. Zoeteman
13 **Man under Vibration. Suffering and Protection** edited by G. Bianchi, K.V. Frolov and A. Oledzki
14 **Principles of Environmental Science and Technology** by S.E. Jørgensen and I. Johnsen
15 **Disposal of Radioactive Wastes** by Z. Dlouhý
16 **Mankind and Energy** edited by A. Blanc-Lapierre
17 **Quality of Groundwater** edited by W. van Duijvenbooden, P. Glasbergen and H. van Lelyveld
18 **Education and Safe Handling in Pesticide Application** edited by E.A.H. van Heemstra-Lequin and W.F. Tordoir
19 **Physicochemical Methods for Water and Wastewater Treatment** edited by L. Pawlowski
20 **Atmospheric Pollution 1982** edited by M.M. Benarie
21 **Air Pollution by Nitrogen Oxides** edited by T. Schneider and L. Grant
22 **Environmental Radioanalysis** by H.A. Das, A. Faanhof and H.A. van der Sloot
23 **Chemistry for Protection of the Environment** edited by L. Pawlowski, A.J. Verdier and W.J. Lacy
24 **Determination and Assessment of Pesticide Exposure** edited by M. Siewierski
25 **The Biosphere: Problems and Solutions** edited by T.N. Veziroğlu
26 **Chemical Events in the Atmosphere and their Impact on the Environment** edited by G.B. Marini-Bettòlo
27 **Fluoride Research 1985** edited by H. Tsunoda and Ming-Ho Yu
28 **Algal Biofouling** edited by L.V. Evans and K.D. Hoagland
29 **Chemistry for Protection of the Environment 1985** edited by L. Pawlowski, G. Alaerts and W.J. Lacy
30 **Acidification and its Policy Implications** edited by T. Schneider
31 **Teratogens: Chemicals which Cause Birth Defects** edited by V. Kolb Meyers
32 **Pesticide Chemistry** by G. Matolcsy, M. Nádasy and Y. Andriska
33 **Principles of Environmental Science and Technology (second revised edition)** by S.E. Jørgensen and I. Johnsen

34 **Chemistry for Protection of the Environment 1987** edited by L. Pawlowski, E. Mentasti, W.J. Lacy and C. Sarzanini
35 **Atmospheric Ozone Research and its Policy Implications** edited by T. Schneider, S.D. Lee, G.J.R. Wolters and L.D. Grant
36 **Valuation Methods and Policy Making in Environmental Economics** edited by H. Folmer and E. van Ierland
37 **Asbestos in Natural Environment** by H. Schreier
38 **How to Conquer Air Pollution. A Japanese Experience** edited by H. Nishimura
39 **Aquatic Bioenvironmental Studies: The Hanford Experience, 1944-1984** by C.D. Becker
40 **Radon in the Environment** by M. Wilkening
41 **Evaluation of Environmental Data for Regulatory and Impact Assessment** by S. Ramamoorthy and E. Baddaloo
42 **Environmental Biotechnology** edited by A. Blazej and V. Privarová
43 **Applied Isotope Hydrogeology** by F.J. Pearson Jr., W. Balderer, H.H. Loosli, B.E. Lehmann, A. Matter, Tj. Peters, H. Schmassmann and A. Gautschi
44 **Highway Pollution** edited by R.S. Hamilton and R.M. Harrison
45 **Freight Transport and the Environment** edited by M. Kroon, R. Smit and J. van Ham
46 **Acidification Research in The Netherlands** edited by G.J. Heij and T. Schneider
47 **Handbook of Radioactive Contamination and Decontamination** by J. Severa and J. Bár
48 **Waste Materials in Construction** edited by J.J.J.M. Goumans, H.A. van der Sloot and Th.G. Aalbers
49 **Statistical Methods in Water Resources** by D.R. Helsel and R.M. Hirsch
50 **Acidification Research: Evaluation and Policy Applications** edited by T. Schneider
51 **Biotechniques for Air Pollution Abatement and Odour Control Policies** edited by A.J. Dragt and J. van Ham
52 **Environmental Science Theory. Concepts and Methods in a One-World, Problem-Oriented Paradigm** by W.T. de Groot
53 **Chemistry and Biology of Water, Air and Soil. Environmental Aspects** edited by J. Tölgyessy
54 **The Removal of Nitrogen Compounds from Wastewater** by B. Halling-Sørensen and S.E. Jørgensen
55 **Environmental Contamination** edited by J.-P. Vernet

CONTENTS

PREFACE

The industrialisation of Europe in the eighteenth and nineteenth centuries was based on the twin economic pillars of iron and coal. The one hundred and fifty years between 1750 and 1900 saw massive exploitation of minerals, expansion of industry and the urbanisation of large populations.

Expansion and economic development has been followed by a period of general decline since the 1920s. This overall decline has affected different regions of Europe in a variety of ways. The contraction of the industries has also been accompanied by the concentration of the industries into larger production units. The impact of this contraction and concentration is characterised by scars on the landscape, unemployment for large sections of the population and a reduction in the quality of the environment. The traditional wealth-producing areas of Europe have become centres of decay and decline.

Some regions have been taking positive action for many years, renovating landscapes, replacing worn out infrastructure and creating the environment in which new industries can develop. These initiatives have produced marked improvements in the quality of life for people in those areas and many lessons have been learned by the professional people involved. Innovative techniques for dealing with technically demanding situations have been developed. Models are available which can guide administrations in areas where reclamation work is yet to begin on a significant scale.

Each generation leaves its brief pattern as a record of the interaction between environment and society. Some patterns are benign, others not so. Unless they are attended to, the relics of industrialisation will mar the environment and quality of life of many future generations.

This book is the result of studies undertaken by Richards, Moorehead and Laing Ltd on behalf of the European Commission. The principal objective was to examine the current 'state of the art' techniques for the restoration of despoiled lands arising from the coal and steel industries. In addition, the results of this work were to be disseminated via technical conferences in Nottingham, Dusseldorf, Milan, Bilbao and Brussels, and by publications, the most important of which is this book.

This book provides technical guidance to aid the process of regeneration in the coal and steel communities of Europe.

ACKNOWLEDGEMENTS

This book is the result of studies undertaken on behalf of the European Commission. The studies were funded by the European Coal and Steel Community in Directorate General XVII and managed by Directorate General XI.

This book was substantially written, compiled and edited by the research contractors, Richards, Moorehead and Laing Ltd (RML). The research team was as follows:

Editors and authors

I.G. Richards	Managing Director, RML
J.P. Palmer †	Director (Environmental Science), RML
P.A. Barratt ‡	Principal Environmental Scientist, RML

Authors

S.M. Blunt	Principal Landscape Manager, RML
A.C. Chapman	Senior Environmental Scientist, RML
E.M. Gallagher	Environmental Scientist, RML
B. Johnson	Director (Engineering), RML
N.J. Ward	Director (Landscape Architecture), RML
K.P. Williams	Senior Lecturer, Department of Minerals, University of Wales, College of Cardiff

† Project Director
‡ Project Manager

B.G. Jones was responsible for producing all figures. B.N. Houldsworth was responsible for text formatting, glossary and index production and, together with I.J. Brown contributed material to some chapters. M.J. Richards headed the word processing team composed of A.E. Hughes, S.E. Pughe and G.M. Rowlands.

Management of the project in Directorate General XI was by N. Hanley and E. Den Hamer.

The project team is grateful for valuable assistance during the project, and also for contributing material for Chapter 1, to the Régions Européennes de Technologie Industrielle (RETI). Particular assistance was given by J-M. Ernecq, I. Cattelat and U. Ferber. We are also grateful to the following organisations for material provided for the book and for assistance with case studies:

- AHU - Buro für Hydrogeologie und Umwelt Gmb
- Bonifica Spa
- British Coal
- Ensidesa
- Etablissement Public de la Métropole de Lorraine
- Etablissement Public Foncier du Nord-Pas de Calais
- Falck Spa
- Groupe d'Etude Habitat-territoire
- Herkules
- IHOBE (Industria-Hondakinentzako Bateango Enularaztegia)
- Ilva Spa
- Internationale Bauausstellung Emscher Park
- International Energy Agency Coal Research
- International Iron and Steel Institute
- Landratsamt Riesa
- Landscape Development, Lothian Regional Council
- Mid Glamorgan County Council
- Scottish Enterprise
- Société de Rénovation et d'Assainissenent des sites Industries
- Strathclyde Greenbelt Company
- TBC Proyectos e Ingenieria, s.l.
- Thyssen Umweltschutz
- TNO (Netherlands Organisations for Applied Scientific Research)
- UK Department of the Environment
- Welsh Development Agency
- Welsh Office
- World Coal Institute

We are also grateful to the following external reviewers:

D.G. Griffiths	Welsh Development Agency
B. Evans	Gwent County Council
J-M. Ernecq	Etablissement Public Foncier du Nord-Pas de Calais
A.E. Ward	Department of the Environment
A. Couper	Lothian Regional Council

Any comments regarding the content of the book should be addressed to:

Richards, Moorehead and Laing Ltd
55 Well Street
Ruthin, Clwyd, LL15 1AF
United Kingdom

Tel +44 824 704366
Fax +44 824 705450

USE OF THE BOOK

This book has been produced for the purpose of transferring knowledge, not only between regions but also between disciplines, in the hope that past mistakes can be avoided and that progress in achievement and understanding within this field is accelerated.

It is unrealistic to attempt to prepare a definitive text dealing with every aspect of every discipline now involved in land reclamation. This book has been written for practitioners and assumes familiarity with these disciplines, but aims to draw out applications of these skills in unfamiliar situations. Much of what has already been achieved by way of reclamation has not been described before, let alone brought together in one publication. The current research effort in Germany, as just one example, is immense and varied. Even a review of the work in German universities and technical institutes is beyond the scope of this book. The book has attempted to review, in particular, the last twenty years, and to indicate where recent research and developments can be applied to good effect on site.

The book is divided into seventeen chapters which guide the reader through the subject from introduction and site investigations, the description and characteristics of former collieries and steelworks, through the application of techniques of treatment, towards final site use and management.

For many, the book will be accessed by referring to the contents and index. In addition, the following chapter-by-chapter summary will aid reference of the book:

Chapter 1 - An introduction to dereliction in Europe caused by the coal and steel industries.

Chapter 2 - Site assessment: covers methods of site investigation for chemical contamination, ecological and archaeological value, geotechnical parameters and preliminary site audits.

Chapter 3 - The investigation and treatment of mineworkings and unstable ground: deals with mining subsidence and instability and the methods available to investigate and improve ground stability.

Chapter 4 - Demolition and site clearance: includes the measures to be taken to avoid wastage of materials and spread of contamination during site clearance.

Chapter 5 - Colliery spoil heap characteristics: the physical and chemical profiles of colliery spoil can affect after use. This chapter discusses spoil characteristics with regard to revegetation and site re-use.

Chapter 6 - Colliery spoil heap stability: concerns tipping methods, modes of tip failure, the investigation of spoil heap stability, and the methods available to treat unstable tips.

Chapter 7 - Colliery spoil heap combustion: examines the reasons for combustion and its effects, including risk. Methods of assessment and treatment are defined.

Chapter 8 - Colliery spoil washing: describes feasibility of coal recovery from spoil materials according to their characteristics, the methods of spoil washing, and the use of coal recovery to offset reclamation costs.

Chapter 9 - Steel industry raw materials and wastes: describes the iron and steel making processes and the physical and chemical characteristics of the materials involved. The testing and assessment of these materials is discussed.

Chapter 10 - Coal carbonisation: examines the range of potential pollutants produced during coke making and coal gasification, and the constraints that these have on the reclamation of such sites.

Chapter 11 - The treatment of contaminated soils: reviews the range of treatments that are available to remove, contain, stabilise or destroy chemically polluted soils, and the practical application of these methods to coal and steel sites.

Chapter 12 - Water quality: discusses the association of water with former industrial sites, the adverse effects of the industries on water quality, assessment of water pollution, methods of control, legislation and standards.

Chapter 13 - Landform and earthworks: landforms are extremely variable at former coal and steel sites. This chapter examines the options and methods available for remodelling landform in relation to site after-use, including consideration of material placement and drainage.

Chapter 14 - The establishment and care of vegetation: considers vegetation of reclaimed land in relation to site after-use, including soil improvement methods, selection of species, planting and sowing, and aftercare.

Chapter 15 - Management of reclaimed land: the post-reclamation management of a site is discussed with reference to management plans, responsibility and the requirements of different sites.

Chapter 16 - A framework for site regeneration: examines the holistic approach to reclamation and the need to plan for regeneration at an early stage, and to organise a programme of reclamation appropriately.

Chapter 17 - Case studies: takes the reader through a variety of coal and steel site reclamations in Europe, giving a regional context to the book.

The chapters are fully cross referenced in the text so that all information relevant to a particular subject will be brought to the attention of the reader. In addition, where more technical or specific detail is provided, this has been placed into 'boxes' rather than the main text. These boxes, and some of the more technical tables and figures, can be referred to according to the amount of detail required. Where references have been cited, these appear as superscript numbers within the text. The reference section lists references alphabetically; reference numbers follow this alphabetical sequence.

1 INTRODUCTION

Chapter contents

1 INTRODUCTION

1.1 Coal and steel in Europe

1.1.1 Coal

Coal was the fuel on which the industrial revolution of the eighteenth and nineteenth centuries was based, and it is still an important primary energy source. However, in many of its former uses it has been replaced by other fuels such as oil and natural gas. In 1989 approximately one third of world electricity generation used coal as a fuel (see Figure 1.1). Within the European Community, coal provides 20% of the primary energy source and is used to produce 40% of the electricity generated.[123]

The major coal producing countries in the European Community are Belgium, France, Spain, United Kingdom, Germany and Greece. Italy, Portugal and Eire also produce some coal, though in smaller amounts. The Netherlands and Denmark produced coal in the past, but their mines have closed and they join Luxembourg as non coal-producing members of the European Community. Figure 1.2 illustrates the main coal fields of the European Community. Box 1.1 describes the formation and geology of coal.

The majority of coal produced in the European Community countries is 'hard coal' *i.e.* bituminous or anthracite coal (see Box 1.1). Major deposits of lignite, or brown coal, are found in Eastern Germany, and lignite is also mined in Greece, Eire, Northern Ireland, Italy and, formerly, Denmark. Table 1.1 shows the hard coal production of each country for 1980 and 1990.

Early mining activity exploited near-surface outcrops of coal. From the fourteenth century onwards underground mining techniques were developed. The depth of mines increased as surface deposits became exhausted and mining techniques improved. Most West-European coal in the past 150 years has been produced from deep mines, often in difficult geological conditions. Production costs in these mines are high and, under pressure from cheaper imports and the decline in coal

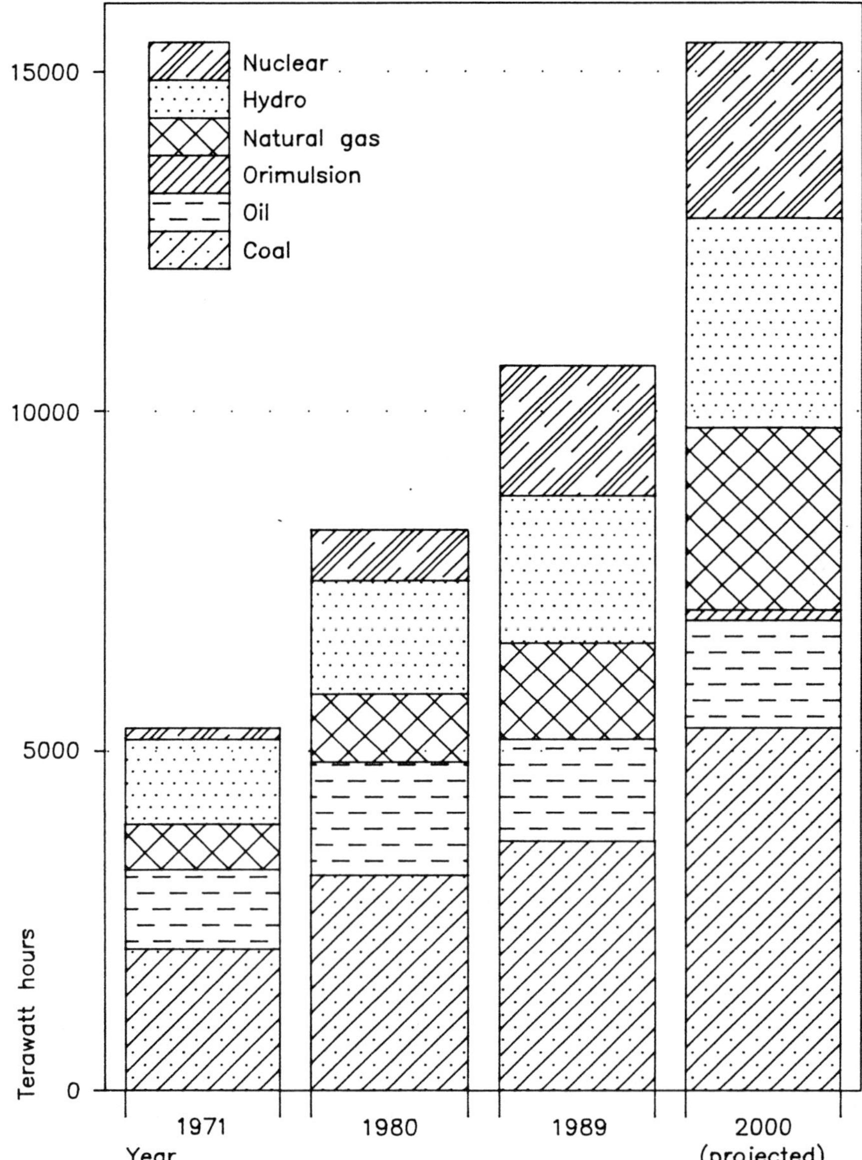

Figure 1.1: World electricity production by fuel (after IEA Coal Research, 1992[123])

utilisation, the number of deep mines in the European Community has fallen considerably. Mine closures have had a severe impact on the communities which have grown up around the mines. These communities are now characterised by high rates of unemployment and social deprivation. There is a need to bring the large tracts of land affected by coal mining back into productive use and in many areas much work has already been done, leading to rejuvenation of local economies through improvements to the environment and incentives for inward investment.

With the development of large earth-moving machinery it has become economical to remove overburden several tens of metres thick overlying coal seams and so mine coal by opencast methods. Opencast coal mining in Western Europe has, generally, involved progressive restoration, so dereliction caused by opencast mining is not a major problem. This is not so in the countries of the former Eastern Bloc. As a result, in the Eastern Länder of Germany there are large tracts of barren land produced by the opencast mining of lignite (see Section 17.7).

Box 1.1: Formation and geology of coal (Tucker, 1981[242])

Coal is formed by the action of heat and pressure on accumulations of partially-decayed vegetable matter, such as peat. Over time and under conditions of increased temperature and pressure, the carbon content of the coal increases and the content of volatile matter decreases, giving a transformation from peat through lignite (brown coal), sub-bituminous, bituminous, sub-anthracite to anthracite coal. Thus, lignite generally formed in the Mesozoic or Tertiary periods, is of a younger age than anthracite, which was formed in the Carboniferous period.

Most European coal was formed in coastal deltas, where progressive sedimentation of particles occurred as the delta was built out into the sea. Coarse sands were deposited over layers of silts, with vegetation then developing on those sands. Thus coal measures typically consist of layers of mudrocks containing marine fossils, or sometimes a marine limestone, overlain by non-marine mudrocks, siltstone, sandstone and then the seam of coal. Coal seams in these formations are generally less than 3m thick. Seat earths, the material at the base of the coal seam, which was originally the soil in which the vegetation grew, may contain nodules of iron ore (siderite). The changing position of the delta over time, and events such as tectonic movements and changes in the sea level resulted in cycles of deposition, producing repetitions of the coal measures sequence.

In some coal basins, such as those of the carboniferous rocks of France, the coal was formed in continental basins, associated with lakes. In these areas rapid subsidence produced thick beds of coal, in some cases hundreds of metres thick.

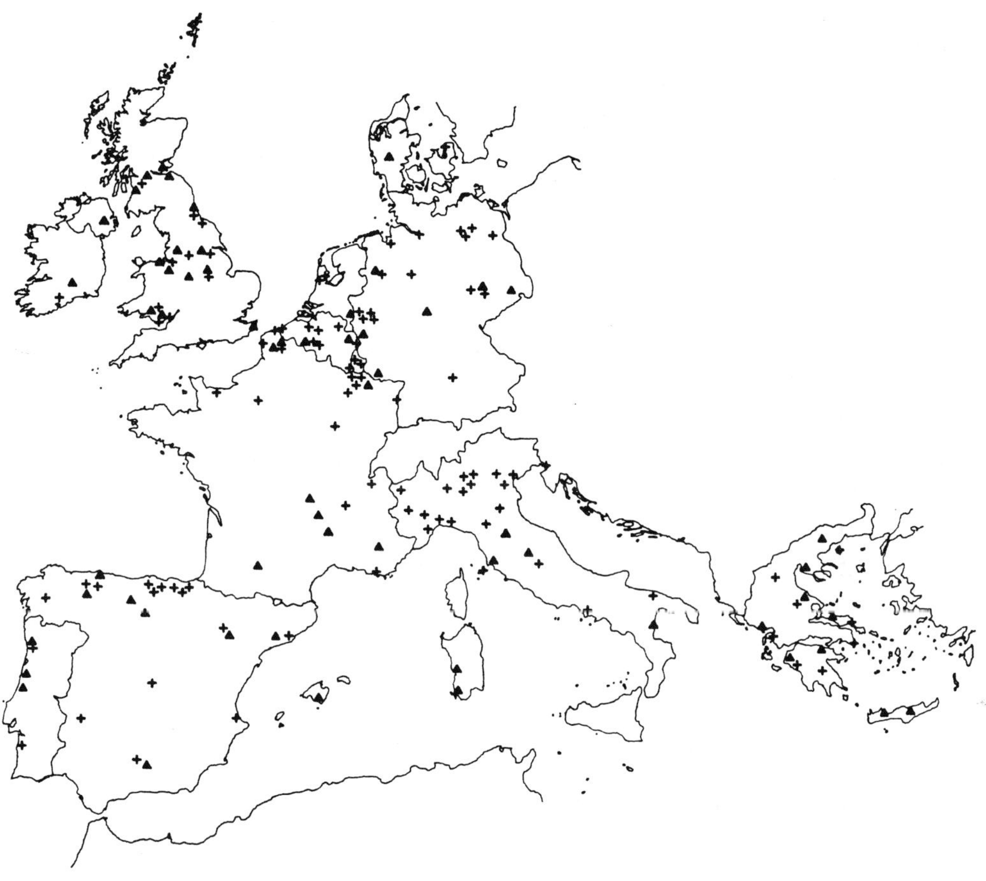

Figure 1.2: Main coal and steel producing areas in the European Community

Table 1.1: Hard coal production in the EC for the years 1980 and 1990 (after IEA Coal Research, 1992[123])

Country	Steam production 1980	Steam production 1990	Coking production 1980	Coking production 1990	Total production 1980	Total production 1990
Belgium	2.5	1.5	4.0	-	6.5	1.5
Denmark	-	-	-	-	-	-
Eire	0.1	-	-	-	0.1	-
France	13.8	8.5	4.2	1.8	18.0	10.3
Germany (Western)	35.4	27.5	55.4	44.2	90.8	71.7
Greece	-	-	-	-	-	-
Italy	-	-	-	-	-	-
Luxembourg	-	-	-	-	-	-
Netherlands	-	-	-	-	-	-
Portugal	0.1	0.2	-	-	0.1	0.2
Spain	8.1	10.5	1.4	0.3	9.5	10.8
United Kingdom	96.0	75.2	9.7	1.6	105.7	76.8
TOTAL	**156.0**	**123.4**	**74.7**	**47.9**	**230.7**	**171.3†**

Figures are in million tonnes of coal equivalent (Mtce)
† Includes a further 46.9 Mtce of brown coal
Steam production: coal for electricity generation
Coking production: coal for production of coke

1.1.2 Steel

Steel is an alloy of iron with small amounts of carbon and, in the case of alloy steels, other metals such as nickel and chromium. Iron ore has been mined in several countries of the European Community, including the United Kingdom, France, Belgium, Germany and Spain. European production of iron ore has now, however, been largely replaced by imported ore from Africa and Brazil. Iron ore is reduced to form pig iron, a form of iron containing approximately 4% carbon and other impurities, in blast furnaces. The use of coke in this process forms a connection between the coal and steel industries which resulted in many iron and steel works being located in the vicinity of coal mines. Steel is made by refining pig iron and/or scrap steel and iron. An increasing proportion of steel is being produced from scrap rather than from pig iron. Section 9.1 gives more details on the processes involved in iron and steel making. Table 1.2 outlines the crude steel production by process for each of the European Community member countries. Steel producing areas in the European Community are shown in Figure 1.2. World steel production has risen four-fold since 1950 (see Figure 1.3). Considerable economies of scale have been achieved in the iron and steel industry by concentrating production at fewer but larger works. This trend, as well as increasing competition from steel-producing nations outside of the European Community and the increasing use of other materials such as plastics in place of steel, has led to the closure of many steel sites in the past three decades, causing dereliction and economic decline of towns once dominated by their steel works. This decline is

Table 1.2: Crude steel production by process, 1991 (after IISI, 1992[124])

	Production million metric tons	Oxygen %	Electric %	Open Hearth %
Belgium	11.3	90.9	9.1	-
Denmark	0.6	-	100.0	-
Eire	0.3	-	100.0	-
France	18.4	71.1	28.9	-
Germany	42.2	77.9	20.3	1.8
Greece	1.0	-	100.0	-
Italy	25.1	41.1	58.9	-
Luxembourg	3.4	100.0	-	-
Netherlands	5.2	95.6	4.4	-
Portugal	0.5	42.3	57.7	-
Spain	12.9	43.5	56.5	-
United Kingdom	16.5	76.1	23.9	-
TOTAL	**137.4**	**67.9**	**31.6**	**0.6**

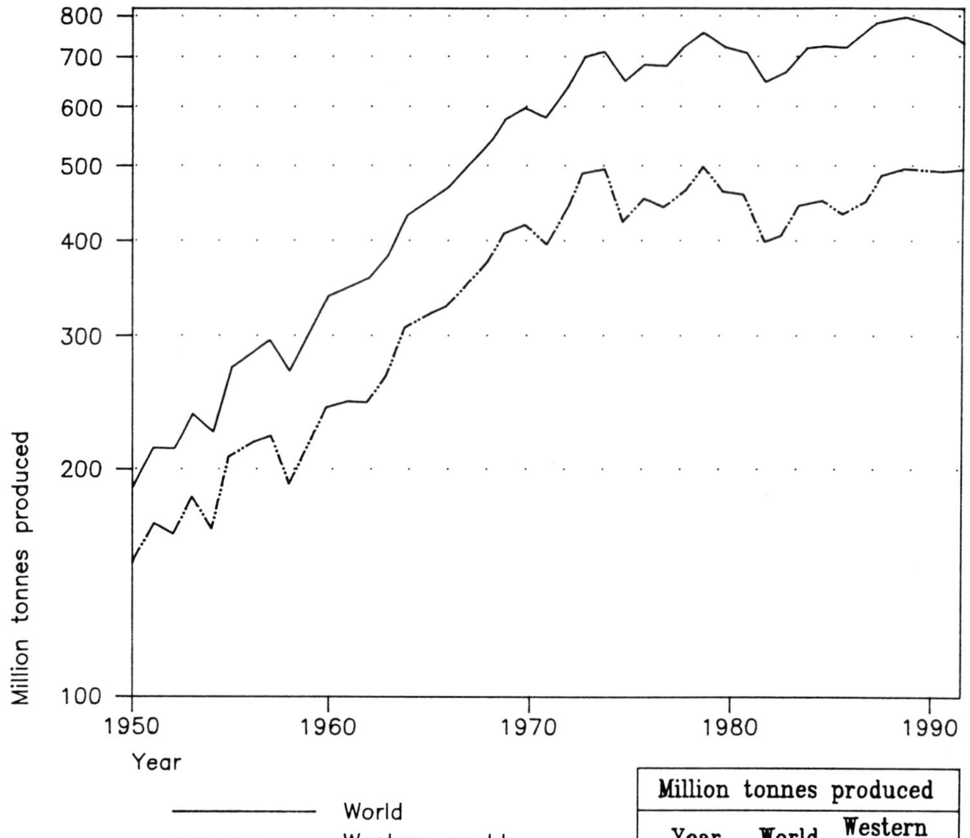

Million tonnes produced		
Year	World	Western world
1950	190	153
1960	336	241
1970	594	419
1980	716	464
1981	707	459
1982	645	398
1983	664	407
1984	710	445
1985	719	451
1986	714	433
1987	736	449
1988	780	489
1989	786	498
1990	770	493
1991	736	492

Average growth rates % per annum		
Years	World	Western world
1950–60	5.9	4.6
1960–70	5.9	5.7
1970–80	1.9	1.0
1980–90	0.7	0.6
1990–91	−4.4	−0.1

Figure 1.3: World crude steel production, 1950-1991 (after IISI, 1992[124])

illustrated by the fall of more than 50% in employment in the European steel industry between 1974 and 1991, shown in Table 1.3.

1.1.3 The European Coal and Steel Community (ECSC)

The European Coal and Steel Community (ECSC) was created in 1950 by the Treaty of Rome, which was signed in 1951 and ratified on 25 July 1952. The ECSC has its headquarters in Luxembourg. At the time of establishment the members of the ECSC were: Belgium, France, Germany, Italy, Luxembourg and the Netherlands.[149] The other six European Community member states (Denmark, Greece, Eire, Portugal, Spain and the United Kingdom) joined the ECSC later.

The main aims of the ECSC were initially to reconcile Franco-German relationships and later, more generally, to promote and increase productivity and social well-being in the steel and coal communities. All barriers against the trade of coal and steel between the member countries were abolished and the ECSC became a free market.

Table 1.3: Employment in the steel industry, 1974 and 1986-1991 (after IISI, 1992[124]). Figures are in thousands employed at the end of the stated year.

	1974	1986	1987	1988	1989	1990	1991
Belgium	64	33	30	30	30	28	28
Denmark	2	2	2	2	2	2	1
France	158	68	58	53	49	47	44
Germany	232	143	133	131	130	126	147
Italy	96	66	63	59	58	57	56
Luxembourg	23	12	11	11	10	9	9
Netherlands	25	19	19	18	18	17	17
Portugal	4	6	6	5	4	4	4
Spain	89	51	45	41	39	37	36
United Kingdom	194	57	55	55	54	51	45
TOTAL	**887**	**457**	**422**	**404**	**393**	**377**	**387**

The use of technology and economic research was seen as a method of improving the status of the members of the ECSC. Thus an obligatory levy system was introduced, whereby 1% of annual production was paid into a fund. This source of funds has made possible and financed the research for and production of this book.

1.2 The role of the European Community in environmental protection

The reclamation of derelict land arising from the coal and steel industries can be set in the wider context of the policy of the European Commission on environmental matters. This Section describes how this policy has developed and some of the principles on which it is based.

In 1974 the growing number of groups protesting about environmental degradation combined their strengths in the form of the European Environmental Bureau (EEB).

The EEB's principal objective was to influence and ensure that an environmental policy was established and set into motion at a Community level. It succeeded and on 22 November 1973 a programme of "action on the environment" was established and the Community's first common environment policy began.

The Community's environment policy is not a set of static rules, but it does have one strict fundamental aim:

> "to help bring economic expansion into the service of Man by preserving the environment in which she/he lives and managing the natural resources upon which economic expansion depends".[80]

The need to manage resources is implicit if we are to safeguard our future economy.

The main objectives laid out by the policy were to:

- prevent and eliminate pollution and nuisances;
- maintain and protect the biosphere;
- manage resources to maintain an ecological balance;
- ensure that development improves the quality of living and working conditions;
- take care that environmental aspects are included in regional planning;
- collaborate with international institutions and others outside the Community to seek common solutions to environmental problems.[80]

Several sets of principles have been defined by the Community in order to attain these objectives. The main ones include:

- the 'polluter pays' principle;
- prevention is better than cure;
- environmental action must take place at the appropriate level: local, regional, national, Community or international.[80]

The second action programme was instigated in 1977, and by 1983, the time of the third action programme, the Community programmes had become more focused, such that the principles which form the basis of the policy could be seen to emerge.[88]

In 1987 the importance of having a dynamic policy on the environment was fully acknowledged and was given a recognised position through "explicit legal and political backing" in the amendments to the Treaty of Rome, known as the Single European Act.

The fifth environment action programme commenced in 1992. This was directed by the terms of the Single European Act, which sought to establish a unified common market. The fifth environment action programme seeks to safeguard the aims of environmental policy to ensure that the economic expansion resulting from the Single European Market does not have negative implications for the quality of the European environment.

It is imperative then that economic development and environmental protection are considered together. It would be meaningless to improve the wealth and social status of the individual and of the Community as a whole if the fundamental quality of the conditions of their very environment, and thus their future survival, is undermined in the process.

1.3 Derelict land arising from the coal and steel industries

1.3.1 The scale and nature of dereliction

Derelict land has been defined as "land so damaged by industry and other development that it is incapable of beneficial use without treatment". This is the definition adopted by the UK Government. It is estimated that there are approximately 200,000ha of such damaged land in the member states of the European Community, its distribution being a function of the intensity of the past industrial activity in given areas. Approximately 40,000ha of this land have resulted directly from the activity of the coal and steel industries.[200] The continued contraction of these industrial sectors, and resultant coal mine and steelworks closures, provides an ongoing challenge to those involved in regional development in the coalfields and steel towns of Europe.

Dereliction is often associated with the visual appearance of a site: unused buildings in a state of disrepair, unkempt vegetation, the presence of deposits of unsightly wastes such as spoil from mining or slag from metal processing. The air of neglect at such sites attracts unwelcome activities such as fly-tipping and theft of stone, bricks or other materials from buildings, causing further deterioration in the appearance of a site. Other properties of derelict sites are less obvious but may be of a more serious nature and present greater technical difficulties. Examples include physical hazards from open mine shafts and unstable ground, and chemical hazards from the presence of toxic waste materials. Derelict land may affect the area surrounding it by pollution of water and air, for example by the wind-blown dispersal of fine-grained particles of dusts from unvegetated spoil heaps or seepage of toxic substances into watercourses from contaminated ground.

1.3.2 Dereliction in individual countries

Introduction

Patterns of industrialisation and the degree of industrial decline, and thus the amount of derelict land, vary from one European country to another. The United Kingdom and Germany became industrialised at an early stage and have seen much restructuring of their industrial base in recent decades. These countries have had to give a high priority to reclaiming derelict land to prevent the economic collapse of regions once dependent on the coal and steel industry. Other countries, which historically have not had a large industrial base, such as Eire and Greece, or in which the decline in traditional industries such as coal and steel has not yet occurred, for example Spain, have not had to give such a high priority to tackling problems of dereliction. Policies relating to derelict land have generally not been developed in such countries and little reclamation has been carried out.

The following sections, provide a precis of the coal and steel industries, and the associated state of dereliction, in each of the member countries of the European Community.

Table 1.4 shows the estimated areas of derelict land from the coal and steel industries in the member states of the European Community. The legislation in each country of relevance to the reclamation of derelict land is summarised in Box 1.2.

Spain

One hundred and sixty-seven coal mines were in operation in Spain in 1990, employing just under 40,000 people. In 1985, when Europe was forced to reassess coal output and decrease production by 40%, Spain, although part of the European Coal and Steel Community (ECSC), was not obliged to restructure its coal industry. Coal production in Spain decreased by only 16% between 1985 and 1993. In 1991 the Spanish government implemented a plan to restructure the industry and significantly reduced its financial aid.

The Spanish steel industry, based on locally produced ores, is an important part of the economy of Asturias and the Basque country (see Section 17.9). Major closures of steel-making capacity, which have taken place elsewhere in Europe, are imminent.

Table 1.4: Derelict land arising from coal and steel production in the European Community (RETI, 1993[199])

Country/ region	**Estimated area of dereliction (ha)**		**Country/ region**	**Estimated area of dereliction (ha)**	
	Coal	**Steel/iron**		**Coal**	**Steel/iron**
Belgium			**Portugal**	□	□
- Wallonia	4715	574	**Spain**		
- Other regions	□	□	- Basque Country	□	660
Denmark	3000	□	- Aragon	600	200
France	□	□	- Cantabria/Murcia	□	100
Germany			- Andalusia	100	□
- Ruhr	3188	724	- Castile y Leon	500	□
- Aachen	680	240	- Asturias	4500	□
- Saar	110	90	- Catalonia	500	□
- Eastern Länder	□	□	**United Kingdom**		
Greece	300	□	- England	5814	605
Eire	□	□	- Scotland	1870	503
Luxembourg	□	548	- Wales	3800	236
Netherlands			- Northern Ireland	□	□
- East	750	□			
- West	200	□			

□ Not known

Portugal

The development of the steel and coal industries in Portugal has been limited. In 1993 one coal mine and two small iron mines were being worked.

Greece

Lignite is mined but this activity is gradually declining, leaving abandoned and despoiled land. The only recorded abandoned lignite site is Alivéri mine which covers 300ha. It closed in 1989 due to exhaustion of reserves.

There is no great pressure to reuse abandoned land in Greece. As a consequence no official organisation or body exists to promote the reclamation of derelict land.

Italy

The steel industry has played a major role in the Italian economy despite the lack of the natural resources necessary for steel production in Italy.

The industry has seen many changes in structure and in its production methods. The general trend was one of increased production from the 1950s through to 1970, followed by decline in the 1970s. The 'Steel Plan' was approved in 1978 which led to restructuring and a change of emphasis in the type of goods produced. The steel crisis in 1980 and another heavy recession led to further legislation and incentives to close down private companies.

By 1986 no further aid was being given to the steel industry. The resulting closure of works had, by 1992, produced 750ha of derelict land.

Box 1.2: Legislation relevant to the reclamation of derelict and contaminated land in the European Community (after RETI, 1993[199] and RETI-CEE-BRGM, 1992[200])

Spain There is no national law relevant to the coal and steel industries; rather there are several regional laws on restoration of mining land.

Portugal No specific laws.

Greece All derelict land is owned by the state. There is no great pressure to reuse derelict land and, as a consequence, no official organisation exists to promote reclamation.

Italy No laws or regulations exist specifically to deal with dereliction itself. The urban planning instruments and laws on the reorganisation of industrial production and of environmental regulations are applied.

Denmark No laws specific to dereliction exist, but the government does recognise the problem of contamination and established the Chemical Wastes Act, 1982. Under this act, regional and local authorities are obliged to carry out surveys to identify sites and localities requiring action in order to prevent chemical contamination of groundwater.

Luxembourg The Minister of Economy encourages the reuse of abandoned land. Sites are cleaned up before they are sold. This treatment is financed by the government and the European Commission with some limited input from individual communes. The potential for contamination is recognised by the government. This is shown by the fact that industries whose use of land may cause contamination are required to provide assurance against civil liability and guarantees that the land will be reclaimed after operations have ceased.

continued ....

Box 1.2 continued

France	No specific laws. Applicable legislation falls into four main categories: (1) Control of industrial facilities when operational and after cessation of operations; (2) Environmental protection; (3) Public health and civil defence; (4) Local and Departmental health regulations.
Netherlands	The Dutch government has a well-developed policy on protection of soil and their aim is that soil should be fit for all types of use. The Interim Soil Clean-up Act 1983 introduced Guideline values for soil clean-up which are widely used in several countries. The Soil Protection Act 1987 incorporates legislation to prevent soil pollution.
Belgium	Only in the Wallonian Region is there a law dealing specifically with derelict industrial land. This is the 1978 Act on the Rehabilitation of Disused Economic Sites in Wallonia. It establishes in principle the obligation for a proprietor of a disused site to clean and restore it.
Eire	No specific laws.
United Kingdom	The Department of the Environment, Scottish Office and Welsh Office periodically carry out surveys of derelict land which record the area and type of dereliction. Legislation relevant to the reclamation of derelict land includes: Derelict Land Act, 1982; Environmental Protection Act, 1990; Water Resources Act, 1991, and Town and Country Planning Act, 1990.
Germany	Land reclamation in Germany is subject to both Federal and Länder laws. The main legislation is included in: (1) the Bundes-Immissionsschutz-Gesetz; (2) the Abfallgesetz; (3) the Baurecht; and (4) the Wasserhaushaltsgesetz. There are other more general statutes on planning, mining and safety.

There are few coal deposits in Italy and what does exist is poor quality lignite. These reserves lie mainly in Sardinia, Tuscany, Umbria and Basilicata, and production from these areas is not sufficient to meet demand in Italy. Even in 1929, the year of peak production, Italy only produced approximately 1 million tonnes of coal when demand was 16 million tonnes.

The coal industry has, nevertheless, produced 3,100ha of abandoned and derelict land (see Table 1.4).

Some examples of coal and steel site decline and abandonment, and plans for future reclamation, are provided in Section 17.6.

Denmark

The exploitation of lignite coal deposits, located in the Central West of Jutland began during the 1940s. These deposits were of poor quality and by the early 1950s production had ceased at all but one mine. This one mine remained open until 1970.

Approximately 3,000ha of land was exploited for its lignite. This land lies abandoned mainly as sand dunes and ponds, planted and naturally revegetated areas. The mining activity has left a legacy of unstable ground and water pollution.

The only steelworks in Denmark is Danish Steelworks Limited. The works were founded in 1940 at Frederiskvaerk, and are still in operation.

Luxembourg

Steel was an important factor in the economy of Luxembourg. The steel industry was based on utilisation of iron ore deposits located in the southern part of Luxembourg and in Lorraine, across the border in France.

Closures of steelworks began in 1975 with the restructuring of the industry necessitated by the decline in demand for steel. Much former steelworks land now lies derelict.

France

Coal and steel production has been an important part of both the history and economy of France. Coal deposits in France lie in three main regions: lignite is produced in the south at Gardanne, Provence, whereas hard coal is found in the Nord-Pas de Calais and Lorraine regions. The French coal industry is in decline, and many mines have closed. In 1993, coal was only produced in Lorraine, and these remaining mines are scheduled for closure by the year 2005.

The French steel industry was based in Lorraine, where it utilised local iron ores. Although the iron ore mines have closed and steel production in France has declined, steel is still produced, and in 1993 France was the third largest producer in the EC.

Examples of dereliction resulting from the closure of French coal mines and steelworks, and the reclamation now being undertaken, are given in Section 17.4.

The Netherlands

The coal industry in the Netherlands dates from the Middle Ages. Coal extraction took place mainly in the south in the Limburg province. The two mining basins in the province, the eastern and western, became the most industrialised areas of the Netherlands. By 1965 the coal mining basins had become the most densely populated areas of the country, with 70% of the population directly involved in the coal industry. At the end of the 1960s competition from natural gas led to the decline and eventual closure of the coal industry.

The Dutch steel industry is based in two regions: Ijmuden and Albasserdan. Two steelworks are in operation. There is also land that was once occupied by steel companies where steel making has ceased, and this land has been reclaimed and is being used for new industries. Contamination on these sites was found to be below levels considered to present significant risk.

Belgium

Belgium was once one of the most important coal producing countries in Europe. The coal and steel industries in Wallonia date from the nineteenth century, and in Flanders from the late 1940s. The industry is less important in Flanders than in Wallonia.

The coal industry in Wallonia began to decline in 1964-65 and in 1993 4,715ha of land formerly occupied by the coal industry lay derelict. In Flanders the coal industry covered an area of 1,320ha in 1950-52. Coal production had ceased by the 1990s, though there are no estimates of the amounts of derelict land it has created.

The Belgian steel industry is still active, notably in the Wallonian region around Mons and Charleroi, but also at Gent, Genk and near Brussels. Some sites of former industrial activity in the coal and steel producing areas are discussed further in Section 17.5.

Eire

The coal and steel industries do not play a major role in the Irish economy. There is one steelworks situated at Cork, and the coal industry is based on small lignite deposits.

No problem of dereliction arising from these industries is recognised.

United Kingdom

Coal deposits are distributed throughout England, Scotland and Wales and there is an opencast lignite mine in Northern Ireland. The main English basins are situated in the East and West Midlands, Yorkshire, Lancashire and the North-East. In Scotland the coal deposits lie in the central belt, within Strathclyde, Central, Lothian and Fife regions. In Wales the deposits are divided into a northern and southern basin.

The coal industry has witnessed fluctuations in the market as has the rest of Europe. Many of the former mines throughout the United Kingdom have now closed, and coal is mined in a limited number of large deep mines, and various opencast areas. In the 1990s a programme of further mine closure and the planned privatisation of the industry is likely to further change this picture.

The steel industry has also undergone many changes over the years. In 1967 most of the steel industry, consisting of 38 steel works, was nationalised to form British Steel. British Steel has since been privatised and in 1993 only four integrated steelworks and a number of smaller units remained.

The Department of the Environment (DoE), the Scottish Office and the Welsh Office periodically carry out surveys of derelict land which record the area and type of dereliction. The government encourages reuse of derelict land and has provided financial aid through the derelict land grant (DLG) to regenerate derelict industrial sites.

Germany

The coal and steel industries in the Western Länder of Germany are situated in 3 main areas: The Ruhr, Aachen and Saar. There are also steel plants in Osnabrück, Braunschweig and Saltzgitter.

The Ruhr is one of the most industrialised regions of Europe, with coal and steel forming the base of its economy. The decline of these industries in recent decades has resulted in large areas of derelict land, much of which is undergoing reclamation. The development agency, the Landesentwicklungsgesellschaft, of North-Rhine Westphalia has been active in promoting reclamation. The steel industry in The Ruhr is structured from private companies, dominated by Krupp, Hosch, Thyssen and Klöchne, and these contribute to the fact that, in 1993, Germany was the largest steel producing nation in the EC. The coal companies regrouped in 1968 to form 'Ruhrkohle Ag'.

In the Aachen area the last coal mine will close in 1997. Only one steel site is still working.

The coal industry in Saar was restructured in the 1950s when it faced a decline in the market for coal. The steel industry saw a similar decline around 1974. Both market declines resulted in the closure of facilities and creation of derelict land.

In the Eastern Länder of Germany opencast mining for lignite has been carried out on a large scale for several decades and lignite supplies 69% of the energy in the region. The major deposits are in Saxony (see Section 17.7).

Approximately 1,200ha of land are occupied by the steel industry, and it is expected that 50% of this land will soon become derelict.

1.3.3 Future trends

In many parts of Europe, where industrial development has created wealth as well as dereliction, considerable advancement in the methods used to reclaim and reuse land have been made. The lessons learned in these countries and the methods which have been developed need to be publicised so that other countries may benefit from this experience.

In some regions industrial closure is imminent, and local solutions to the dereliction and economic decline which may be produced by such closure are often unclear. National and European funding for reclamation and redevelopment are only part of the solution. An appreciation of the constraints to and opportunities for reclamation, and a firm understanding of the mechanisms involved are essential if former industrial land is to be brought back into beneficial use.

1.4 The stimulus for reclamation

This brief introduction to the coal and steel industries has indicated the scale of the problems in Europe caused by the decline of these industrial sectors. Large communities have lost their economic base and large areas of land have been left, at best degraded and unused, and in many situations sources of danger. Whilst the steelworks, mines and other enterprises were active the environmental degradation they caused was sometimes obscured by the industrial activity, and in many situations condoned or accepted as the price of economic survival. The industries and their inherent problems developed over a long period; closure and run-down on the other hand, has been rapid. The rapidity of this change in circumstances has magnified the degradation and pollution which has resulted.

Within the European Community there are centres of experience where the ability and desire to rehabilitate dereliction came early, triggered sometimes by disaster, as at Aberfan in the UK, in 1966 (see Section 6.1), or else by severe industrial contraction, as in the Ruhr in Germany. Years of experience in landscape rehabilitation and industrial regeneration have resulted in the development of skills and techniques which are able to convert weaknesses and threats into strengths and opportunities. The dirt and grime of the original industrial landscape have given way to a cleaner and advanced technological environment. New landscapes have created new lives for communities and individuals. The benefits of long running reclamation programmes can be measured on a macro, regional scale and at the level of the individual. There is clear evidence in regions of the United Kingdom and Germany of the social and economic benefits which flow from this work.

Less fortunate regions, burdened with yet more declining enterprises based on coal and steel, and with a legacy of dereliction, should be encouraged by the progress made by pioneer regions in the field of derelict land reclamation.

The reclamation of derelict land in the coal and steel industries embraces many different aspects and has become increasingly complex as environmental awareness and regulation have developed through the last few decades. Figure 1.4 provides an indication of how, in the United Kingdom, the engineer has been joined in the design team by a great many others, and illustrates that the early emphasis on public responsibility and the safety of the community has developed into a holistic approach to the environment. Most significantly there has also been a change in emphasis with regard to responsibility. Total responsibility for derelict land lay in the public sector in the 1960s and 1970s, but with the implementation of acts and regulations dealing with pollution, the private sector, commercial enterprises and individuals are having to accept ever greater levels of responsibility for the consequences of their activities.

1.5 Reclamation strategy

Sites occupied by the coal and steel industry frequently cover large areas of land and the concentration of a large number of sites in a town or region means that, following the decline in these industries and the closure of mines and steelworks, large tracts of land become unused and derelict. The scale of the derelict land problem in industrial regions of Europe calls for a strategic approach, where the reclamation of derelict sites in a particular geographical area is carried out in an integrated manner. Post-reclamation land-uses of different sites can then

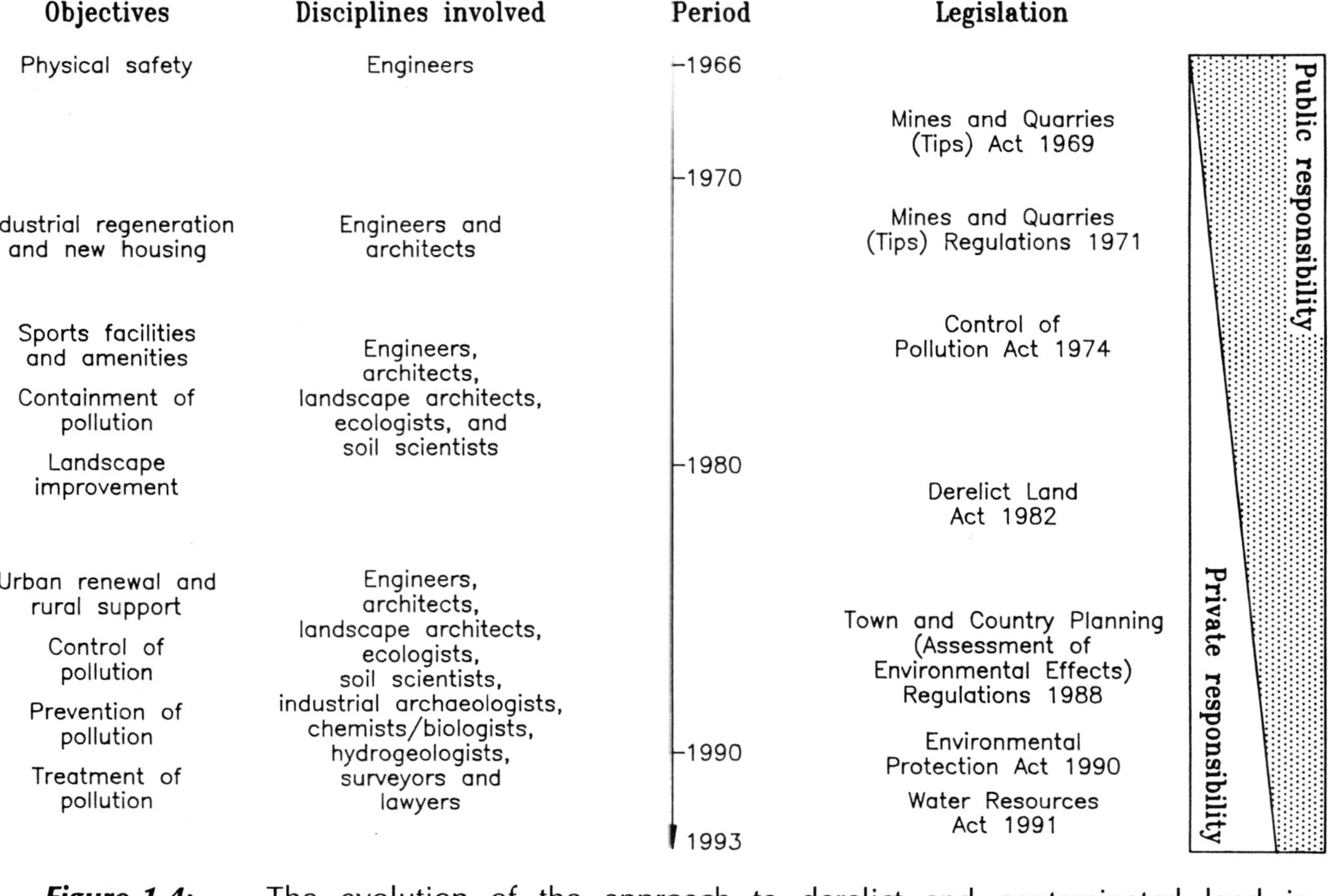

Figure 1.4: The evolution of the approach to derelict and contaminated land in England and Wales

complement each other and the infrastructure required for these developments be provided.

A reclamation strategy should make explicit the objectives of reclamation at any particular site. The overall aim is often to rejuvenate economically depressed areas, but this may be achieved through the objectives of, for example, improvement in appearance, provision of land for industry or housing, or removal of hazards and amelioration of pollution. The latter are, increasingly, dictated by statutory requirements and regulatory authorities. For example, a land owner may be required to take action to stop the pollution of ground or surface waters from a contaminated site, or to provide deterrents to the fly-tipping of refuse or other public nuisance on derelict land.

Development of a reclamation strategy for an area allows a rolling programme for reclamation and redevelopment to be implemented in a rational way, with resources appropriately directed to solve the most pressing environmental problems or development requirements first.

1.6 The experience of reclamation in Wales

Wales provides an example where reclamation of derelict industrial land has played a key role in the economic rejuvenation and environmental rebirth of a region severely affected by early industrialisation and a protracted economic decline.

Wales is an extremely diverse and complex region, and this complexity derives from its geography and geology, soils and topography, climate, history and its people. Some of the finest landscapes in Europe can be found in Wales alongside what were, in the early part of this century, some of the densest and most deprived urban areas in Europe. The industrial areas of Wales developed from about 1750 on the twin economic pillars of iron and coal, supplemented by all the other elements which are needed to create an industrial society of immense wealth and energy. Little attention was paid by anyone at any level of government or in any other facet of society to the environment in which most people lived and worked in the industrial areas.

Complacency was rudely shattered on 21 October 1966 when a slide of colliery spoil engulfed a school and adjoining houses at Pantglas, Aberfan, a village in South Wales formed around an active colliery (see Section 6.1 and Photograph 1.1). The number of deaths totalled 144, including school children. The disaster affected everyone in Wales, and many other people too. The realisation that an industry which was the economic base of the region could cause such devastation was an emotional shock. At the highest levels of government it was resolved that such an event should not occur again under any circumstances and this was reflected in opinions throughout the country.

Photograph 1.1: The Aberfan disaster: colliery spoil from tips on the hillside has swept away houses and the local school. In this photograph the operation to clear away the spoil has begun (source: Report of the tribunal appointed to inquire into the disaster at Aberfan on October 21 1966[5])

Many of the elements of the United Kingdom's central government are administered in Wales through the Welsh Office which is a separate government department answerable to a Secretary of State. Many aspects of government policy and control change in degree or emphasis when they are applied in Wales, and this includes the clearance, treatment and redevelopment of derelict land.

After the Aberfan disaster the government embarked on what has been the biggest derelict land clearance in Europe, with over £250 million having been spent by 1993 and a supplementary £1,000 million spent on development.

Many colliery spoil heaps in South Wales, set on steep hillsides above towns and villages, were potentially unstable, threatening landslides as had occurred at Aberfan. The first priority in reclamation was therefore safety, but it was immediately apparent that carrying out works to improve safety could create land which would be available for the development of new industries and housing in an area where land suitable for modern development was virtually non-existent.

The earliest reclamation schemes involved:

- controlling ground and surface water which were the main causes of instability in coal tips;
- reducing the angle of tip faces to increase stability and enable machines to move across the new landform;
- developing vegetative cover, usually grass, though with some tree planting, on the new landform;
- a very nominal and generally inadequate management and maintenance programme for vegetation.

From the outset the Welsh Office looked to district authorities, that is local government, as the promoters of schemes. The Welsh Office also established a principle which is still applied today. This principle was that the whole programme should be administered by a very small team of specialists within the Welsh Office, who looked to other organisations such as local government and private consultants to design schemes, and to private contractors to carry out the work. The administration of the land reclamation programme has now been taken over by the Welsh Development Agency (WDA), established in 1976 to develop a new industrial base for the Welsh economy. The main tasks of the WDA have been to regenerate Welsh industry, clear industrial dereliction and improve the environment. The scale of the work carried out has been such that the WDA confidently predicts that all sites of major dereliction will have been removed by the year 2000. At central government, Agency and local government levels in Wales it is recognised that it is in the best business interests of Wales to improve the quality of the environment. This must be the most important lesson to have been learned. Reclamation and regeneration has created new jobs in new technologies, tourism, health care, housing and leisure, and has increasingly involved a partnership between the public and private sectors.

Although the most severe and technically difficult problems lay in the narrow industrial valleys of the South, the initiative was applied throughout Wales. Over the years one of the key roles played by the Welsh Development Agency has been as a 'clearing house' for good ideas and problem solving. Economy in design, materials and construction are common features of reclamation schemes in Wales and much of this is due to the co-ordinating and overseeing role of the Agency's land reclamation team.

Wales is a small country, and this smallness has helped the Agency to provide local control. Agency staff are convinced that because the majority of the reclamation projects were within 1 to 2 hours drive from their office this gave them an effective control over, and familiarity with, the varied problems presented by the many projects being undertaken.

Reclamation schemes have also become more sophisticated. Design teams invariably contain engineers, landscape architects and environmental scientists, ecologists and botanists for example, but one frequently finds geologists, hydrogeologists, mining engineers, and industrial archaeologists involved. One of the lessons learned is that many interesting industrial artifacts were lost in the early years of the reclamation programme. Later schemes have taken into account the industrial history of the sites. Some notable sites have been retained and refurbished as museums and interpretative centres.

The sites reclaimed under Welsh Office and Welsh Development Agency programmes by 1991 have been used as follows:[253]

- 130 sites have been used for industrial development;
- 70 sites have been used for housing;
- 50 new or improved roads have been built;
- 50 sites have been used for playing fields;
- 12 country parks have been created.

From the earliest days of the reclamation programme the emotional stimulus of the Aberfan disaster has been a driving force. The main thrust of the reclamation programme has been to improve the environment of Wales for the sake of improvement, knowing instinctively at first but with ample proof now, after twenty-seven years of experience, that new landscapes mean new lives for the people.

2 SITE ASSESSMENT

Chapter contents

2 SITE ASSESSMENT

2.1 Introduction

Site assessment is the process by which all relevant information concerning a site is compiled and evaluated to enable the most appropriate reclamation proposals to be produced. This chapter describes the basic principles of site assessment, providing a framework for detailed assessments of specific aspects. The detailed assessment of specific issues are dealt with in later chapters (see Table 2.1).

Site assessment usually includes:

- a walkover survey;
- a desk study, with preliminary investigations;
- detailed investigations and surveys;
- analysis of the information collected.

The site assessment process should identify:

- risks to people or to the environment from the site in its present state;
- constraints on the future use of a site, such as poor ground conditions or areas of contamination;
- the opportunities presented by a site, such as ecological or wildlife value or the presence of structures of historical importance;
- structures which could be put to beneficial use;
- materials which could be put to beneficial use.

2.2 Preliminary planning

Site assessment and future site use are closely related, and a particular proposed site use will require the investigation of aspects of the site that are specific to the after-use concerned. For example, reclamation for

Table 2.1: Chapter location of site assessment methods

Chapter number	Assessment described
2	Ecology, landscape, structures, archaeology, contamination
3	Mine workings
4	Pre-closure site audit
6	Tip stability
7	Spoil combustibility
8	Feasibility of coal recovery
9	Expansion of slags
12	Water pollution
14	Suitability of materials for revegetation

public open space, or amenity woodland, will require the investigation of soils and water for their ability to support growth of suitable plants, whereas reclamation for industrial development requires the investigation of ground conditions for foundation design and to establish the presence and extent of buried foundations. Both uses would require the investigation of contamination and its potential detrimental effect on subsequent users of the site or other targets such as groundwater. Whilst it is important in the early stages of site assessment to maintain flexibility over reclamation proposals some indication of proposed site use will assist the site assessment process and enable appropriate assessments to be made without wasting time and resources.

Reclamation objectives and site after-use proposals can be expressed in the form of a 'masterplan'. A masterplan is a document which can be a combination of drawings and text and which provides a framework for detailed site assessment and subsequently for the design of reclamation works.

2.3 Desk study and preliminary investigations

A desk study involves the collection and collation of all relevant information relating to the site and its surroundings. Sources of this information include:

- maps and plans;
- aerial photographs;
- industry records;
- mining records;
- results of earlier investigations;
- utility companies.

Organisations such as local authorities should be consulted, and local knowledge should be utilised as much as possible, for example by interviews with people living near the site. Past employees of a former industrial facility are particularly useful as sources of information.

The information obtained in the desk study will include:

- planning policies related to the site and the surrounding area;
- land ownership and other rights over land, including rights of way;
- current land uses, of the site and surrounding land;
- current infrastructure (roads, railways, utility services);
- information on geology, hydrology, hydrogeology, soils and climate;
- positions of shafts and mine workings;
- all former uses of the site and surrounding land;
- layout of plant and former process activities;
- waste disposal practices and licences issued;
- industrial archaeology including any designation or listing of particular site features;
- any reports on the ecology of the site.

Visits to the site should complement the desk study and provide information on the current status of the site and surrounding land. The following aspects, which serve to highlight the complexity of derelict industrial sites, should be considered during the visits:

- current land use;
- character of the surrounding landscape;
- the visual impact of the site;
- the presence of buildings, with an assessment of their structural soundness, their potential for reuse and their historic value;
- the presence and condition of other structures *e.g.* walls, culverts, bridges;
- the nature of materials at the surface, particularly whether they impede investigative excavations at the site;
- the presence or absence of vegetation, the nature of the vegetation and an assessment of its ecological and landscape value;
- deposits of waste materials, for example colliery spoil or slags and other materials such as stone, bricks or hardcore;
- the existing landforms and the constraints or opportunities they present;
- soil or soil substitute resources available on site;
- signs of contamination, for example unusually coloured materials, odours, lack of vegetation, presence of tanks or drums which may have held, or still hold, hazardous materials;
- surface hydrology;
- the presence of utilities (such as water or gas mains, electricity cables, sewers).

In general the presence of contaminating substances derives from activities and processes which were or may still be carried out on a site. Much information on the location of contamination will be gained from a pre-closure site audit (see Box 4.5). Frequently in the past a pre-closure site audit has not been undertaken, and access to the site for the purposes of site audit was only possible after closure and often after demolition had already taken place. In many cases sites have been levelled-off, removing all significant landmarks, and leaving imprecise records of the works which may be misleading. At coal carbonisation sites, such as coking works and gas works, the presence of above-ground by-product storage and process facilities provides invaluable information for those required to design and undertake a site investigation. Closure of the works and the demolition and/or removal of process structures increases the difficulties in making an investigation appropriate, well targeted and cost effective.

The preparation of a reclamation scheme will generally involve a team of people from different disciplines. It is important that all members of the team are familiar with the site at an early stage. To bring together and brief the team at or before the preliminary investigation stage is likely to enhance the efficiency of the site assessment and reclamation design process.

A critical appraisal of the information gathered by the desk study and preliminary investigations will enable identification of likely constraints on, and opportunities for the reclamation and reuse of the site. For example it may be possible to predict likely areas of contamination, make an estimation of quantities of spoil, or assess the potential for re-use of buildings. This analysis will enable development of the reclamation proposals to proceed in a well informed way, and to include the preparation of broad budget costings as part of a masterplan.

2.4 Detailed investigations

2.4.1 Introduction

Detailed investigations should aim to identify and quantify the constraints or opportunities which, on the basis of past use, are present in any particular area of a site and which will impinge on the proposed use of the site. Typically the team of investigators will include civil, structural and geotechnical engineers, environmental scientists, landscape architects and topographical surveyors, although more specialised disciplines will be required in some cases.

The desk study and preliminary investigations will have identified aspects in need of detailed investigation. Some of these aspects will only require investigation if certain site uses are planned, whilst others will require investigation whatever the site use. The latter may include the stability of tips, surface water and groundwater pollution arising from contaminated ground and air pollution and ground instability caused by burning colliery spoil or unstable mine workings.

2.4.2 Topographical survey

An accurate topographical survey is essential baseline information. The establishment of control stations for the site before any demolition takes place is invaluable. There is then a clear and unambiguous correlation between the site as it was in its former use and its condition when clearance is completed. The features which should be included in a topographical survey are shown in Box 2.1.

Box 2.1: Features included in a topographical survey

A topographical survey should record the following:

- permanent and seasonal watercourses;
- drainage systems - open and closed;
- sinks and issues of water;
- tracks, paths and boundaries;
- shafts, adits and tunnels;
- buildings and other structures;
- major vegetation zones;
- positions of trial pits, boreholes and soil probes;
- positions of other sampling points;
- access roads;
- railways and canals;
- spoil heaps;
- slag heaps;
- lagoons;
- contours.

2.4.3 Ecological survey

The detailed investigations will include further study of the matters considered in the preliminary investigations. These include assessment of the existing vegetation present on the site (see Box 2.2) and its ecological value (see Box 2.3).

2.4.4 Visual assessment

Box 2.4 discusses the assessment of the visual characteristics of a site. Assessment of the visual characteristics of a site requires a rational, professional approach, and involves description, analysis and evaluation of the landscape.

Visual assessment is important because it equips the design team with information which is needed to take full advantage of the visual assets of the site, and to cater for its visual problems.

Box 2.2: The value of existing vegetation

Most derelict or abandoned sites have some vegetation, either that which has colonised abandoned areas or that which survived through the operational life of the site. This vegetation is worthy of examination since existing vegetation:

- can indicate the characteristics of the site and guide revegetation;
- provides a habitat for common or uncommon species of plants and animals;
- provides a reservoir from which plants and animals can recolonise the reclaimed areas of the site;
- may possess a tolerance of hostile site conditions, which can be utilised by propagation;
- lends visual maturity to the site;
- may provide a buffer between nearby residents and major reclamation works;
- may protect spoil materials from erosion or instability;
- retention reduces expenditure on the establishment of new vegetation.
- may include invasive species such as Japanese Knotweed (*Fallopia japonica*), which may be spread by reclamation earthworks, and for which control measures will need to be specified at the design stage.[201, 212]

Where examination suggests that existing vegetation can play a useful role within the reclamation objectives for the site, it should be protected from disturbance and integrated with the design of the landform and revegetation works. The existing vegetation may require additional management works such as species enrichment, control of invasive weeds or the improvement of fertility, in order that reclamation objectives are fulfilled. Conversely, the objectives of reclamation may be modified to take in the conservation of valuable existing vegetation.

Box 2.3: Ecological assessment

Ecological assessment is the process of assessing the 'ecological value' of an area of land. In the context of site assessment prior to reclamation work, the aim of ecological assessment is to assess what value the site has so that the information may aid decision making about how the site is treated. The process of ecological assessment is broadly the same throughout Europe although some of the criteria for assessment may be given different emphasis in different countries. Typically an assessment will involve:

- gathering of background ecological information on the site and on similar sites;
- field survey and annotation of maps with different habitat types and species of interest;
- evaluation of the ecological value of the site and its component parts;
- placing of the site in a local, regional and national context in terms of its value.

The most frequently used criteria for placing value on a site are as follows:

- diversity;
- rarity;
- area;
- threat of human interference;
- amenity value;
- education value;
- recorded history.

An ecological assessment will also provide information on the nature of the materials on site. A botanist will be able to make assessments of the contamination and nutrient status of the materials on the basis of the vegetation composition. Such information will aid in deciding if suitable soil-forming materials are available on site and if any materials should be set aside as a 'seed bank' for a reclamation scheme. Similarly recommendations may be made about the protection of existing species, measures to assist their spread and opportunities for habitat creation.

Box 2.4: Visual assessment

It is important to consider the visual qualities of the surroundings of a site, as well as the site itself, so as to:

- study the appearance of the site and its features from outside its boundaries;
- appreciate the character of the landscape into which the site must be integrated and identify the local 'sense of place'.

The assessor should seek to analyse what gives particular character to the setting and the site, distinguishing the positive and negative aspects. Areas of common character should be identified on a plan. This gives a framework for action to build on positive aspects and to deal with the negative ones.

Visual assessment frequently examines the following:

- the 'visual envelope' of the site, *i.e.* the area which can be seen from the site, or which provides views of the site;
- noteworthy views out of the site;
- views of the site and its features, together with assessment of its visual impacts, considering how many people can see it, from how far away;
- the nature of all views, *i.e.* panoramas;
- description of landform, land-use patterns, vegetation (see Box 2.2), the built environment and any special natural or manmade landscape features, both on and off site;
- identification of remnants of the pre-industrial landscape within the site and any other features which may be worth preserving, to add character and maturity to a scheme;
- description of the emotional responses to the site and its setting e.g. does it give a feeling of enclosure or exposure and is it thus stimulating or oppressive?

2.4.5 Structural integrity of buildings

Where buildings or other structures exist on a site an assessment of their structural integrity should be made by a structural engineer to determine whether they are in a safe condition and whether they can be reused. Box 2.5 indicates the types of defects which can be caused by differential movements beneath buildings and the possible causes of those movements.

2.4.6 Ground conditions

Investigation of the nature of materials below ground level must be undertaken at former industrial sites. These ground investigations may be used to obtain information on the following:

- the location and condition of underground structures, foundations, cavities, shafts, mine workings *etc.*;
- a geotechnical description of ground conditions to determine foundation requirements;
- the nature and location of contaminated ground;
- groundwater levels, flow and contamination;
- stability of spoil heaps and waste tips;
- potential of materials as a growth medium for vegetation.

This investigation is multifaceted and to avoid duplication of effort it is important that there is coordination between the different disciplines. A variety of methods, both invasive and non-invasive, are available to the practitioner for the investigation of sub-surface conditions. These methods are discussed in detail in the following section.

Box 2.5: Structural assessment

Buildings may show signs of distress from relative movements of the building fabric. Distress can manifest itself in the following defects, which tend to be concentrated around or radiate from an area of maximum stress:

- tapered cracks indicating sagging or hogging modes of failure;
- inclined cracks usually associated with movement at the extremity of the building;
- vertical and horizontal cracks especially at relatively weak areas around openings and at damp proof course level;
- sloping lintels;
- bulges and lack of verticality in walls;
- slopes in floors;
- doors and windows jamming in distorted frames.

The possible causes of movements include:

- differential settlement due to uneven consolidation of clay soils;
- subsidence of foundations;
- heave due to frost, a rising water table or expansive chemical reactions;
- instability of sloping ground.

Subsidence may be caused by:

- mining activities below or adjacent to the building;
- settlement of filled ground;
- water movement under the building from natural sources or from broken services;
- water abstraction from ground below the building;
- shrinkage of clays;
- collapse of solution cavities;
- excavation adjacent to the building;
- earthquakes.

2.5 Techniques of ground investigation

2.5.1 Trial pits

The excavation of trial pits is a relatively inexpensive method of ground investigation, which can provide valuable information on subsurface characteristics.

Trial pits are excavations carried out from the ground surface. The advantage of trial pits is that they allow direct visual examination of the material below ground level and the taking of large samples of solid materials and groundwater. Box 2.6 shows the types of information which can be obtained from a trial pit. Sampling and chemical analysis are discussed in Sections 2.6.2 to 2.6.5.

Site personnel should not enter trial pits without first ensuring that it is safe to do so. Precautions such as shuttering are generally necessary to prevent collapse of pits of more than 1m deep. Before entering a pit, or any other enclosed space, monitoring of the atmosphere should be carried out to ensure that toxic gases or vapours are not present. Reliable and accurate gas monitoring equipment should be used for this purpose.

Machine dug trial pits can extend to 6m below ground level although 3-4m is often sufficient for observation of ground conditions and the collection of samples. The depth of unsupported trial pits is frequently limited by the collapse of the sides of the trial pit.

Trial pits can easily be extended lengthways into trenches to determine the extent of any visible materials of interest. If concrete or other hard materials are present, either at the surface or at depth, these may require breaking with a percussion breaker.

The excavation of trial pits causes considerable disturbance of the ground. To avoid settlement following reinstatement, trial pits should be refilled and compacted in layers, with the addition of imported material where

Box 2.6: Trial pit logs

Trial pits can be used to obtain a wide range of information and should be recorded by photographs of the exposed ground profile and by a written log. The trial pit log should include observations on the following:

- the nature of the materials present according to depth e.g. colour, soil type, whether fill materials or natural ground;
- subsurface structures e.g. foundations of former buildings, pipes, ducts;
- location and rate of flow of groundwater seepages;
- standing levels of groundwater, at specified times after pit excavation;
- the location of samples taken.

necessary. Where heavily contaminated materials are excavated these should never be left exposed at the surface after infilling of pits. Care should also be taken to avoid the introduction of contaminated materials into previously uncontaminated strata when backfilling trial pits. This is particularly important when those strata contain groundwater which could become contaminated and thus cause pollution over a much wider area. If necessary contaminated materials should be removed from the site rather than used to backfill the trial pit. Generally, trial pits should not be left open if the site is unattended, but should be backfilled as soon as possible after inspection.

2.5.2 Boreholes

Introduction

Boreholes are the most common method of investigation of ground conditions beyond the reach of trial pits. They can also be used for shallow investigations when the disruption caused by trial pits would be unacceptable.

Cable percussion

Cable percussion, also known as shell and auger, is the most frequently used technique in soils and weak rock (see Figure 2.1). The drilling tool, an open-ended tube, is driven into the ground by repeated lifting and dropping. When full of soil or other subsurface material the tool is brought to the surface and emptied. Samples may be obtained in this way. In cohesive soils such as clays and silts the tube is known as a clay cutter. A core of undisturbed material may be obtained by discarding the outer portion of material brought up by the clay cutter. In sands and gravels the tube has a flap valve to retain loose material during lifting. This tube is known as a shell or bailer and its use requires the borehole to contain sufficient water to cover the lower part of the shell. In dry ground conditions water may have to be added to the borehole as a lubricant. Chisels are used to break up rocks or other hard layers. Flush screw-thread steel casings are driven into the ground in wet, very soft, or loose soils to line the borehole as it advances. Shell and auger boring is a fairly simple and economic technique which can provide accurate information on ground conditions. Borehole depths of 20 to 30m are readily achievable using this technique, providing there are no obstructions, and depths in excess of 50m are possible in favourable conditions.

Rotary techniques

The alternative to percussive methods are rotary techniques such as rotary augers and rotary drilling. Rotary augers (see Figure 2.2) are used in soils but are not suitable for thick deposits of granular materials or for soils containing cobbles or boulders. The equipment used is more expensive than for cable percussion, and augering is a less suitable technique for obtaining accurate information on ground conditions as materials from different strata become mixed as they are carried to the surface by the auger. However, rotary augering is far quicker than percussive methods so it is frequently used when ground conditions are already known and rapid installation of groundwater or gas monitoring

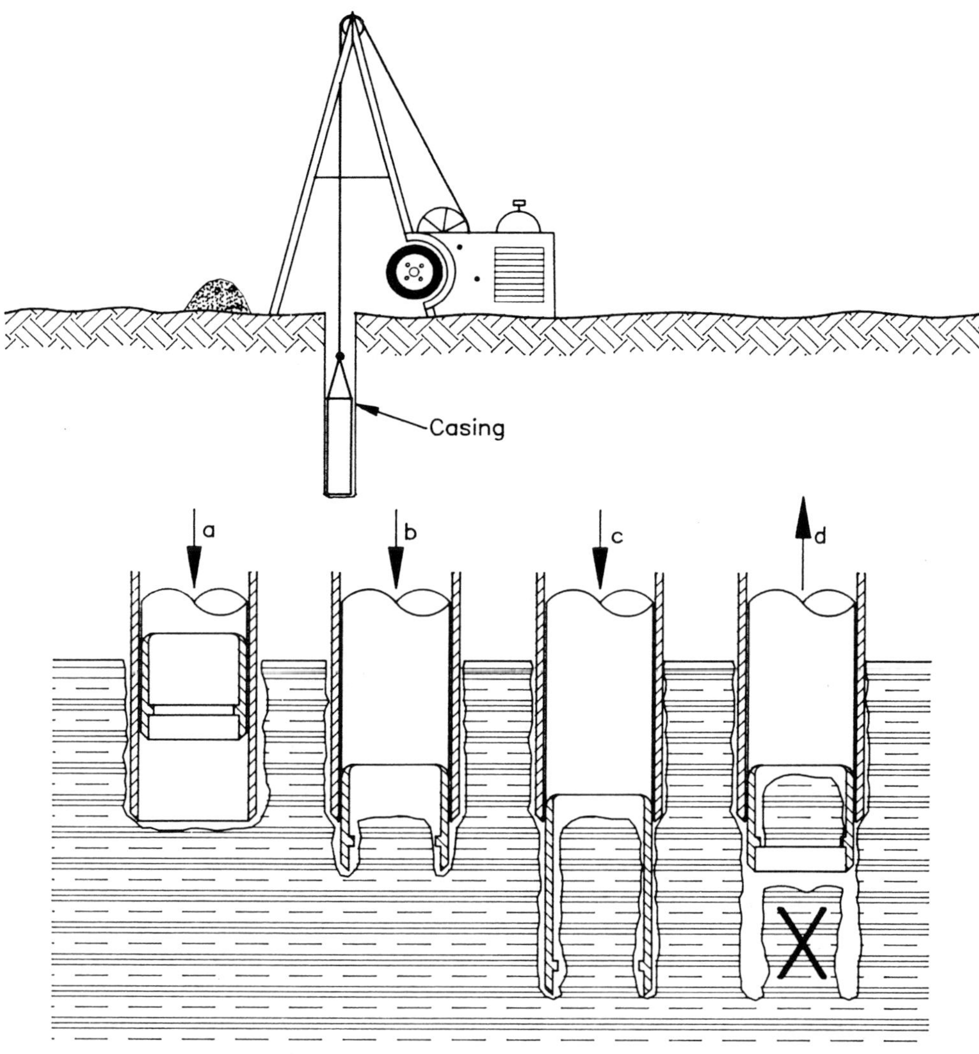

a Clay cutter on first downward stroke.
b End of first stroke. Clay cutter has penetrated base of borehole.
c Repetition of strokes a and b causes clay cutter to penetrate further.
d Retaining ring eventually shears soil at base. Clay cutter is raised to the surface for emptying. The soil at X is disturbed.

Clay cutter action

Figure 2.1: Cable percussion (after Weltman and Head, 1983[255])

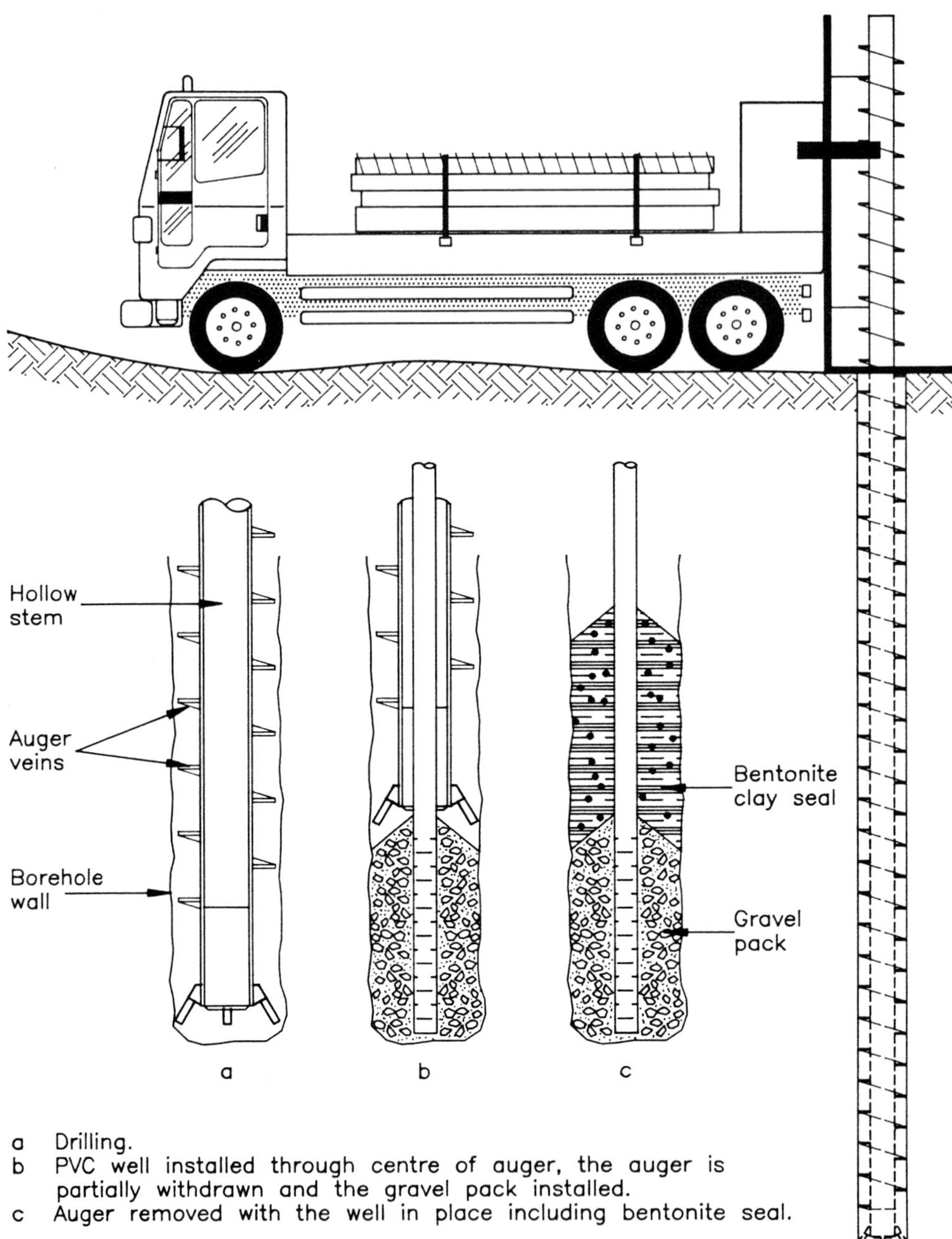

Figure 2.2: Rotary augering and installation of PVC well

boreholes is required. Undisturbed samples may be obtained using hollow-stemmed augers.

Rotary drilling (see Figure 2.3) is used in hard rock strata and includes rotary core and rotary percussive techniques. Rotary core drilling is used to obtain core samples and provide detailed information on rock strata. Rotary percussive drilling methods are used for probe drilling or the investigation of mine workings (see Section 3.3) and are more economical than core drilling although they are of limited use in obtaining samples. Sometimes the two drilling methods may be used together by penetrating to the required depth using rotary percussion and then using rotary coring to obtain core samples. A drilling fluid, such as water, or air, foam or a drilling mud, is required to cool and lubricate the drill bit and flush the fragments of rock to the surface.

Monitoring wells

Boreholes can be fitted with stand pipes or piezometers for measurement of groundwater levels, obtaining samples of groundwater or for monitoring gases. Design of wells for groundwater monitoring should ensure that only water from the stratum of interest is allowed to enter the well. For example, when monitoring groundwater in natural sands beneath fill materials, slotted pipe should be used within the sands but not through the depth of the fill material above. In contrast, in order to monitor gases produced by the fill materials, slotted pipe must be used within the fill.

Boreholes in contaminated land

When boreholes are used at contaminated sites, care is needed to ensure that the boreholes do not allow contamination to spread into underlying uncontaminated strata. Where a confined aquifer is present beneath contaminated ground, boreholes should not extend into the aquifer. Where boreholes are used for sampling of soils and groundwater in order

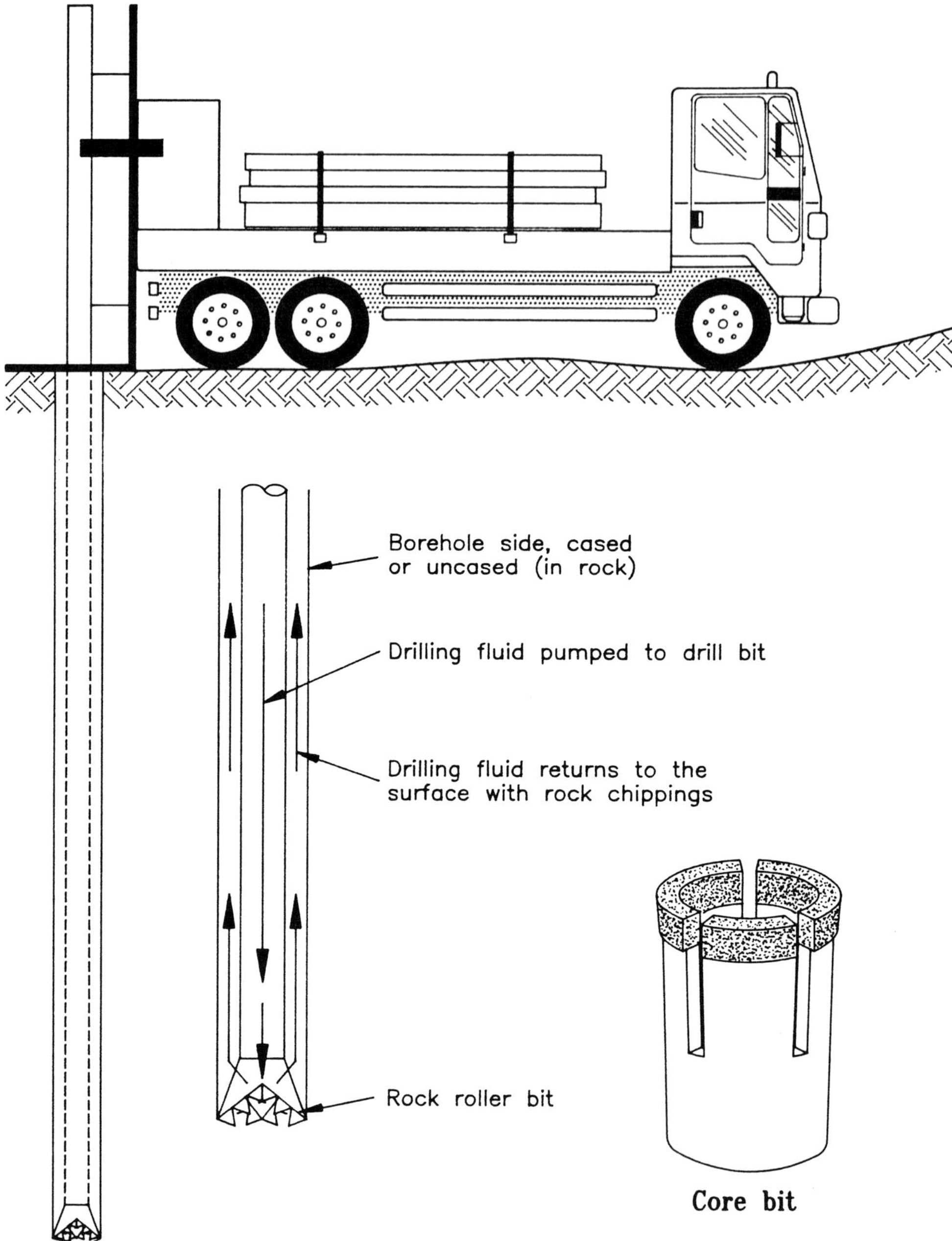

Figure 2.3: Rotary drilling (after Weltman and Head, 1983[255])

to assess contamination the procedures shown in Box 2.7 should be followed.

Geotechnical tests

During borehole drilling, tests can be carried out to determine geotechnical characteristics of the ground. Penetration tests, where a sampling tube is driven into the soil for a set distance by repeated blows of a drop hammer, give an indication of the strength of the stratum. Penetration tests are frequently carried out during cable percussion drilling. The permeability of strata may be measured by pumping water out of a borehole and measuring the rate at which it fills (rising head test) or by pumping water into the borehole and measuring the rate of fall (falling head test). A variety of laboratory tests may be carried out on samples collected during the ground investigation to provide information on the geotechnical properties of materials. Such tests include moisture content, liquid and plastic limits, bulk density, particle size and, for undisturbed samples, shear strength and compressibility.

Box 2.7: Measures to avoid introducing contamination whilst boreholing

The following precautions should be taken to avoid cross contamination whilst boreholing:[204]

- equipment should be cleaned before moving from contaminated to less contaminated areas;
- the use of lubricants should be avoided;
- air filters should be fitted to the exhaust of the drilling rig and to the air compressor where air flushing is used;
- drilling fluids should be used only where absolutely necessary, when water of mains quality should be used, a sample of which should be taken for analysis, and the quantity used recorded;
- packing media used around stand pipes or piezometers should be inert and free of impurities.

2.5.3 Soil probes

A variety of small diameter probes have been developed in response to a demand for small, mobile rapid machines which also have the advantage of causing less disturbance than traditional trial pits and boreholes. Probes may be the preferred method of ground investigation where contamination makes it desirable to minimise contact of personnel with materials in the ground, or where the contamination is volatile and disturbance caused by trial pits is likely to result in loss of volatile compounds from materials sampled.

Probes may be used for obtaining information on soil type, sampling of soil gas and obtaining samples of soils and groundwater. Depths of several tens of meters can generally be achieved. Information on soil type can be obtained from cone penetration testing. A probe fitted with a cone tip is driven into the ground at a constant rate and continuous measurements made of cone tip resistance and side friction. The relationship between these two measurements is dependent upon soil type. Adaptations to the standard cone penetration equipment allow measurements of porewater pressure, permeability, conductivity, pH, redox potential and temperature.[247]

Soil gas surveys have been one of the most frequent uses of soil probes. The total concentration of volatile compounds *e.g.* benzene and associated compounds at coal carbonisation sites, can be rapidly measured using portable instruments such as flame ionisation detectors. On-site gas chromatography can be used to obtain information on the nature of the volatile compounds present. The collection of soil samples for subsequent laboratory analysis is, however, necessary to obtain soil concentrations of specific compounds.

2.5.4 Geophysical techniques

Geophysics provides a range of non-invasive techniques for ground investigation which rely on differences in physical properties between different materials. Techniques include ground probing radar, electrical resistivity, seismic refraction, magnetometry and electromagnetics. Further information about these techniques is given in Box 2.8. Such techniques can be used for locating buried foundations, underground voids, tanks, shafts and mine workings and, in some cases, areas of contamination. The techniques are generally fairly rapid but require specialist interpretation. Results can be ambiguous, requiring confirmation by excavation or other techniques. Geophysical techniques do, however, provide a means by which large areas can be covered quickly, directing future ground investigations to the areas most likely to be of interest.

Box 2.8: Geophysical techniques

Ground probing radar
Reflection of electromagnetic pulses of 100-1000 MHz from sub-surface features. The equipment is highly mobile and can be towed at speeds of several km/hour. Depth penetration is poor, sometimes as little as 1m in water-saturated or clay materials.

Electrical resistivity
An array of electrodes is moved across the site to map changes in resistivity, such as may be caused by areas of contamination.

Seismic refraction
Measures travel times of seismic waves, typically induced by sledge hammer blows. Can give information on strata to 20-30m depth, but is ineffective when overburden is dry or unconsolidated.

Magnetometry
Measures changes in magnetic field caused by ferrous objects or differing magnetic susceptibility of geological units.

Electromagnetics
A transmitter and a receiver are moved across the site, while maintaining a constant distance between them. Variations in subsurface electrical conductivity, due to variations in groundwater levels, or the presence of metallic objects, are mapped. Interference may be caused by buildings, power lines, electrical storms and fences.

2.6 Investigation of contamination

2.6.1 Nature of contamination

Contaminated land has been defined by the NATO Committee for Challenges to Modern Society as:

> "Land which contains substances that, when present in sufficient quantity or concentrations are likely to cause harm directly or indirectly to humans, the environment or on occasions to other targets."[223]

Contamination of the ground is associated typically with waste materials, some of which will have been used as fill to raise ground levels, or have been simply disposed of on site. Examples include metals in ashes, slags or flue dusts, and cyanides and sulphates in spent oxide from the purification of gases after coal carbonisation (see Section 10.2).

Contaminants may migrate from the wastes into surrounding uncontaminated materials. The extent of such migration is dependent on the nature of the waste materials and on the mobility of the hazardous substances in the surrounding ground. Contamination may also arise where mobile substances, particularly liquids, have been introduced into the ground through leaking pipes or tanks or through spillages during the course of on-site operations or site demolition. Unlike waste materials which can be identified visually, contaminated natural ground may be visually indistinguishable from uncontaminated ground. The source of mobile contamination may be located outside the boundaries of the site under investigation. The nature of the contamination typically found at coal and steel sites is discussed in Chapter 5 (colliery spoil), Chapter 9 (iron and steel) and Chapter 10 (coal carbonisation), whilst methods for treating contamination are described in Chapter 11.

2.6.2 Sampling strategy

There is always a degree of uncertainty associated with investigations of contamination as it is impractical to excavate the whole of a potentially contaminated site and analyse all of the materials therein to determine the concentration of contaminants present. Critical decisions have to be made with regard to the number and spacing of samples taken so as to obtain information on the contamination status of a site which provides the degree of certainty required. The degree of certainty will be increased by basing these decisions on information, obtained during the desk study, concerning the location of contaminating activities and waste disposal practices. The degree of certainty in contamination investigations is often expressed as the chance, or statistical probability, of finding a 'hot spot' of contamination, that is, a concentrated point source. The statistical probability is dependent on the sampling pattern and the frequency of sampling which are discussed in the following paragraphs.

Sampling pattern

There are three basic approaches to the pattern of sampling:

- **Judgemental**: samples are deliberately taken at certain locations selected on the basis of prior knowledge of contaminant distribution. Such sampling is very unlikely to produce samples which are representative of the site as a whole but it is an efficient way of obtaining information on the concentration of contaminants in an area known to be heavily contaminated, or the extent of migration of a contaminant from a known source.
- **Systematic**: sampling locations are defined by a grid system, generally a square grid (see Figure 2.4 (a)). This is easy to set out on site and is generally the method chosen where there is little prior information on the location of contamination. However, if the pattern of contamination coincides with the pattern of the grid, the samples obtained will not be

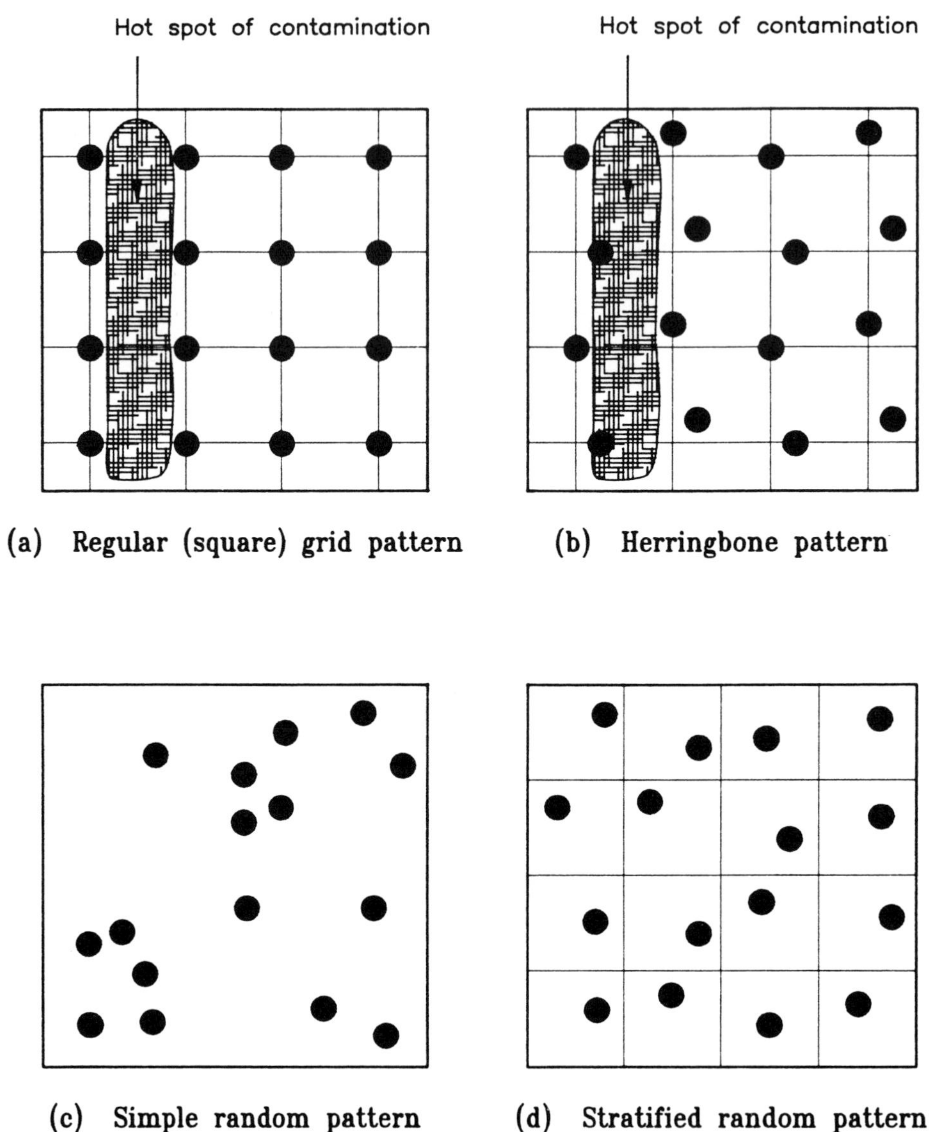

Figure 2.4: Sampling patterns for contaminated land

representative of the site as a whole. For example, elongated 'hot spots' of contamination which are parallel to the grid lines and fall between them will be missed altogether. The risk of this can be considerably reduced by using a herringbone rather than a simple square grid pattern, as shown in Figure 2.4(b).[91]

- **Random**: mathematically determined random sampling allows statistical analysis of results to be undertaken, although the variation between individual data points after sample analysis, especially with regard to chemical analytical data, is often so large as to make meaningful statistical interpretation impossible. In its simplest form, where sample points are placed randomly over the whole site, random sampling is inefficient in that unless a very large number of samples are taken there may be substantial areas where no samples are taken at all (see Figure 2.4(c)). More sampling locations are thus required to give the same probability of locating a 'hot spot' of contamination than with systematic sampling. By dividing the site into a number of areas (*e.g.* equal-sized squares) and placing a sampling location randomly within each area, less sampling points are needed than for simple random sampling. This is known as stratified random sampling and is shown in Figure 2.4(d). An element of judgemental sampling may be introduced by varying the relative sizes of the areas according to prior knowledge of the distribution of contaminants across a given site.

Frequency of sampling

A rational method of deciding sampling frequency is based on the following:

- the maximum size of a 'hot spot' which, if it is not discovered during the site investigation, would not cause unmanageable problems during later development of the site;
- the required degree of certainty of finding such a 'hot spot'.

For the herringbone sampling pattern the number of sampling locations needed to ensure a 95% probability of hitting a target 'hot spot' is given by the equation:

$$N = \frac{kA}{a}$$

where: N is the number of sampling points
A is the total site area
a is the area of the target 'hot spot'
k is a constant which depends on the shape of the target as follows:

circular target	$k = 1.08$
plume-shaped target	$k = 1.25$
elliptical target	$k = 1.80$

An assessment thus has to be made of the likely shape of the target in order to calculate the sampling frequency.[91]

Formulae for sampling frequency which reflect "current practical experience with soil investigation rather than considerations relating to the statistical reliability of pronouncements regarding the soil pollution of a site" are given in the Dutch draft standard for exploratory site investigations.[175] For sites where pollution is not suspected intensive sampling and analysis for a wide range of possible pollutants is necessary to have a reasonable chance of finding previously unknown pollution. Sampling should be done according to a systematic sampling pattern, with the number of sampling locations for near surface samples, n, given by:

$$n = 10 + 10A$$

where A is the total site area in hectares. Where the preliminary desk study has suggested that pollution is present but that its distribution is homogeneous the number of sampling locations is given by:

$$n=5+A$$

Where contamination is thought to arise from known point sources, four sampling locations per point source are recommended, with at least one groundwater observation well per point source. Where the locations of the point sources are not known, the recommended number of sampling locations is given by:

$$n=4+\frac{A}{a}$$

where a is the estimated area of contamination in hectares.

Depth of sampling

It is generally necessary to know how the concentrations of substances vary with depth, so at each sampling location several samples from various depths will need to be obtained. The British Standard Draft for Development, "Code of Practice for the identification of potentially contaminated land and its investigation", DD175,[43] recommends that at least three samples are taken at each sampling point, one to represent the surface and near surface layers, the second to represent the greatest depth of interest and the third at a random intermediate depth. However, more samples may be necessary to achieve the objectives of:

- obtaining representative samples of all types of waste materials present at a given surface location;
- obtaining samples of natural materials in order to enable an assessment of the migration of contaminants.

The Dutch Draft Standard for exploratory surveys,[175] recommends sampling the top soil (or the top 0.5m), then taking three samples of the subsoil one for each interval of 0.5m down to 2m below ground level. If groundwater is present at 2-5m below ground level a sample of soil should also be taken from just above the groundwater table. Fewer sampling locations are thought to be necessary for samples at depth, as shown in Box 2.9. These recommendations apply to sites where pollution is not suspected. For sites where pollution is suspected the draft standard recommends that samples are taken from each suspect soil stratum and from the adjacent strata.

Multi-stage surveys

It may be beneficial in some situations to carry out an investigation of contamination in more than one stage. For example, the initial ground investigation may be followed by the preliminary design of reclamation proposals, and then by further ground investigations to clarify whether certain options are more acceptable than others. Similarly, where it is proposed that contaminated materials be removed, further investigation will be needed to define the quantities involved. However, first stage surveys involving very low sampling frequencies, provide little reliable information on the overall contamination status of a site.

A staged site investigation will have the advantage of focusing a contamination assessment on a particular area or areas of contaminated ground, but does have the disadvantage of being spread over a longer

Box 2.9: Recommended number of sampling locations on sites where contamination is not suspected[175]

Samples to 0.5m:	$n = 10 + 10A$
Samples from 0.5m to 2.0m:	$n = 3 + 3A$
Groundwater observation wells:	$n = 1 + 1A$

Where n is the number of sampling locations and A is the size of the site in hectares

period of time. A focused investigation is likely to provide more accurate information about critical areas of a site and therefore be more cost effective. In addition the results of the first stage investigation will be useful to those people preparing preliminary programmes and budgets for reclamation work.

The hazardous waste site investigation programme of Baden Württemberg[115, 164] recommends three stages of site investigation after the initial historical research. A preliminary investigation, with limited costs, consists of initial systematic measurements and examinations of the site and resources at risk, using existing or easily established measuring and sampling locations and passive bio-monitoring of flora and fauna. This may be followed by a detailed investigation, designed to give a picture of the location, extent and quantity of contaminated materials. Further in-depth studies may then be carried out to assess options for clean up or containment of contamination. Between each level of investigation an evaluation is carried out to assess whether further investigations are necessary.

2.6.3 Sample collection

Solids

Methods of sample collection should endeavour to ensure that the sample taken is as representative as possible of the material being investigated. Bulking of samples from different depths or from different locations is not recommended as it is the range of values, in particular the maximum concentrations, rather than the average concentration, which is of relevance to risk. However, when sampling a particular material from a trial pit, several samples of the material may be taken at the same depth and these combined to produce a sample which is representative of the material at that location. In the laboratory samples should be sub-sampled by coning and quartering.

Mixing of samples in the laboratory to reduce the number of analyses which have to be carried out is suggested in the Dutch draft standard for exploratory investigations.[175] Samples which differ in visual appearance or by smell should not be mixed together. It is also preferable for mixed samples to be made up of neighbouring samples from the same depth. Portions of the original, separate samples should always be retained in case required for further analysis.

Samples from trial pits are taken either from the sides of the pit or from the material held in the bucket of the excavator. The former method is preferred as the place from which the sample has been taken can be more accurately defined. Any surface dusting of material from other parts of the trial pit should be removed before sampling. A long-handled tool should be used for sampling from the sides of the trial pit at depth in circumstances where it is inadvisable for the investigator to enter the trial pit.

When taking samples from the bucket of the excavator care should be taken to avoid materials which have fallen into the bucket from the sides of the trial pit as this may mislead the investigator as to the depths at which contamination occurs. If possible samples should be obtained from the centre of large lumps of material.

The method of sampling used should be recorded in the log for each trial pit. Sampling tools should be made of stainless steel and be easy to clean between each sample.

The volume of sample required will vary with the analysis required but one litre is generally sufficient. Containers should be made of materials which do not react with the sample, or allow escape of substances present in the sample. Thus, polyethylene and many other plastics should not be used where organic contamination is suspected, but are acceptable when carrying out analysis for metal contaminants only. Glass containers should be used when organic contamination is suspected.

Sample containers should be robust and air-tight on closure, with large openings to enable easy filling and emptying. Samples should be clearly labelled, with site name, date, sample location and depth.

Liquids

Samples of liquids found in trial pits may be obtained by a variety of methods, such as lowering a container into the pit. Samples may be obtained from standpipes installed in boreholes and backfilled trial pits by use of a bailer or various hand or mechanical pumping techniques. A period of at least several days should elapse between installation of such standpipes and collection of samples, to allow equilibrium of water in the standpipe with surrounding groundwater.

Containers for liquids should not react with sample constituents, or allow substances to escape. Thus glass bottles with PTFE seals should be used where volatile organic compounds are suspected. Plastic bottles will be suitable when it is known that the analysis of organic compounds will not be required. Containers should be rinsed several times with the liquid being sampled before filling and sealing. Liquid samples are far more likely than solid samples to undergo change after collection, for example by reactions between substances in solution, precipitation and dissolution and adherence to container walls. The storage of samples at low temperature is recommended to arrest chemical and biological activity in the sample. Various chemicals can also be added to preserve particular substances. For example, acids can prevent the precipitation of metals, and sodium hydroxide can prevent breakdown of cyanides. When taking water samples, particularly when low concentrations of contaminants are of interest, a selection of different bottles, each containing different preservatives, may be required. The analysis suite required should therefore be known in advance of taking liquid samples.

2.6.4 Analysis

Analysis of the samples after collection is often the most expensive part of investigations into contamination of land. Careful thought has to be employed in deciding which samples to analyse and for what constituents. Information obtained in the desk study is likely to be of great value in this respect. The marginal costs of sample collection are small, and so it is advisable to collect a large number of samples, and then analyse only a selection of these. Analysis of further samples can be carried out later in the light of initial results. Samples should always be kept until the reclamation proposals have been finalised, in case further analysis is required.

The analysis suite should be chosen on the basis of:

- the substances which are thought likely to be present, on the basis of the past uses indicated by the desk study;
- the substances which are thought likely to cause a hazard, given the proposed use of the site.

Measurement of pH, a basic parameter of soil and water conditions, should be carried out in nearly all site and soil investigation situations. It is inexpensive and is a good general indicator of site and soil conditions. Screening analyses, which indicate the presence of a group of substances but not the concentration of individual compounds, can be used to gain maximum information for minimal analysis expenditure. Further analysis is often then required to identify the particular substances present, but this can be directed at samples shown by the screening analysis to contain high concentrations of the class of compound in question. For example, analysis for total sulphur can be followed by analysis for total sulphate, sulphide and elemental sulphur, and then subsequently water-soluble sulphate if total sulphate concentrations are unusually high. Similarly, analysis for total cyanides can be followed by analysis for free cyanide and thiocyanate. Measurement of electrical conductivity gives an indication of the concentration of soluble ions and samples with high conductivity should be followed by analysis for cations and anions.

Analysis for organic compounds is particularly complex and can be expensive. An initial screen by solvent extraction and gravimetric determination is often carried out, but these methods generally do not measure volatile compounds, as these are lost during evaporation of the solvent. Analysis of the head space, *i.e.* the space at the top of the sample container, for volatile compounds is a more appropriate screening technique for this type of contamination. For example, a flame ionisation detector, which measures total flammable hydrocarbons, can be inserted into the air space in the sample container. Techniques such as thin layer chromatography can be used to separate organic compounds into different types of compounds, but the identification of individual compounds generally requires techniques such as gas chromatography (GC) or high performance liquid chromatography (HPLC), often involving lengthy extractions and run times.

There are standard methods of analysis for many substances, but not for all those likely to be present in contaminated land at coal and steel sites. It is important that the person specifying the analysis suite and interpreting the results has a good understanding of the methods used, their advantages and limitations. Good communication with those carrying out the analysis is essential. In particular, when reporting results laboratories should give sufficient information about the methods used and how they have been reported to enable a correct interpretation to be made.

Generally the errors introduced in the sampling process, in terms of the samples being representative of the material in question, are much greater than those at the analytical stage. It is thus often better to analyse a large number of samples by a reasonably accurate method than a small number by a very accurate and costly one. However, for groundwater samples in particular, very low concentrations of contaminants may be of interest.

The limit of detection of the analytical method employed should therefore always be lower than the lowest concentration which could be of concern.

Analysis should be subject to rigorous quality control procedures to ensure that samples are not lost, correct analytical procedures are carried out and results are reported correctly. Standard reference materials *i.e.* materials as similar as possible to those being analysed but containing a known concentration of the analyte, should be used to verify analytical procedures. A reliable laboratory may be chosen on the basis of its membership of an accreditation scheme which conforms to the European standard for the competence of calibration and testing laboratories, EN 45001.

2.6.5 Interpretation of results

Results of analysis must be interpreted to make an assessment of the contamination status of a site. This requires decisions to be made on what concentrations are to be regarded as signifying contamination. There are three different approaches to this problem:

- comparison with 'natural' concentrations in unpolluted soils;
- comparison with concentrations in surrounding soil;
- consideration of the risks associated with the presence of the substance at different concentrations.

These approaches may be applied to individual sites, on a case by case basis, or they may be used to derive standards for soil concentrations of individual substances which are then used when assessing large numbers of sites.

Whilst any increase in soil concentrations of hazardous substances above 'natural' levels may be considered to be undesirable, in practice, comparison of soils from heavily industrialised urban areas, where the majority of land contains elevated concentrations of potentially hazardous substances, with concentrations found in soils of unpolluted rural situations would lead to unjustified alarm and misdirection of resources.

Comparison with local background soil concentrations is a more practical approach, which allows areas containing high concentrations of contaminants to be identified and resources directed at treatment of these, to give a gradual improvement in the contamination status of an area over time.[118] This approach of developing standards specific to a local area is also necessary where the background concentrations are naturally high, for example in the south-west of England where arsenic concentrations in soils may naturally be as high as several hundred mg/kg.

The third type of approach considers the risks associated with the presence of the substance in question and attempts to define concentration values below which the risk is negligible, and above which the risk is unacceptable. Evaluation of risk implies consideration of the system:

$$\text{Contaminant} \xRightarrow{\text{Route}} \text{Target}$$

In this model the target may be people, animals, plants, water resources or building materials.

Routes by which people may be exposed to contaminants include:

- direct skin contact;
- inhalation of dust or gases;
- ingestion, directly or via food or drinking water.

The degree to which routes of exposure operate depend, amongst other considerations, on the use of the site. Therefore, the risk assessment approach frequently takes into consideration the intended use. Assessment of risks to humans and animals from toxic substances in soils involves determining acceptable dose levels from toxicity data and then relating soil concentrations to dose levels by consideration of exposure routes. A large degree of uncertainty is involved in this assessment, due to the paucity of toxicity data, especially for chronic exposure to low levels of toxic substances, and the many unknowns affecting the routes of exposure.

Approaches to the interpretation of soil concentrations in the Netherlands, Germany and the United Kingdom are discussed here.

Dutch approach

In the Netherlands the Interim Soil Clean-up Act of 1983, introduced the Soil Clean-up Guideline,[227] which contains the most widely used set of concentration values for assessment of contaminated sites. The 'Dutch List', as it is known, consists of three sets of concentration values for a wide range of substances in soil and groundwater. The three types of value are explained in Box 2.10.

The principle of the Dutch approach is that contaminated soils should be restored to 'multifunctionality' *i.e.* the soil should be suitable for a wide range of possible future uses, such as crop production, grazing, groundwater recovery, ecological functions, and construction or development.[41] This principle can however be relaxed when

Box 2.10: Dutch Soil Clean-up Guideline values for soil concentrations

The 'Dutch List' as it is known, consists of three levels of concentration values set for a range of substances in soil and groundwater. The three levels have the following significance:

- A-value — based on average background concentrations, implying unpolluted soils
- B-value — implies pollution present and further investigation is required
- C-value — implies significant pollution present and clean-up required

environmental, technical or financial circumstances make remediation impossible in practice. Where multifunctionality of soils is not restored, any hazardous effects must be controlled by isolation of the site with control and monitoring procedures to ensure the effectiveness of the isolation system. A soil is considered to be multifunctional if it conforms to the A-values of the Dutch List. These values were originally defined by reference to average concentrations of unpolluted soils. However, in 1987, in the Environmental Program 1988-1991, the Dutch Government published a provisional list of reference values for acceptable soil quality, revising many of the original A-values.[248] They are an attempt to define 'no-effect' levels of substances in soils. The preparation of the reference values considered standards set in other policy areas, such as standards for drinking water, surface water and food. In setting standards for metals and fluoride concentrations, 'clean' rural areas were considered, and modelling of soil solution/solid phase relationships and toxicological risk assessment was carried out for organic compounds. Because the bioavailability of substances is dependent on soil type the values are given as formulac which dcrivc soil conccntration from clay and organic mattcr content of the soil.

The 1983 Interim Soil Clean-up Act is to be incorporated into the Soil Protection Act 1987.[227] This Act provides for a wider policy on prevention of soil pollution. The Soil Clean-up Guideline is to be revised and renamed the Guideline for Soil Protection. This will contain the revised A-, or reference values and revised C-values, but no B-values. The new C-values which are to be renamed 'intervention values', have been set at the level thought to represent a serious threat to Man and the environment. The derivation of these values is described in Box 2.11. The presence of contaminants at concentrations which exceed the C-values indicates a need for clean-up, in that there is a potential serious threat to Man or the environment. However, the new Guideline for Soil Protection recognises that further investigation is needed to determine actual threat and therefore whether or not clean up should have a high priority. This involves assessment of the extent of actual exposure, which is primarily determined by soil use.[29]

Box 2.11: Derivation of Dutch C or 'intervention values'[29]

Separate values were initially derived for the two aspects, Man and the environment.

(i) A "serious threat to the environment" was taken to be "irreversible and irreparable damage to the species composition", defined as when fifty per cent of the species present experience adverse effects as a result of the concentration of one or more contaminants. C- values were then derived following a search of ecotoxicological data for plants, soil fauna and micro-organisms.

(ii) Soil concentrations constituting a "serious threat to Man" were derived from consideration of carcinogenicity/toxicity data and exposure pathways. The carcinogenicity/toxicity data were used to set maximum tolerable risk (MTR) levels equivalent to the maximum daily intake of a compound which can be taken orally without experiencing adverse effects on health, or the quantity of a carcinogen which corresponds to a risk of one additional case of a lethal tumour in 10,000 lifelong exposed individuals. The soil concentration which could lead to a dose equivalent to the MTR was then calculated from consideration of exposure pathways.

The lowest of the two C-values (one for ecotoxicology and the other for human-toxicology) was chosen as the new soil C-value, provided the uncertainty attached to this C-value was not much greater than that of the other C-value. Soil C-values were then corrected for organic matter content and clay content, factors which influence bioavailability, using the formula derived for the reference A-values.

Groundwater C-values were calculated by reference to the soil C-values using soil-water partition coefficients and assuming an equilibrium between soil and water partitioning. Because of the uncertainties involved in this process the values obtained were then decreased by a factor of 10.

German approach

In Germany policy on contaminated land is the responsibility of the Länder although a federal law on soil protection is in preparation. Several studies have been carried out on soil concentrations in specific areas to determine regional background levels which are then used for assessment of contaminated land. Risk assessment procedures are also used for individual sites and to produce soil standards, such as those of Eikmann and Kloke.[139] These were published in a communication of the VDLUFA (the German association of agricultural research institutes) and are to be made legally binding in Saxony. The publication gives reference levels for hazardous substances in soils. Three concentration ranges are given:

- **A**: safe range, concentrations found in natural, unpolluted soils;
- **B**: tolerable range, concentrations which, on the basis of present knowledge, do not cause damage to people, animals, plants, or the soil ecological functions;
- **C**: toxic range, concentrations which cause discernable damage to people, animals, plants and the ecosystem. Clean up, or prevention of contact with contaminated soil is considered urgent.

These three ranges are divided by two soil values:

- BWI: between A and B ranges, it does not vary with site use;
- BWIII: between B and C ranges, varies with site use.

Additionally, within the B range there is another soil value, BWII, which signifies a contaminant concentration at which consideration should be given to clean up or changing the use of the area within an appropriate timescale, though any risks are not acute. The interval between BWII and BWIII represents a safety range where clean-up is to be considered but is not urgent.

Values have been set for BWI, BWII and BWIII for a wide range of metals, benzo-a-pyrene (a polyaromatic hydrocarbon), PCBs, furans and dioxins, for various urban and agricultural uses. These include children's play areas, domestic gardens and allotments, sports and playing fields, parks and recreational areas, industrial and commercial areas, with and without impermeable ground cover, agricultural areas, and non-agricultural ecosystems. For each type of use the target group which is to be given protection (*e.g.* children, people of employment age, groundwater or plants), the assumed route of uptake, the soil area and soil depth is given. Soil areas to which the BW values apply are generally those that do not have a good vegetation cover. The depth of concern varies between 0.1 and 0.5m for the various uses.

Guidance on the assessment of the aggressiveness of soils towards building materials is provided in the German standards, DIN 4030 (June 1991) "Evaluation of liquids, solids and gases aggressive to concrete" and DIN 50929 (September 1985), "Probability of corrosion of metallic materials when subject to corrosion from the outside".

UK approach

In the UK 'threshold' and 'action' trigger values have been published for a limited range of substances for various types of end-uses including, in order of decreasing sensitivity: domestic gardens and allotments; parks, playing fields, open space; landscaped areas, buildings and hardcover.[121] In this system there are two types of trigger concentrations. The 'threshold trigger' concentration is the concentration below which the risks are considered to be negligible and the site may be regarded as uncontaminated. The action trigger concentration is the concentration above which some form of remedial action is essential. At concentrations between these two values professional judgement is required to evaluate the need for treatment. 'Action' trigger concentrations for metals have not been specified except for reclamation of former metalliferous mine sites to pasture and grazing which is the subject of a separate document.[122] Trigger values are intended to be used to assess the suitability of a proposed new use of the land, not whether an existing use should be allowed to continue.

More specific guidance on the reclamation of particular types of site is published occasionally by the UK Department of the Environment. This guidance includes that for coal carbonisation sites.[86]

Information on acidity levels and sulphate concentrations which affect concrete is given in Building Research Establishment Digest 363, "Sulphate and acid resistance of concrete in the ground" (July 1991).

2.7 Assessment of industrial archaeological value

The assessment of former coal and steel sites should include an assessment of the possible historical and archaeological nature of a site and its various features. A specialist industrial archaeological input should preferably be sought on all projects (see Section 4.7).

A key task for the specialist is to establish the overall significance of the site *i.e.* whether it is of local, regional, national or even international importance. This will allow rational judgements to be made on the emphasis to give to conservation. Assessment of the industrial archaeological value of a site needs to examine criteria such as scarcity, representativeness, scientific interest, engineering interest, state of preservation and completeness. The amenity recreation/tourism implications also need to be considered.

Many EC countries are now drawing up a 'monuments protection programme' on an industry-by-industry basis; iron and steel sites for example are classified in order of interest and specific site features may be listed for preservation. Such programmes will greatly aid the assessment of sites. Official lists such as any national archaeological record database, lists of scheduled ancient monuments, and regional archaeological sites and monuments records should be examined to see whether they include any features on the site. Specialists, volunteer groups and societies often maintain valuable databanks with lists of important features and they may also have published "standards for the survey and recording" of such sites. In the USA there is a standard "Guidelines for Inventories of Historic Buildings and Engineering and Industrial Structures" when work is carried out at three levels:

- Level I: file search, written overview and bibliography;
- Level II: reconnaissance survey to given standards;
- Level III: intensive survey with simultaneous recording and evaluation.

Levels I and II can often be carried out by voluntary groups but Level III is usually done by contractors. All sites should be studied to Level I, the more historically interesting sites to Level II and III.

2.8 Reclamation proposals

The final stage in the site assessment process brings together all the different issues which have been covered in the investigation phase and puts forward initial proposals for reclamation. At this stage decisions between any conflicting objectives will have to be made to produce the most effective solution for the site and to allow detailed design of a reclamation scheme to proceed. The design of treatment and amelioration measures is discussed in Chapters 11, 13 and 14. Chapter 16 places site investigation and reclamation in the wider context of planning, funding and local support, illustrating the need for an integration of good design, public acceptability and clear objectives in the fulfilment of a successful scheme.

The relationship between the activities carried out during site assessment and those concerned with the reclamation and management of former coal and steel sites is shown in Figure 2.5, which is a flow chart of the principal stages involved.

Site available for reclamation

Is the end-use defined ?

No

Define end-use at earliest possible time. If this is not clear at least produce a list of desirable options, given the nature of the site and local planning or policy context

Yes

Reclamation brief

Appoint design team

Walkover site

Has site been cleared of buildings ?

Yes

Obtain demolition records

No

Carry out pre-demolition site audit

Gather site information:
- processes, layouts, mine plans *etc.*;
- waste disposal plans;
- geology, soils, hydrogeology;
- previous studies on the site;
- ecological information;
- archaeological information.

Develop site management considerations

Formulate site investigation proposals including: engineering, land-use and visual assessments

Site investigation needs

Obtain permissions to enter site

Carry out site investigation

Reconsider end-use proposals

Management needs

Consider information on services and utilities

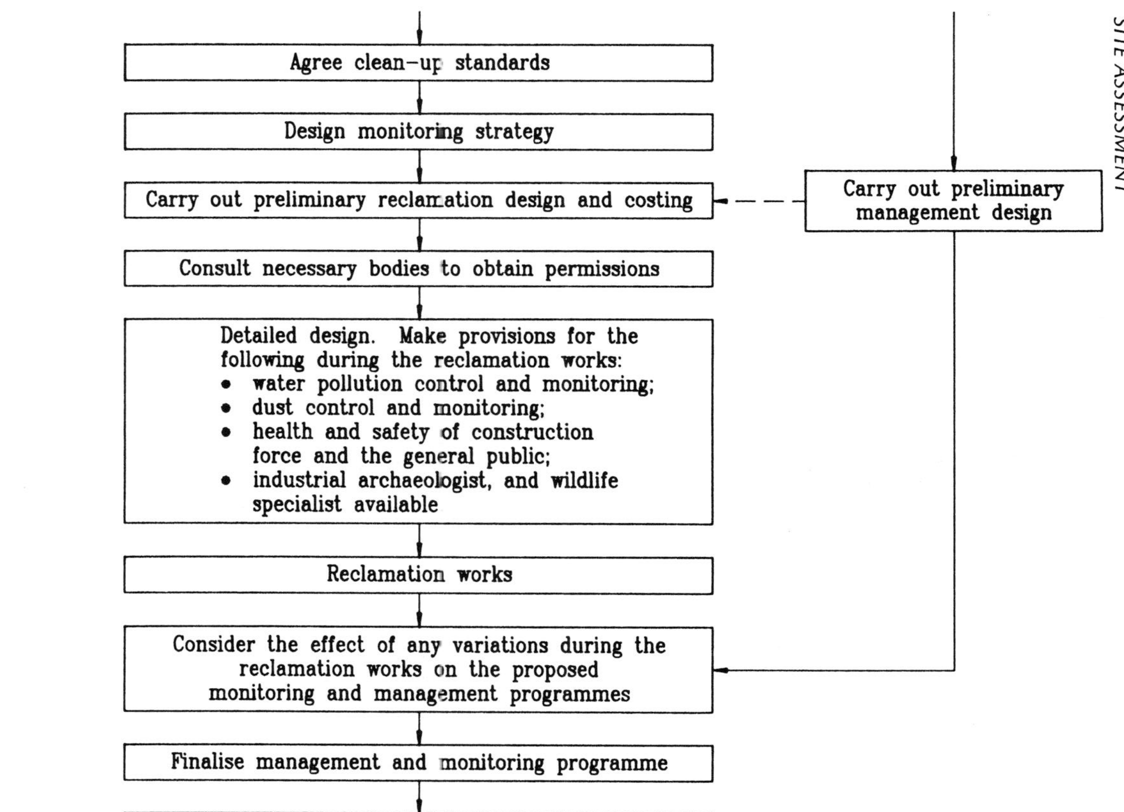

Figure 2.5: Flow chart of the principal stages in the investigation, reclamation and management of a former coal or steel site

3 INVESTIGATION AND TREATMENT OF MINE WORKINGS AND UNSTABLE GROUND

Chapter contents

continued...

3 INVESTIGATION AND TREATMENT OF MINE WORKINGS AND UNSTABLE GROUND

3.1 Introduction

Mining instability occurs as a result of collapse of the underground workings, or from the collapse of shafts or adits which lead to the workings. This gives rise to general or localised subsidence of the ground surface. The term subsidence is generally used to refer to vertical displacement, but a degree of horizontal movement may also accompany the downwards movement of the surface.

Apart from subsidence due to underground mining, subsidence may also occur for a variety of other reasons, such as: soil compaction, soil shrinkage, lowering of the water table, collapse of natural underground cavities, earthquakes or volcanic activity.

Most early workings for both coal and iron ore were carried out at surface outcrops. Underground mining from bell pits, drifts or adits did not generally take place until around the fourteenth century. The use of room and pillar methods (see Section 3.2.3) was not introduced until around the seventeenth century. Modern mining for coal and iron ore may involve opencast or underground mining techniques.

Modern opencast mining sites worked in accordance with good practice are reinstated in a controlled manner to an agreed programme of restoration. Nevertheless, subsidence of the backfilled material often continues for many years after restoration as a result of consolidation settlement.

Modern underground mining techniques for coal, and sometimes iron ore, generally involve total extraction by longwall methods. The resulting ground subsidence is usually predictable and rapid. However, extraction of coal from modern mines using room and pillar methods is still practised in some areas.

Mining subsidence can have a significant effect on buildings, infrastructure and utility services, and will therefore result in a safety hazard. Photographs 3.1 and 3.2 show examples of typical subsidence damage to buildings. Damage occurring as a result of subsidence varies depending on the method of mining, the degree of movement involved and whether any structural precautions were taken to counteract the effects of anticipated subsidence.

When planning a development on a site which may be affected by mine workings, investigations should be carried out to establish the nature and extent of any workings and what effect any future subsidence may have on the proposed development (see Section 3.4). The investigations should be designed and carried out so as to enable the design team to determine the treatment which may be necessary to stabilise shafts, adits or shallow workings, and any special precautions required in the design of foundations buildings and other structures.

3.2 Causes and effects of mining subsidence

3.2.1 Factors affecting subsidence

The extraction of coal and other minerals by underground mining inevitably involves the risk of mining subsidence. Different mining methods and the type of subsidence which may be produced, and their effects on structures, have been well researched and are discussed in the following sections.[114, 126, 250, 251, 257]

The effect of mining subsidence varies considerably depending on factors such as:

- the type of mining carried out;
- the depths of the workings;
- the nature and condition of the overlying strata;
- the effects of changes in site conditions;
- the sensitivity of buildings, infrastructure and utility services to subsidence damage.

Photograph 3.1: Example of external structural damage caused by mining subsidence (source: British Coal)

Photograph 3.2: Example of internal structural damage caused by mining subsidence (source: British Coal)

3.2.2 Bell pits

Bell pits were used in early shallow workings for coal and iron ore. The technique involves sinking a shallow shaft of about 1m diameter down to the seam. Depths of bell pits varied depending on the extent of overburden and the thickness and quality of the seam, but rarely exceeded 12m. Mining proceeded radially from the base of the shaft until the area became too large to be naturally or artificially supported, or problems of ventilation and water ingress occurred. The pit was then abandoned and a new pit started nearby. Bell pit diameters of 8 to 20m have been recorded. In areas of intensive mining, shafts have been found as close together as 8 to 10m.

Surface evidence of the existence of bell pits can take the form of cones of mine waste around the top of the shaft, or numerous depressions along the outcrops of major seams. The pits were often partly backfilled or simply left to collapse after abandonment. Areas containing bell pits present a risk of continued mining subsidence due to the collapse of voids, or due to the poor bearing capacity of the disturbed ground (see Figure 3.1).

The collapse of a bell pit generally leads to the formation of a crater at the surface. In areas where intensive mining has been carried out by this method many such craters may be present side by side within a small area. Where the ground surface has subsequently been regraded the signs of former shallow mining activity may no longer be visible. However, areas of disturbed ground may be present below the surface which could lead to subsidence damage of surface structures as a result of differential settlement. The effect of the sudden collapse of a single bell pit is normally fairly localised, but could lead to subsidence damage of nearby structures. The timing of a collapse cannot be predicted.

3.2.3 Room and pillar workings

In water-free conditions or in situations where water could drain naturally, room and pillar methods were developed. These methods became more widespread with the introduction of steam pumps.

(a) Plan of site mined using bell pits

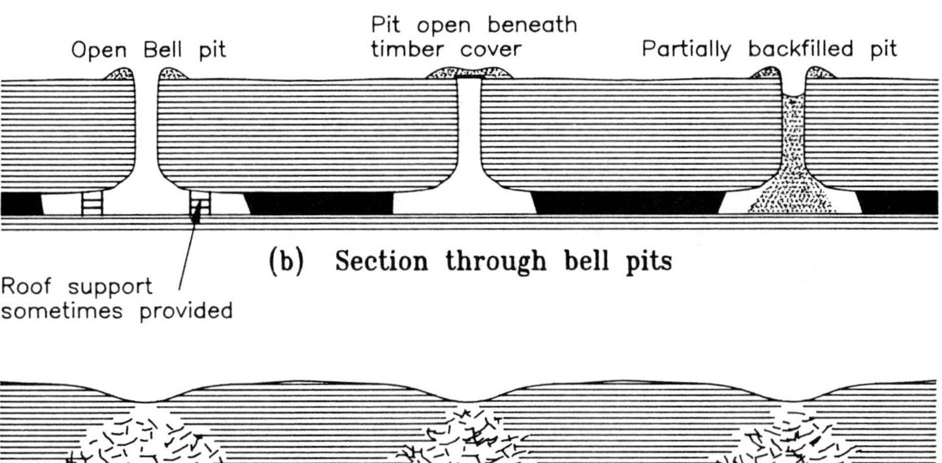

(b) Section through bell pits

(c) Section through bell pits after collapse

Figure 3.1: Bell pits

Unworked pillars of coal or iron ore were left in place to support the roof of the workings.

Regional variations occurred in the room and pillar layout. For early mines the room widths were usually determined on a trial basis. Room widths for later mines tend to vary between 2 and 5m, with pillars initially 10 to 12m square. After initial extraction pillars were often reduced in size or removed completely as the workings were abandoned. For deeper workings in the UK, pillar widths normally exceed 10% of the working depth to ensure pillar stability.

Areas of room and pillar workings present a risk of mining subsidence due to: roof failure, pillar failure, floor heave, or a combination of these effects (Figure 3.2). Voids resulting from a collapse may cause surface effects such as craters or 'crown holes' to appear. Sometimes the collapse of a group of pillars may occur, leading to localised subsidence.

Pillar collapse may occur naturally or may be triggered by external factors such as:

- surface loadings including those due to embankments, spoil heaps or building foundations;
- undermining by total or partial coal extraction;
- seismic shocks occurring either naturally or due to the collapse of longwall workings.

The condition of room and pillar workings is difficult to establish and the timing of any collapse cannot easily be predicted. Formation of a crown hole is likely to lead to differential settlement and subsidence damage of nearby structures. Areas of disturbed ground due to former collapses may also lead to reductions in allowable bearing pressures and possible subsidence damage.

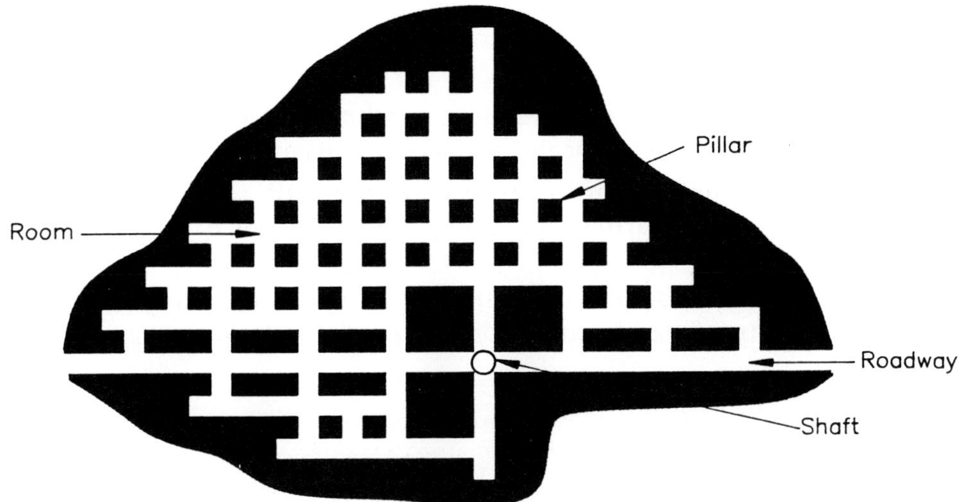

(a) Plan of room and pillar workings

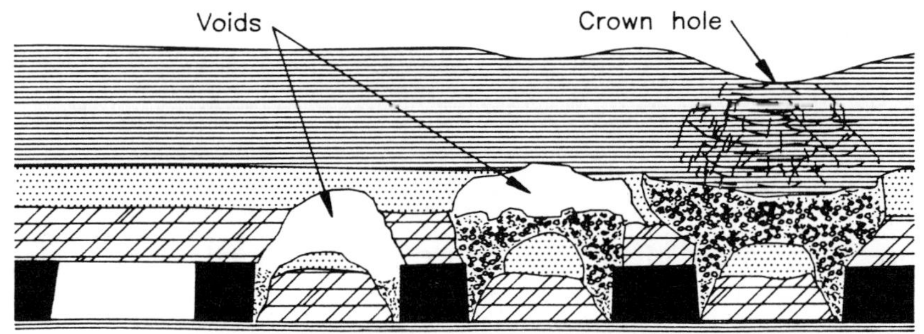

(b) Section showing effect of collapse of rooms

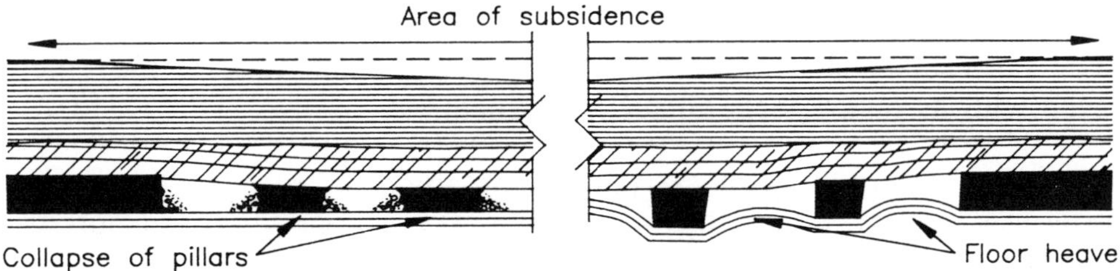

(c) Section showing effect of collapse of pillars and effect of floor heave

Figure 3.2: Room and pillar workings

3.2.4 Drift mining

Drift mining was carried out in situations where seams had a shallow dip, and the outcrop was either at the surface or beneath a shallow soil cover. Coal was originally obtained from the adit or 'drift', or from connectors between drifts. Such drifts were comparatively short in length.

Subsidence may occur due to the collapse of the drifts, or due to the poor bearing capacity of the disturbed or previously collapsed ground.

3.2.5 Longwall mining

Longwall mining involves the working of a single coal face with access from roadways at right angles to the face at each end of the seam. The working face is temporarily supported, the support being advanced as the face is worked, allowing the roof to collapse into the space behind the workings (Figure 3.3). Modern workings typically involve face lengths of up to 300m.

Subsidence from longwall mining occurs in the form of a wave moving parallel to and at the same rate as the line of the advancing coal face (Figure 3.3). Most of the subsidence is transmitted to the surface in a relatively short period of time depending on the depth of the seam and the characteristics of the overlying strata, but residual subsidence may occur up to 2 years after mining has taken place. The subsidence produced is fairly regular and can be accurately predicted.

The surface subsidence develops with various components of movement each with its own effects on surface structures:

- **Vertical or displacement subsidence:** This may have little effect on structures providing it is uniform but may have a significant effect on rivers and lowland drainage by inhibiting the rate of flow and increasing flood potential;
- **Tilt or differential subsidence:** This may affect building structures of which tall structures would be the worst affected.

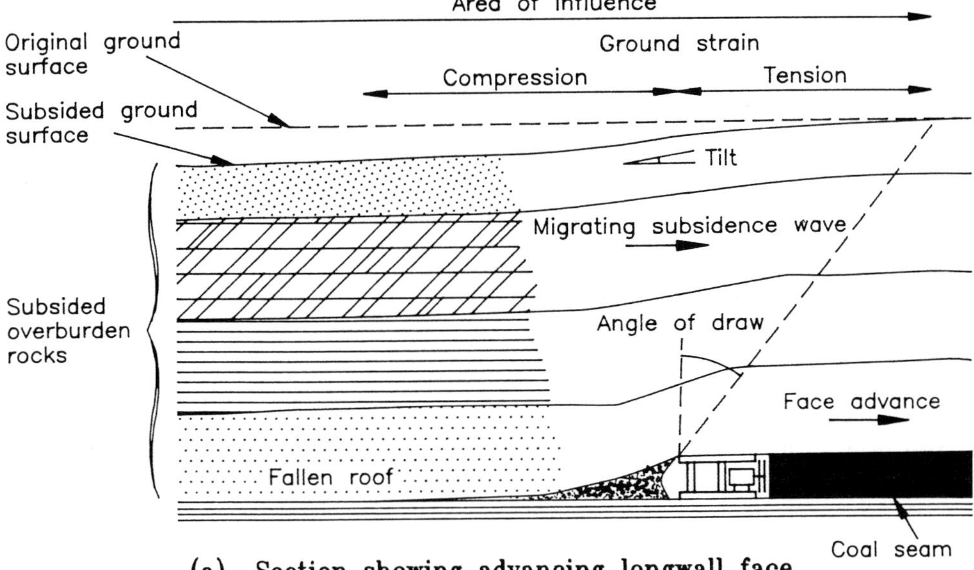

(a) Section showing advancing longwall face

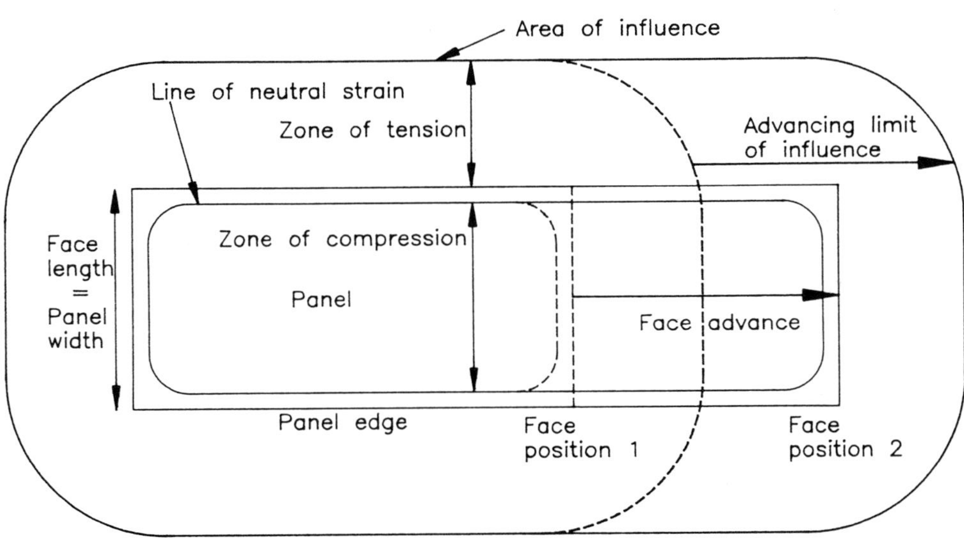

(b) Plan showing advancing longwall face

Figure 3.3: Longwall workings (after Waltham, 1989[250])

Tilt can have a significant effect on canals, land drains, pipelines and sewers causing blockages and flooding. Bridges and long buildings may also be affected;

- **Ground strain:** This is developed as tension over the crest of the subsidence wave and compression over the trough, with horizontal displacement reaching a maximum in the wave centre. Ultimate strain, which is the sum of tension and compression, is typically several mm/m, and causes the majority of building damage, with long terraces of houses often being the worst affected.

Provision is generally made for repairing any subsidence damage resulting from new mining operations. Subsidence effects due to longwall mining can be predicted accurately by subsidence engineers.[171, 257]

3.2.6 Shafts

Shafts provide vertical access to the workings. Shaft size and depth varies, but modern shafts may be as much as 8m diameter and greater than 1000m deep. Shafts are often circular, but can also be oval, square or rectangular. Linings of timber, brick, stone or concrete are generally provided except in strong rock were a lining may not have been considered to be necessary. Most disused shafts have been wholly or partially filled and may be concealed at the surface.

Subsidence can occur due to the sudden collapse of the shaft and a crater is generally formed at the surface as material slips into the shaft (see Photograph 3.3 and Figure 3.4). The size of the crater depends on the diameter of the shaft, the depth to bedrock and the angle of repose of the surrounding material.

Shaft collapses will occur due to:

- deterioration and collapse of the shaft lining;
- settlement or degradation of the fill material within the shaft;
- the collapse of staging within a partially-filled shaft.

The shaft collapse may occur naturally or can be triggered by factors such as:

- marked changes in groundwater levels due to extremes of weather, or due to soakaways or burst water mains;
- dumping of materials near or above shafts;
- vibration from plant or traffic;
- blasting and seismic shocks;
- mining subsidence.

In some cases collapse may occur slowly leading to a gradual loss of support and worsening damage to nearby structures. However, there is a risk of sudden collapse occurring in some instances which may have a catastrophic effect on structures, roads and utility services and will thus present a serious safety hazard.

Photograph 3.3: Collapsed mineshaft (source: Richards, Moorehead and Laing Ltd)

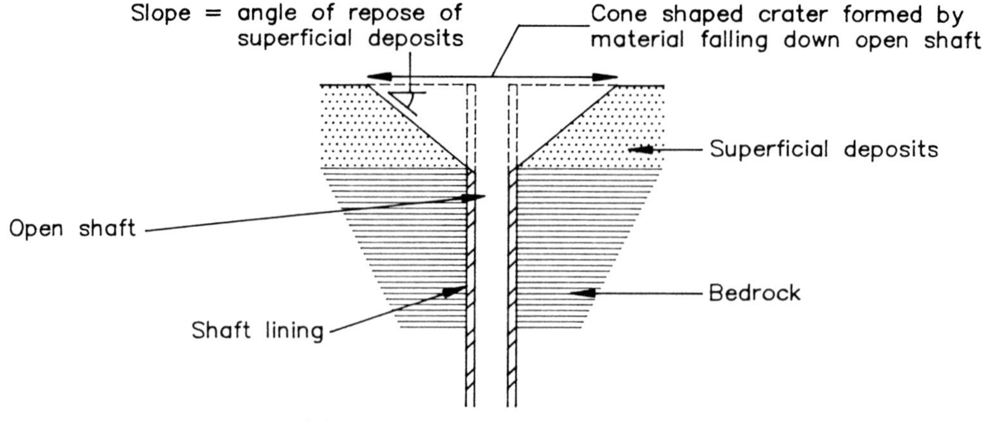

(a) Collapse of shaft lining

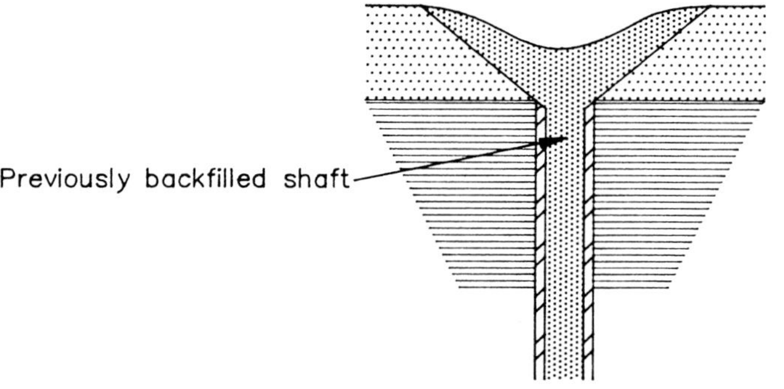

(b) Settlement of shaft fill material

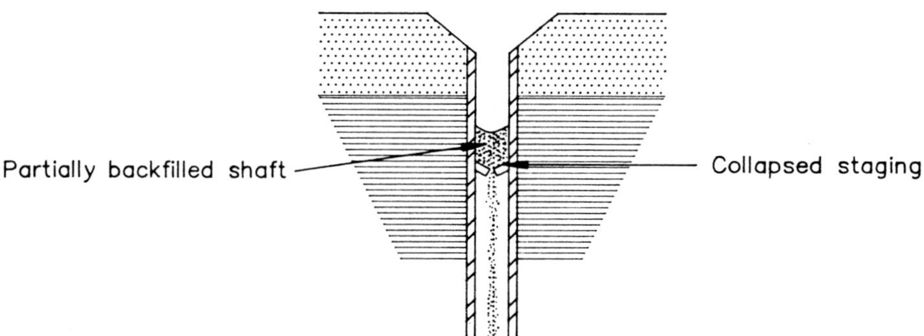

(c) Collapse of staging within a partially filled shaft

Figure 3.4: Shaft collapses

3.2.7 Adits

Adits provide horizontal or inclined access to the underground workings. Adit size varies, but a brick, stone, concrete or timber portal is normally provided and the adit is lined until competent rock is reached. The mouth of a disused adit may be concealed if the portal and roof near the entrance has collapsed.

Subsidence may occur due to the sudden collapse of the ground around the entrance to the adit. Collapse will take place due to:

- deterioration and collapse of the adit portal or tunnel lining;
- collapse of the fill material and stopping placed within the adit.

An adit collapse may be triggered by factors similar to those which could cause a shaft collapse (see Section 3.2.6).

Subsidence caused by the collapse of an adit is generally fairly localised and is usually less serious than a shaft collapse. However, subsidence damage may still be caused to nearby structures.

3.2.8 Opencast mining

Opencast mining involves the removal of the overburden to allow coal to be worked by surface mining methods. For coal workings the open pit is normally progressively filled as work proceeds, and the site is then restored. In some instances opencast coal workings have taken in and reclaimed former derelict areas from older underground workings (see Section 17.2.2).

Subsidence inevitably occurs on opencast sites due to settlement of the often considerable depth of backfilled material which is usually placed without being compacted. Such settlement of restored opencast sites is likely to continue for several years.[53]

3.2.9 Subsidence damage to structures

The effects of subsidence damage to structures and civil engineering works is discussed in the following paragraphs.

Buildings

The response of building structures to mining subsidence can be assessed, but is difficult to predict with accuracy. The likely behaviour of a structure will depend on several factors. The most important of these are:

- the size, shape and orientation of the structure in relation to the underground mine workings;
- the foundation design and type of superstructure;
- the methods of construction and quality of materials used;
- the age of the building and standard of maintenance and repair.

The first signs of subsidence damage may be the appearance of cracks, doors and windows becoming ill fitting, or floor tiles becoming loose and heaving. Damage may become progressively worse leading to severe cracking in some cases and possibly to eventual collapse. In the event of the collapse of a shaft or shallow workings, damage will occur more suddenly.

Roads

Subsidence damage to roads may occur as follows:

- distortion of horizontal and vertical alignment;
- fracturing and distortion of road foundations;
- undulations in the running surface;
- damage and displacement of kerbs, channels, flagging and fences;
- disruption of drainage and consequential flooding due to surface settlement;
- consequential damage *e.g.* water action from fractured mains.

The formation of local changes in gradients on high speed roads may result in a safety hazard. Disruption to drainage may cause problems from ponding in some cases.

Railways

Subsidence damage to the running track may occur as:

- distortion of horizontal and vertical alignment;
- disturbance to the track bed;
- undulations in the track bed;
- disruption of drainage with consequential flooding.

Restrictions may be imposed on traffic until remedial works are carried out.

Bridges

Bridge piers or supports can experience movement towards or away from each other which may affect the bridge bearings. Tilt or twist may have a serious effect on bridge decking. Bridge arches may also be affected and flattening of the arch may occur in some cases.

Canals and rivers

Canals and rivers affected by subsidence may require remedial works to:

- raise bank levels;
- raise weir levels;
- seal leaking banks;
- raise bridge levels;
- repair locks and culverts.

Drainage

Subsidence may result in a loss of gradient in drainage pipes with resulting blockages or lack of capacity. Cracked or broken pipes may also occur in some cases.

Utility services

Underground water, gas, electricity and telephone utility services may be disrupted as a result of subsidence damage. Overhead electricity lines and pylons may also be affected.

Colliery spoil heaps

Mining subsidence can have a destabilising effect on some tips (see Chapter 6).

3.3 Investigation of mine workings

3.3.1 Requirement for investigation

On development sites where there is a possibility of past mining having taken place, an archival search should first be carried out of available data. If the information collected suggests that mine workings are present, a mining investigation will be necessary. A mining investigation will also be necessary in cases where damage has occurred to structures and mining subsidence is the suspected cause. Figure 3.5 shows a flow diagram for the investigation of a site potentially affected by mine workings. Such an investigation should be planned by a specialist consultant. The following sections describe the various stages of the investigation and the methods which are employed.

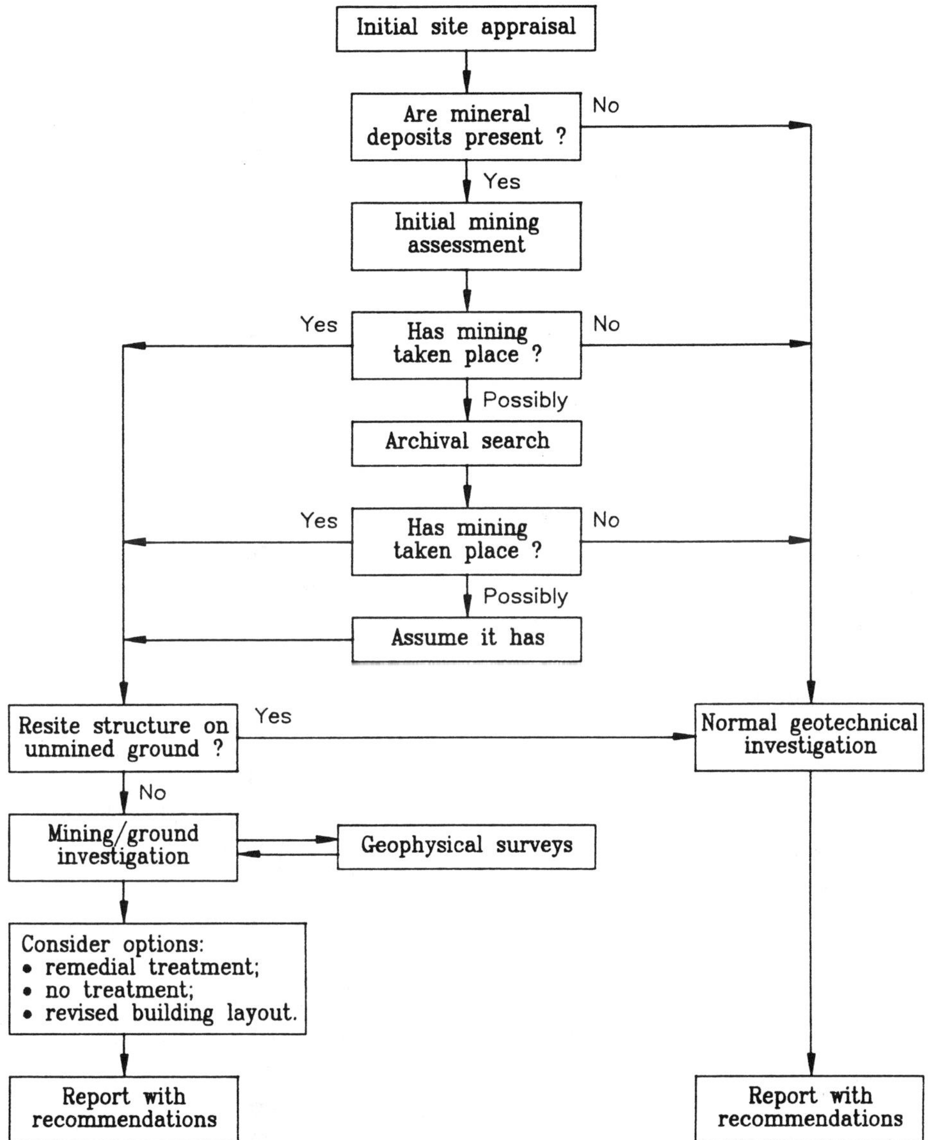

Figure 3.5: Flow diagram showing the general procedure for assessment and investigation of a site potentially affected by mine workings (from Healy and Head, 1984[114])

3.3.2 Archival search

Sources of information vary from site to site. The following are likely sources of information:

- mining company records;
- local authority records;
- geological survey records;
- old topographical maps;
- aerial photographs;
- local mining specialists;
- local inhabitants.

A lack of information on past mining activities does not necessarily mean that no mining has taken place. In areas with a long history of industry and development it is reasonable to assume that any accessible coal seams or mineral veins have been worked.

3.3.3 Walk-over survey

A visit should be made to the site to verify, as far as possible, the information collected in the archival search. Any visible evidence of mining should be noted. Such evidence may appear as spoil heaps, crown holes, shafts, surface workings, and remains of mine structures.

3.3.4 Desk study

From the available information it may be possible to assess the extent, type and date of any mining activity and to determine the presence or possible location of any shafts, and whether there is a history of subsidence on the site.

Plans should be produced to show the following information:

- the location and depth of mineral deposits, details of whether these deposits have been mined and the mining methods employed;
- the locations or approximate locations of known or suspected shafts;
- visible signs of mining activity such as spoil heaps, crown holes and bell pits. Details of any ground subsidence and settlement damage to buildings;
- zones of possible development constraint due to mining activity should be plotted *i.e.* parts of the site where development cannot be carried out without investigation and necessary remedial treatment.

3.3.5 Mining investigation

The object of the mining investigation will be to obtain all or some of the following information, depending on the individual circumstances:

- site geology, including details of ground strata;
- geotechnical properties for foundation design;
- presence or absence of shallow workings and the condition of any workings present;
- the location and condition of any shafts and adits;
- details of the groundwater regime.

3.3.6 Investigation techniques for mine workings

Direct investigation methods will normally need to be employed using boreholes, drillholes, trial pits or trenches, or a combination of these techniques. Indirect investigation methods may be useful to provide preliminary information and reduce the amount of direct investigation required. Investigation techniques are discussed in Section 2.5.

Indirect methods

A number of geophysical methods are available, but specialist knowledge is required to interpret survey results as anomalies and patterns can be difficult to discern.[157]

Geophysical methods which may be appropriate include the following (see also Box 2.7):

- electrical resistivity;
- magnetometry;
- electro-magnetic surveys;
- seismic surveys;
- ground probing radar.

Geochemical investigation by the sampling of soil gases can be used to indicate the presence of coal workings or a shaft.[213] However, methane may be present in the ground for other reasons *e.g.* as a result of the anaerobic degradation of waste materials, but it is possible to differentiate between methane from different sources by analysis of other gases. The simultaneous presence of carbon dioxide and volatile organics *e.g* volatile fatty acids (VFAs) is thus indicative of landfill gas rather than mine gas (see also Section 11.2.4).

Box 4.3 provides more details on methane associated with abandoned mine workings.

Direct Methods

Direct investigation methods are more expensive than indirect methods, but generally provide more reliable information. The methods usually employed are:

- **Trial pits and trenches**: Trial pits and trenches can be used to investigate shallow ground conditions, to determine the effects of subsidence, and to locate shafts, adits and other features.

- **Boreholes:** Drilling is the method most commonly used for investigating underground workings, and rotary-percussion drilling is the quickest and least expensive method using water flush or sometimes air flush (see Photograph 3.4). Monitoring of water returns provides information on rock type, state of fracture and whether voids are present. Rotary-core drilling will be appropriate where core samples are required. The use of close circuit television downhole cameras can be used to check the condition of underground voids. The depth and spacing of boreholes should be subject to advice by a specialist consultant and will vary depending on site conditions and the type of development proposed. A depth of 30m and a staggered grid spacing of 5m or less is appropriate for investigating shallow workings beneath a proposed building development. Where the strata dips steeply, inclined boreholes are likely to be required.

Photograph 3.4:
Investigation of shallow workings using rotary-percussion drilling (source: NKC Geotech Ltd)

3.4 Investigation of shafts and adits

3.4.1 Shafts

Investigation of shafts by trial pitting and trenching is the method most commonly used when the depth to rockhead is shallow *i.e.* up to 5m. Drilling will be required where the depth to rockhead is greater or where the shaft is located close to existing features which cannot be disturbed. The most efficient method is to drill in a spiral pattern working outwards from the assumed shaft location (Figure 3.6(a)). Drilling will be required at close centres of around 1.5m, or greater for larger shafts. Once the shaft has been located the following information should then be obtained by further investigation involving drilling and trial pitting within and adjacent to the shaft:

- shaft size and shape;
- depth to rockhead and details of superficial deposits;
- type and condition of fill material if present, and whether the shaft is fully or partially filled;
- in an open shaft, the locations of roadways and culverts;
- the type and condition of the shaft lining if present, and details of any fill material behind the lining;
- details of any existing plug, cap or staging;
- details of groundwater levels and seepages, and any existing drainage provisions;
- results of gas monitoring.

3.4.2 Adits

Investigation of adits is generally carried out by excavation. Drilling is appropriate in some cases *e.g.* where the adit has subsequently been covered with a significant depth of fill material, or where the adit is

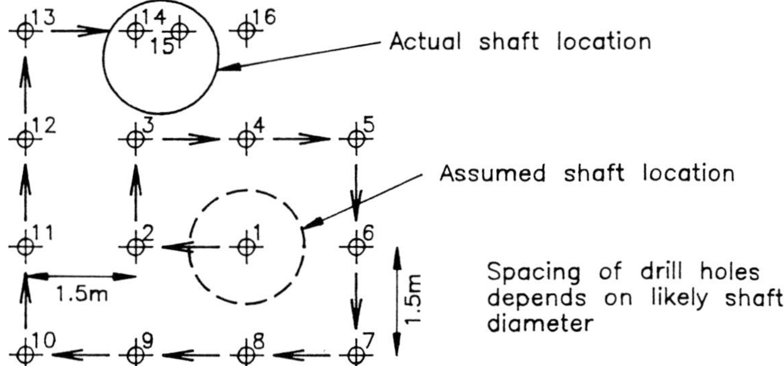

(a) Plan showing sequence of exploratory drill holes in a spiral pattern

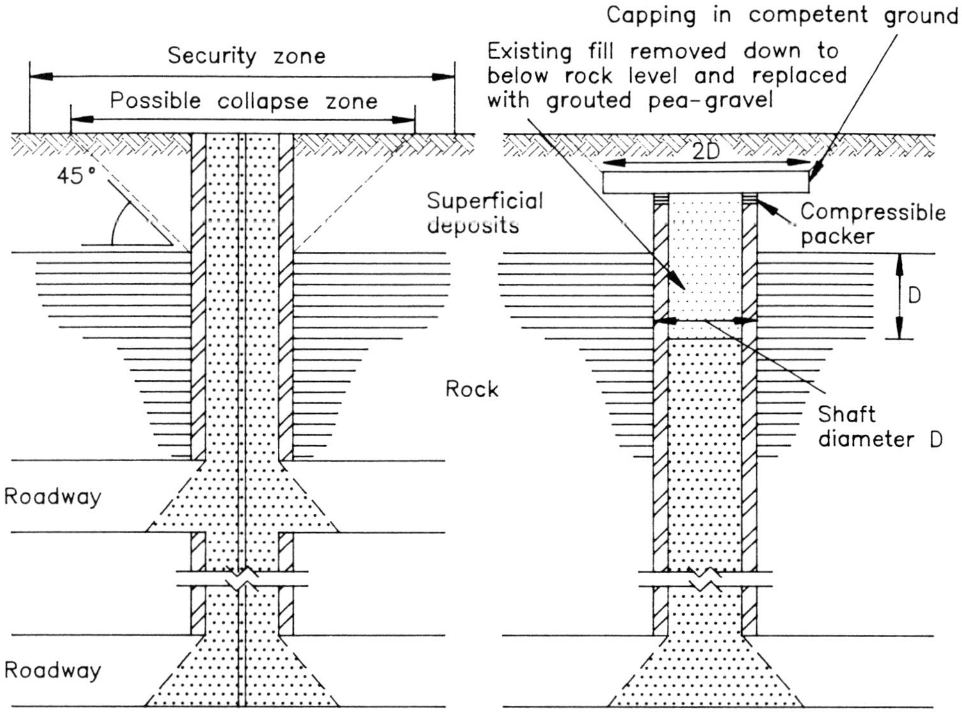

(b) Drilling and progressive grouting of existing shaft filling (assuming filling is suitable)

(c) Plugging with grouted pea-gravel in conjunction with capping in competent ground

Figure 3.6: Shaft location by drilling and shaft treatment by grouting

located close to existing features. Once the adit has been located the following information should be sought:

- adit size and shape;
- inclination of adit;
- type and condition of adit portal if still present;
- type and condition of tunnel lining where provided;
- depth to rockhead and details of superficial deposits;
- condition of strata overlying the adit;
- details of any existing stopping or filling;
- details of groundwater levels and seepages, and any existing drainage provisions;
- results of gas monitoring.

3.4.3 Safety precautions

Extreme care is required when investigating concealed shafts as these can be subject to sudden collapse when the fill or support staging is disturbed. Adits or shallow workings may also collapse during investigation, although this is generally less serious than the collapse of a shaft. A fenced security zone should be established around the investigation area. This area should extend beyond the potential collapse zone, which is assumed to be a cone projecting upwards with sides at an angle of 45° from the point where the walls of the shaft intersect rock head (Figure 3.6(b)).

All personnel involved in the investigation should be equipped with harnesses and safety lines secured to anchorage points at least 5m outside the collapse zone.[172] Plant and equipment should also be tethered to anchorages at least 10m outside the collapse zone. Suitable anchorages are steel posts concreted into boreholes.

There is a risk of explosion if mine gas is present, therefore continuous gas monitoring should be carried out during the investigation of shafts, adits and shallow workings. Sources of possible ignition should be prohibited within the security zone *e.g.* smoking, naked flames, spark ignited engines, or electrical apparatus which is not flameproof. Specialist guidance should be sought from a mining ventilation engineer regarding precautions required to deal with gas emissions during shaft treatment.

3.5 Treatment of shallow workings, shafts and adits

3.5.1 Options for treatment

Treatment methods for shallow workings, shafts and adits are well documented.[72, 114, 172, 250]

Construction in areas of shallow workings *i.e.* less than 30m depth, generally requires some form of treatment to consolidate the workings, followed by the use of appropriate foundations to support structures.

Treatment is generally carried out to fill open or partly filled underground cavities and prevent upwards migration of voids which may lead to subsidence damage. Fully collapsed workings are likely to require treatment in order to improve the bearing characteristics of the ground.

Depending on the site conditions, treatment is not normally necessary where there is cover of 20m of competent rock above the workings. A minimum rock thickness of 15m or 10 times the seam thickness, whichever is greater, is a rule sometimes applied in the UK. However in unstable conditions treatment of workings to depths in excess of 50m may be required.

The main treatment options are as follows:

- excavation through the shallow workings and backfilling with suitable material compacted in layers. This method is normally only suitable for shallow depths up to 5m below ground level;

- partial grouting either to improve the bearing characteristics or reduce the risk of void migration;
- full grouting of mine workings and possibly overlying strata to improve overall ground-bearing characteristics.

Where grouting is to be carried out the grout pressure should be carefully controlled in order to avoid the risk of ground heave occurring at the surface.

In some cases treatment may be required to deal with shallow workings discovered beneath existing structures.

3.5.2 Grouting of open workings

Grouting is generally carried out using a mixture of cement and pulverised fuel ash (PFA), or cement, PFA and sand in varying proportions depending on the circumstances. Pea gravel infilling may be used to limit grout travel within open workings. Grout is injected via grout tubes of approximately 25mm diameter inserted into drill holes of 50 to 75mm diameter drilled on a grid pattern (Figure 3.7). Grout pressures should not normally exceed 10kN/m^2 per metre of depth. Grouting continues until the specified pressure is reached. The grout tube is then partially withdrawn until the pressure drops and further grout can be injected. The grouting procedure continues until the grout reaches the top of the drillhole. Where the specified pressure cannot be reached, re-drilling after 24 hours and further grout injection may be necessary. Records should be kept of the grout take at each drill hole.

Grouting is normally carried out to a boundary located well outside the proposed building area. In order to limit the quantity of grout injected into workings, a perimeter wall of grout is provided around the edge of the treatment area using viscous grout injected into pea gravel. Perimeter drill holes may be 75 to 100mm diameter to accommodate the pea gravel,

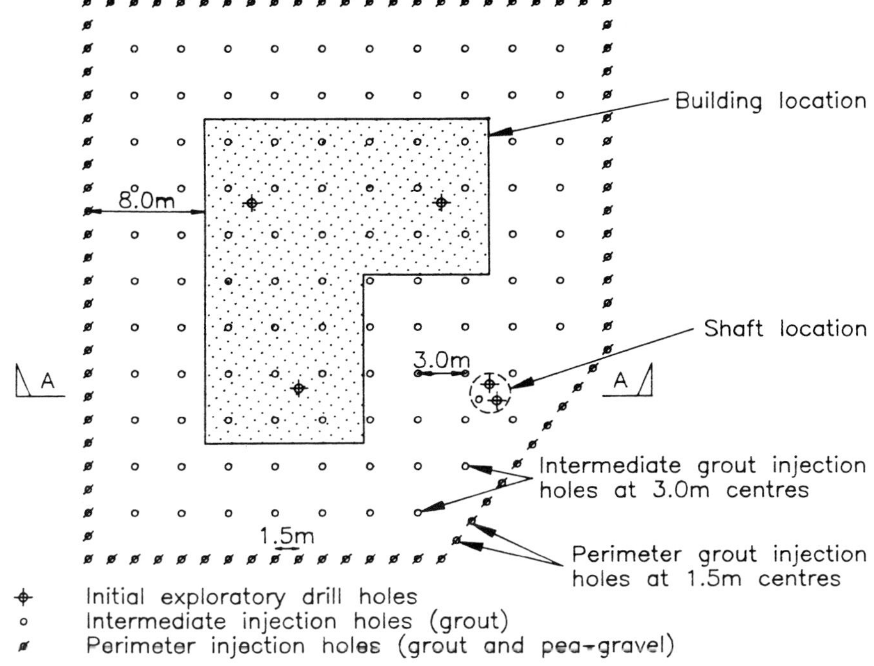

Plan showing grout hole layout

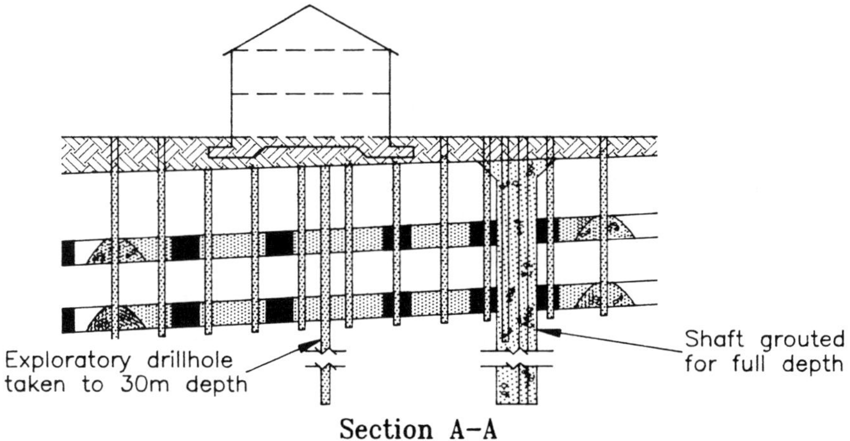

Section A–A

Spacing of grout injection holes and distance from building line to perimeter grout injection holes depends on the depth and condition of the workings.

Figure 3.7: Grouting of open shallow workings

and spaced at 1.5 to 3m centres. Typical drillhole centres for infill grouting are 3 to 6m, although closer spacing may be required in areas of high grout take.

3.5.3 Grouting of collapsed workings

Grouting of collapsed workings is carried out in a similar manner to that described for open workings, although a perimeter wall is not normally required. Treatment of collapsed workings is less straight forward than for open workings in that void locations are more difficult to predict. Grouting is carried out from primary holes at approximately 6m centres. If required, secondary infill holes may be drilled at 3m centres (Figure 3.8). If voids are discovered these may require individual treatment. Re-drilling and further grouting will be necessary in areas where the specified pressure cannot be reached or the grout take is excessive.

3.5.4 Treatment of abandoned shafts

The choice of shaft treatment method depends on a number of factors including:

- available funding;
- depth to rockhead;
- size of shaft;
- condition of existing lining;
- whether the shaft is open or filled;
- location and proximity to existing structures;
- proximity to urban areas;
- whether future access is required by authorised persons or wildlife *e.g.* protected species such as bats.

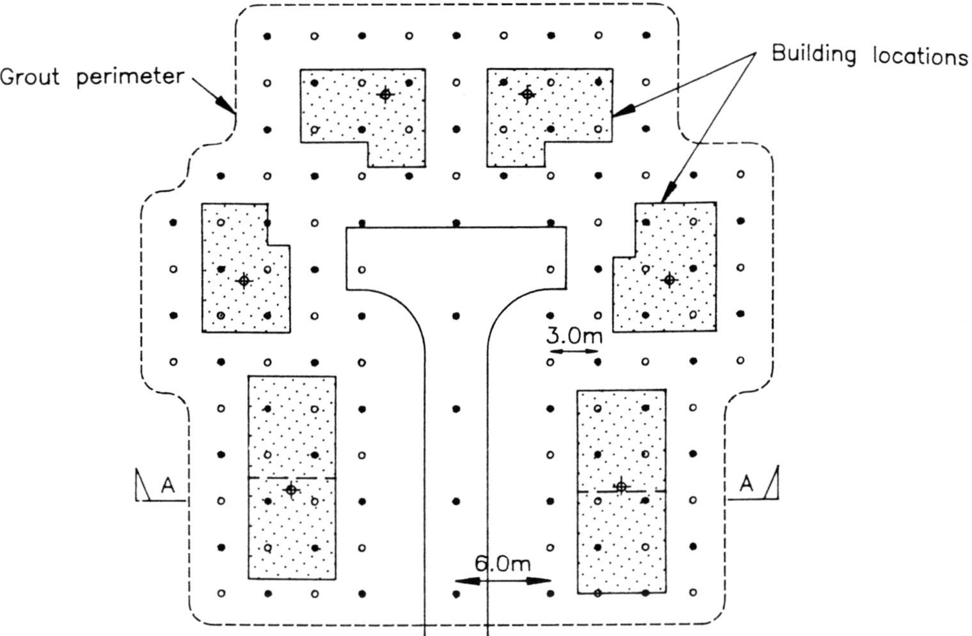

Plan showing grout hole layout

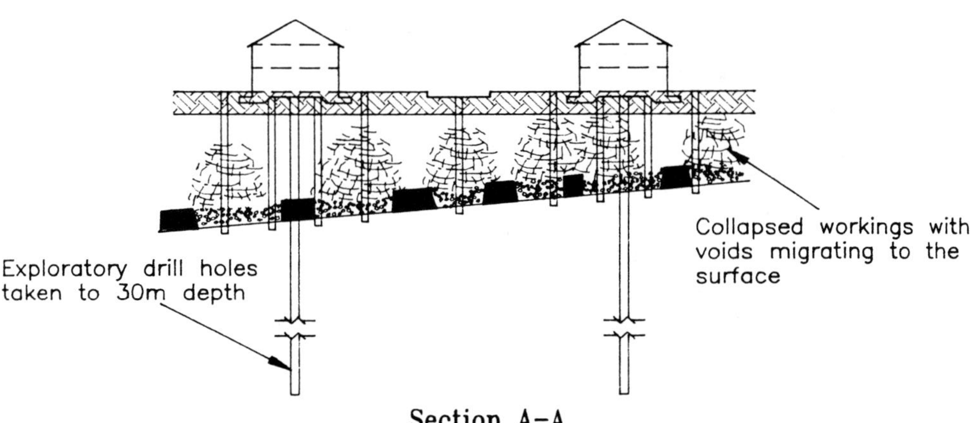

Section A–A

Figure 3.8: Treatment of collapsed shallow workings

Methods commonly used to ensure the security of shafts are as follows:

- capping using reinforced concrete;
- plugging with reinforced concrete;
- filling and plugging with grout;
- drilling and grouting;
- provision of prefabricated steel or concrete covers;
- filling in conjunction with security fencing and warning signs;
- security fencing and warning signs only.

Capping

Capping with reinforced concrete is the most common method of shaft treatment (see Photograph 3.5). The cap should ideally be located at or below rockhead (Figure 3.9(a)) and the width of the cap should normally be at least twice the shaft diameter. The cap thickness and reinforcement requirements will depend on the shaft diameter, the depth of overburden

Photograph 3.5: Capping of a mineshaft (source: Richards, Moorehead and Laing Ltd)

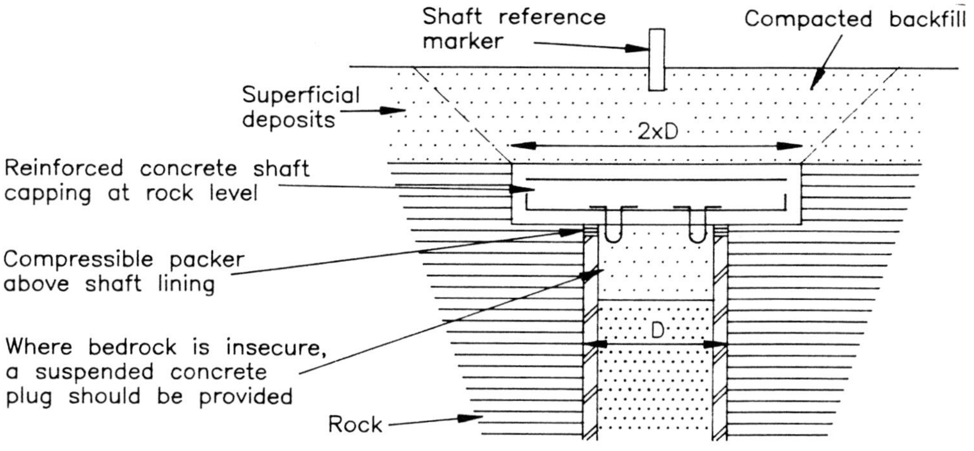

(a) Typical shaft capping detail

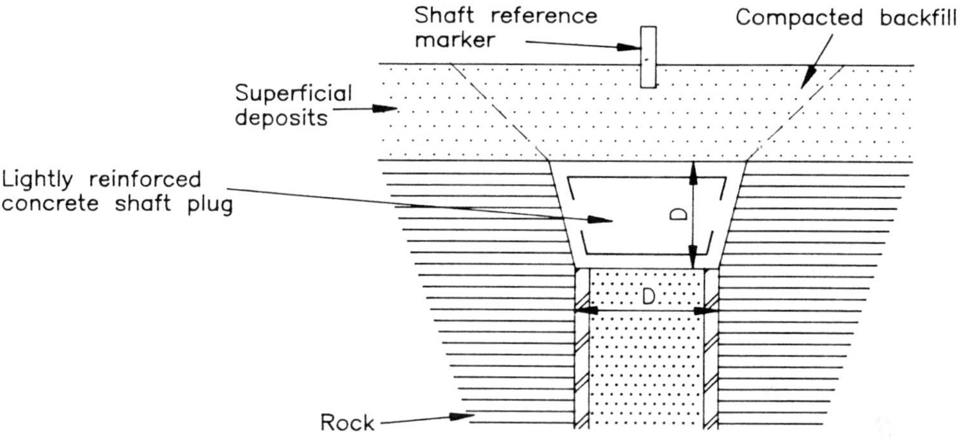

(b) Typical shaft plug detail

Notes:

1. Capping or plug to be designed to support overburden plus imposed loading of $33kN/m^2$ at cap level.
2. Concrete to be at least grade $30N/mm^2$ made using sulphate-resisting cement.
3. Where mine gas is present, surface of capping or plug to be provided with membrane or clay seal. Alternatively, gas venting with a flame arrestor is to be provided.

Figure 3.9: Shaft capping and shaft plugging (after UK National Coal Board, 1982[172])

and any surcharge loading. Where accumulation of gas may occur venting should also be provided. A provision for drainage may be necessary in some circumstances.

If personnel access is required, an access shaft of precast concrete manhole rings with a lockable cover may be provided down to cap level. In some cases a grilled access may be required *e.g.* for wildlife such as bats, or to maintain existing ventilation levels. Where the shaft capping is to be covered, a location marker such as a concrete post or stone plinth should normally be provided at the surface. When the shaft cannot be capped at bed rock due to excessive depth or limited access then it is acceptable to cap the shaft in competent ground at a higher level provided the existing shaft lining below cap level is in a stable condition. Compressible packing 50 to 75mm thick should be provided between the underside of the cap and the top of the shaft lining. In some cases it may be necessary to fill the shaft and grout the fill down to just below bedrock level (Figure 3.6(c)).

Plugging

Plugging using reinforced concrete may be appropriate for small shafts up to 3m in diameter in situations where access is limited (Figure 3.9(b)). The plug would normally be installed at rock head and the rock should be trimmed back to provide a stable keyed surface. The provision of gas venting or personnel access would be as described earlier for capping.

Filling and plugging with grout

When the depth to rockhead is excessive, or when access is limited, it may be necessary to fill the shaft with granular material and provide a plug by grouting the fill down to just below rockhead. For shafts which are already filled, unsuitable fill should to be removed down to rockhead and replaced with granular material prior to grouting.

Drilling and grouting

Drilling and grouting may be required in situations where buildings are to be located above a shaft (Figure 3.7), although building close to shafts, whether treated or untreated, is not normally recommended. Grouting may also be required for a shaft discovered to be present beneath or close to an existing structure. The existing shaft filling should be investigated by drilling prior to grouting operations.

Prefabricated steel or concrete covers

This method is appropriate for existing shafts which are to remain open and where bedrock is close to the surface, or where the existing lining is in a stable condition. Personnel access or grilling may be provided through the cover as necessary.

Filling, in conjunction with perimeter fencing and warning signs

Shaft filling and fencing is an appropriate form of treatment in open country away from populated areas. Filling of large shafts would normally be carried out using suitable granular material tipped into the shaft in a controlled manner by means of a conveyor. Free draining single size fill material should be used in the lower part of the shaft where existing mine drainage systems have to be maintained. Where shafts are already filled, investigations should be carried out by drilling to determine whether the shaft is fully or only partly filled. If voids are discovered within the shaft below a platform or obstruction it will be necessary to fill the void using an appropriate fill such as pea gravel.

The shaft filling is likely to be subject to settlement, and may require topping up at a later date. As the shaft may not be properly stabilised by filling alone, the provision of security fencing and warning signs is also recommended.

Security fencing and warning signs only

The provision of security fencing and warning signs is only acceptable as a short term safety measure, and should not be considered as a long-term solution. The fencing should be located outside the cone of failure which may be formed should the shaft collapse.

3.5.5 Treatment of adits

The choice of adit treatment method depends on:

- available funding;
- location and proximity to existing structures;
- condition of existing portal;
- condition of existing lining, if provided;
- whether the entrance to the adit has collapsed;
- depth to rockhead and the condition of strata, overlying the adit;
- requirements for drainage or gas venting;
- whether future access is required for authorised persons or wildlife *e.g.* protected species such as bats.

Methods normally used to treat adits are as follows:

- stopping with filling and grouting;
- stopping only;
- grilling;
- excavation and backfilling;
- drilling and grouting.

Stopping with filling and grouting

Adits are normally treated by stopping and filling to prevent subsequent collapse leading to subsidence at the surface. A stopping of brickwork, concrete or grouted fill should be provided which is capable of withstanding the pressure from infill material or grout. The stopping should be located at a suitable distance down the adit to ensure an adequate depth of rock cover of at least 10 times the tunnel height above the remaining length of untreated adit. The length of adit between the stopping and the entrance is then backfilled with suitable fill. The fill may be progressively grouted either from within the adit or from the surface. The entrance to the adit should be faced with a wall of brick or concrete. Provision should be made through the filling for drainage or gas venting as necessary.

Stopping only

Assuming the adit is considered to be stable, in some circumstances it may be appropriate to close the adit by the provision of a brick or stone wall at the entrance. A lockable door may be included within the wall if access by personnel is required.

Grilling

Where continued access is required to the workings for birds and mammals, or where existing ventilation needs to be maintained, a steel grille offers the best solution providing the roof of the adit is stable. A lockable gate may be included in the grille.

Excavation and backfilling

This method of treatment may be used for shallow lengths of adit where the depth below ground level is less than 5m.

Drilling and grouting

Drilling and grouting may be used to stabilise lengths of open or collapsed adits in situations where access is difficult due to the proximity of nearby structures.

3.6 Methods of minimising subsidence damage

3.6.1 New structures affected by contemporary longwall workings

Where mining subsidence from longwall workings is expected, special precautions should be taken in the design of the building foundations and superstructure.[126, 240, 257] Buildings should be designed, as far as possible, to be flexible. Shallow, flexible raft foundations with a smooth underside constructed on polyethylene placed over a 150mm thick layer of sand provides good protection against tension and compression strains at the ground surface (Figure 3.10(a)). The raft foundation should be constructed close to the ground surface to avoid problems due to the effects of side thrusts. An alternative method is to use reinforced strip or pad foundations laid on a layer of sand. A space should be provided at the ends of the foundation trenches which is then filled with compressible material. The construction of basements should be avoided. Large structures should be divided into independent units by means of construction joints passing through the structure and its foundations. For multistorey structures, foundations consisting of pad and beam construction, or cellular rafts are required. Where large movements are expected, provision should be made for jacking to allow the structure to be re-levelled. Buildings should not be constructed close to the surface locations of geological faults.

Special flexible building structures have been developed for use in mining areas *e.g.* the CLASP system used in the UK for many school and office buildings.[29] This system involves the use of a pin jointed steel frame with diagonal bracing between column locations which incorporates compressive springs capable of resisting wind loading but designed to distort when ground settlement occurs.

Drains should be provided with ample falls to ensure that they will continue to operate after any reductions in gradients. Pipes should be bedded in granular material, and double joints should be provided to

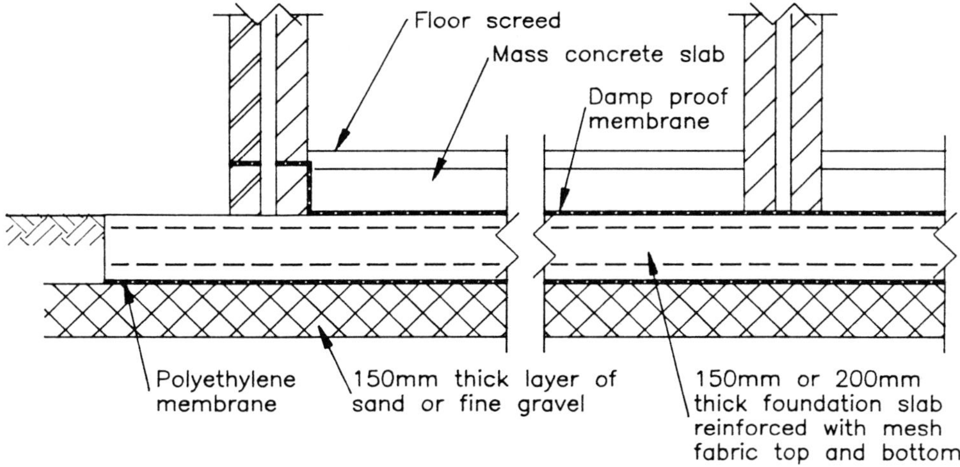

(a) Section showing typical flat raft foundation for housing subject to subsidence from longwall workings

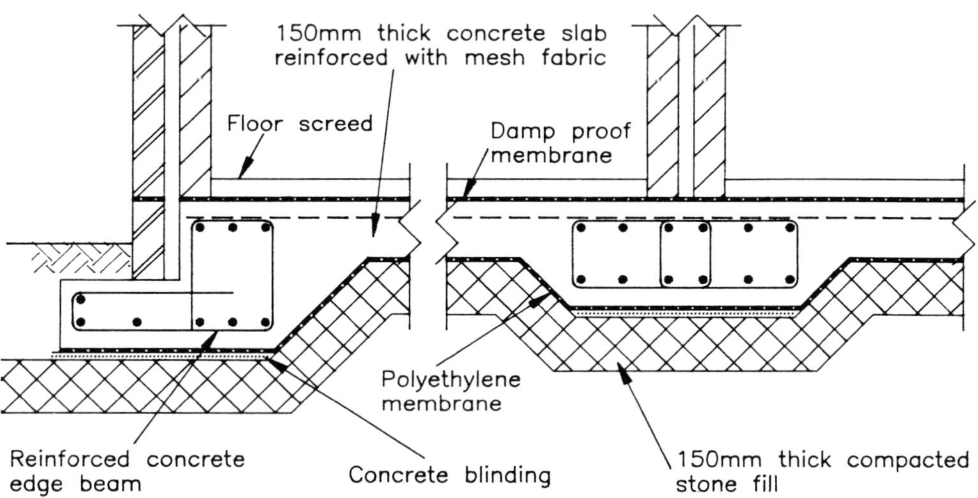

(b) Section showing typical semi-raft foundation for housing subject to differential settlement

Figure 3.10: Raft and semi-raft foundations (after Tomlinson *et al.* 1978[240], redrawn by the permission of the Controller of HMSO)

allow articulation to occur at manhole positions and entry to structures. Manholes, should be constructed of precast concrete rather than brick for greater strength.

Roads should be of flexible macadam construction rather than more rigid concrete construction. Bridges should be designed to articulate with movement joints provided within the structure. Jacking points may be required to allow re-levelling to be carried out.

Service installation should be designed to accommodate ground movements.

3.6.2 Existing structures affected by contemporary longwall workings

Subsidence damage to existing structures can be limited by the use of appropriate preventative measures. Trenching, for instance, may be provided around buildings to reduce damage from horizontal strains, and damage to long buildings can be reduced by splitting them into smaller units with the incorporation of movement joints.

Buildings of special importance can be protected by the provision of special foundations incorporating jacking facilities. Temporary internal propping may also be provided in some cases. Surface subsidence effects can be reduced by using partial extraction methods, with support pillars being left in place beneath sensitive sites. The use of backstowing to reduce surface subsidence is sometimes carried out by backfilling the workings behind the longwall face.

Drains should be modified or relaid to provide adequate falls with flexible connections. Service installations may need to be modified, and frequent inspections should be carried out to check for damage.

3.6.3 New structures affected by abandoned mine workings

Different techniques are required to protect structures to be built above abandoned shallow workings. Following a mining investigation, treatment would normally be carried out to stabilise the workings beneath the proposed development and prevent the collapse of any nearby shafts and adits. Raft or semi-raft foundations would then be provided to minimise the effect on the superstructure of possible differential settlement. The foundations should be designed to span or cantilever over possible areas of reduced bearing capacity. A typical semi-raft foundation for use in low rise buildings is shown in Figure 3.10(b). The building type, the location on the site and the foundation solution adopted should be chosen to suit the mining conditions and the extent of stabilisation treatment carried out. Where possible, major structures should be located outside the area known to be affected by shallow workings. Alternatively the foundations to major structures should be taken below the level of the workings using deep excavation or piling. However, care should be taken when using piling to ensure that the piles are founded in competent strata below the level of any workings. Structures should if possible, be sited away from shaft locations even when the shaft may have been treated.

3.6.4 Existing structures affected by abandoned mine workings

Where existing structures have been affected by the collapse of abandoned workings, investigation and treatment will be required to stabilise the workings and deal with any shafts and adits. Underpinning of the existing foundations may be required followed by remedial works to the superstructure. Investigation and treatment of the workings may be constrained by the need to avoid further disruption to the existing structure.

3.6.5 Highway construction over abandoned mine workings

Where investigation of shallow workings indicates that there is a risk of instability, it is normal practice to stabilise the workings by grouting. An alternative approach which has been adopted is to use a geotextile reinforcement incorporated into the road formation which is designed to span over any voids caused by crown holes in order to keep the deformation of the road surface within acceptable limits.

4 DEMOLITION AND SITE CLEARANCE

Chapter contents

4 DEMOLITION AND SITE CLEARANCE

4.1 Introduction

Steelworks are generally larger and more complex than collieries because steel making is a multi-stage, multi-material process where scale of production has brought economic benefits. The size of a deep coal mine is determined largely by the practicalities and economics of the underground conditions, as well as ownership of the minerals. Where these conditions are particularly difficult and complex, individual collieries occupy a relatively small area of land compared with steelworks and are usually grouped in clusters the pattern of which is determined by geology and topography.

Steel making is a complex process which has had a long history of development, and as a result many steelworks have experienced long periods of stability interspersed with periods of rebuilding and development as new processes and techniques have been taken up. Sometimes several generations of steelworks may be present side by side on the same site.

Some of the earliest ironworks and steelworks were located close to their raw materials and can be found established on coal fields and close to both iron ore mines and limestone quarries. In these circumstances topography and other physical constraints have limited the scale of the development or have been overcome only at great expense. More recent and modern steelworks of large size, being sensitive to the cost of importing raw materials and the export of finished products, tend to be located in large areas of flat land well served by road, rail and waterborne transport systems (see Photograph 4.1).

The mining of coal has changed and developed over the 200 years in which coal has been extensively mined in Europe, but abandoned collieries are unlikely to exhibit complex and multiple stages of development in size or working methods at the surface.

Photograph 4.1: A steelworks site in Asturias, Spain (source: Richards, Moorehead and Laing Ltd)

Steel making has always involved large numbers of people and therefore old and new 'steel towns' are large centres of population. By contrast the combination of smaller size and larger number of coal mines found on the coalfields has resulted in mining communities being more 'village-like' in scale producing very different patterns of development and communications.

Although the two industries are closely linked, the marked differences in character between them mean that the issues involved in the demolition and site clearance of a steelworks are different from those concerning collieries (see Box 4.1).

In the past, short-term low cost strategies were often thought to be the best approach to the removal of the old to make way for new industries. It was common for demolition to be carried out during the decommissioning of the site. Machinery may have been sold off intact,

Photograph 4.2: View of an abandoned colliery site in South Wales, UK (source: Richards, Moorehead and Laing Ltd)

but often the demolition work was carried out for the scrap value of the buildings and machinery. Little or no regard was paid to the efficient recycling of materials or the adequate treatment of contaminated ground, or safe disposal of wastes. Generally few records were kept of the work being carried out. Sites were levelled with underground voids being backfilled in an uncontrolled manner. Below-ground features such as old foundations, tanks, flues and culverts frequently received little or no treatment. These demolition methods have subsequently given rise to problems where unforseen ground conditions were encountered during site redevelopment (see Photograph 4.2).

Wherever possible an integrated and disciplined approach should be adopted and the demolition and site clearance works should be carried out as part of a reclamation scheme specifically designed to suit the proposed end-use of the site.

Box 4.1: Steelworks and colliery site characteristics

Site characteristics	Steelworks	Collieries
Processes	Many and varied	Few
Raw materials and products	Many	Few
Sites	Complex	Simple
Wastes	Several in large quantities, several in smaller quantities	One at some sites, many where coal carbonisation has been carried out
Contaminants	Many and varied	Few at some sites, many where coal carbonisation undertaken
Degree of contamination	Severe	Slight, other than where coal carbonisation undertaken
Buildings structures and machinery	Many	Some
Buried structures	Many	Some
Shafts, adits and shallow workings	May be present	Will be present
Industrial archaeological features	May be present	May be present
Recovery and re-use of materials	Likely to be possible	Likely to be possible

4.2 Colliery site features

Features normally associated with colliery sites are listed in Box 4.2.

Buried foundations and other underground structures will be present, and may need to be removed in the parts of the site which are to be redeveloped.

Shafts, adits and shallow workings will be present, some of which may already have been stabilised. Others may still require investigation and treatment.

Spoil heaps and lagoons may be subject to instability problems and can require special treatment (see Section 6.6). Spoil heaps and fill material with a high coal content are susceptible to accidental or spontaneous ignition, and special precautions will be required if the waste on site is shown to have high coal content (see Chapter 7).

Contamination may also be present in certain parts of the site. This is particularly likely in areas where coal carbonisation and by-products storage or disposal was undertaken (see Section 10.1). Fuel and oil spillages are likely to have occurred in some areas *e.g.* around storage tanks or in railway sidings, and asbestos is likely to be present in insulation or construction materials, or as a result of the on-site disposal of waste materials containing asbestos. Asbestos is invariably found in steam-raising plant areas including pithead baths.

In addition to the health and safety aspects of abandoned colliery and coal processing sites, there may also be chemical contamination features that affect underground structural developments at a site. High sulphate levels for example, will be present in some cases and this will give rise to sulphate attack of concrete (see Section 10.5.2).

Deep mines were invariably required to be drained of groundwater by pumping, and pumping usually ceased at the time of abandonment. As

Box 4.2: Features associated with colliery sites

Structural features at colliery sites include:

Headframes and winding gear, ventilating fans and drifts, coal processing plant including crushing, screening and washing plant and associated conveyors. Coking plant with associated by-products areas, and also brickworks, which were sometimes present at the colliery site itself or on nearby land.

Buildings will include offices, laboratories, workshops, boiler houses, motor rooms and buildings housing processing plant, ventilating plant and pumping arrangements.

Underground structures will include buried foundations, basements, retaining walls, tanks, culverts, fan drifts, drains, buried pipework and underground services.

Shafts and adits will be present together with shallow underground workings, and in some cases opencast workings.

Other site features may include coal blending and stockpiling areas and railway sidings. Extensive spoil heaps may be present together with open and buried lagoons. Access roads are likely to be still in existence and there may also be railway lines, embankments, cuttings, canals and wharves.

the water level rises in a closed mine system it compresses mine gases which accumulate in the underground workings, and these gases may find their way to the surface (see Box 4.3). Gas and water migration is a major problem on some sites.

Some colliery sites will contain historic features worthy of recording or retention. Proper consideration should be given to this even before the colliery becomes disused (see Section 4.7).

Box 4.3: Emission of methane from abandoned workings

In active mines methane is channelled from within the mine to the surface where it is utilised at the colliery site for heating or power generation, or it may be piped away for industrial use.

Methane production varies depending on differences in geology, methods of mining and coal seam properties. Methane is emitted from the coal face, from the coal as it is worked, and also from the surrounding strata which have been disturbed.

The rate of gas production is linked to the rate of coal extraction, but methane will continue to be emitted from the workings after coal production ceases, and will find its way to the surface via mine entrances or via fractured strata above the workings. The potential for emission of residual methane from old workings increases in periods of low barometric pressure, or as a result of flooding of the workings, or subsidence of the overlying strata.

Therefore at closed or abandoned mines there is the potential for gas migration, with the associated risk of explosion if a build up of gas in an enclosed space is accidentally ignited.

At mines which have a history of methane production measures should be taken during abandonment to ensure that gas is safely vented to the surface. Where the potential for gas production is unknown, gas monitoring should be carried out during the site assessment in order to establish whether any remedial measures may be necessary.

Remedial measures can include the following:

- provision of gas vents to shafts and adits;
- provision of gas diffusion trenches or boreholes adjacent to buildings considered to be at risk;
- the grouting of underground voids and fractured strata to prevent or reduce gas migration;
- modification to existing buildings such as the replacement of suspended ground floor slabs with solid slabs, and the venting of enclosed spaces within buildings;
- provision of gas-proof membranes beneath new buildings;
- provision of passive or active gas venting systems within new buildings;
- sealing of service ducts and drainage entry points to prevent the ingress of gas;
- for high risk situations gas detection and alarm systems should be provided within buildings.

Following the implementation of remedial measures, ongoing monitoring should be carried out to check that these measures have been effective.

4.3 Steelworks sites features

Features normally associated with steelworks sites are listed in Box 4.4. Buried foundations and other underground structures will be present, and may be complex, extensive and massive. A long-standing or disused site may have been occupied by earlier phases of development for which the foundations still remain.

Iron or coal workings can be present as either former surface or underground workings which could give rise to subsidence. These workings will require investigation and treatment (see Chapter 3). Contaminated ground will exist in discrete parts of the site *e.g.* coking and by-products areas, waste tips, rubble from flues and chimneys, and

Box 4.4: Features associated with steelworks sites

Structural features at steelworks sites include:

Blast furnaces, open hearth furnaces, electric arc furnaces, ore preparation and handling plant and associated conveyors, coking and by-products plant, gas holders, casting machines, rolling mills, finishing mills, pickling and plating plants. Power generating plants, chimneys and cooling towers may also be present.

Buildings will include: offices, laboratories, workshops, boiler houses, motor rooms and buildings housing furnaces and processing plant. Many of these buildings are very large, and may contain overhead cranes and lifting gear.

Underground structures will include massive foundations some of which may be piled. Other underground features may include cellars, pits, tanks, flues, culverts, drains, buried pipework and underground services.

Iron ore or coal workings may be present on or near the steelworks site.

Other site features can include steel stockpiling areas, scrap reception and handling areas, coal stockpiling areas, slag processing plant, and slag heaps. Open lagoons and buried lagoons can be present and access roads are likely to remain, along with railway lines, railway sidings, canals and wharves.

old slurry lagoons. Asbestos may also be present in considerable quantities from insulation or construction materials or as a result of the on-site disposal of waste materials containing asbestos, and must be identified at an early stage so that precautions can be taken to avoid dust inhalation and resulting adverse health effects during and after demolition and site clearance.[71]

Some steelworks slag may be unsuitable for use as fill material for construction purposes, as under certain circumstances some slags can be subject to expansion (see Section 9.2). Where high sulphate levels are present this will lead to sulphate attack of concrete unless new concrete structures in contact with the ground utilise sulphate resisting cements or protective coatings (see Section 10.5.2).

Waste heaps at steelworks may be subject to underground fires which could lead to subsidence damage and health and safety hazards (see Section 7.3). Where underground fires have occurred on steelworks waste heaps this is generally due to the presence of colliery waste which may have been ignited by hot material from the furnaces.

Slag heaps can contain fused material or may have a high scrap iron content, both of which could cause problems during earthmoving activities.

Features of historical interest may be present on the site, as discussed in Section 4.7.

4.4 Site investigations prior to demolition and site clearance

At operating sites or where production has only recently ceased, information on previous site usage and contamination status should normally be available from site staff, or from records. Pre-closure site audits are an extremely useful aid to the identification of structures, process areas, waste disposal areas, and potentially contaminated materials that exist on site (see Box 4.5). It is far easier to obtain useful information about such sites whilst they are still operational, and a well-targeted environmental audit of this sort can provide an excellent foundation to a future development path. Such audits should also highlight historic equipment, buildings or processes.

Where materials are to be reused or recycled during demolition and reclamation works (see Section 4.6) a preliminary site audit will identify at an early stage those materials that are uncontaminated, and those that may require separation and/or treatment due to contamination.[140]

For sites which have been abandoned for some time, or which have inadequate or non existent records, in-depth investigations will be required to provide detailed information regarding the previous history of

Box 4.5: Factors to be included in a pre-closure site audit

Environmental audits of operational industrial facilities are a useful method of accounting for the movement of materials on and off site, and are often used as a management tool to aid the setting of environmental objectives and to minimise waste production and risk at a given site.

For a site identified for closure, an audit, involving existing documentation and members of staff, can provide useful information to aid site redevelopment. A pre-closure audit should address the following:

- location and quantities of waste deposited on site;
- records of accidents, particularly those concerning spillages or leakages of chemicals;
- location of process plant and storage areas;
- environmental data on, for example, groundwater quality or atmospheric emissions;
- locations of features of archaeological importance;
- structures of potential use during subsequent reclamation works e.g. buildings for materials storage, hardstanding, tanks for liquid storage and/or treatment;
- recycling potential of structures and materials on site.

the site and the likely problems to be overcome during the demolition and site clearance works (see Box 4.6).

At sites where some demolition and site clearance works have already been undertaken, the following information should be obtained:

- previous site history;
- scope of previous demolition works;
- method of dealing with toxic and contaminated materials;
- treatment of underground basements, tanks, culverts, drains, flues;
- treatment of buried foundations;
- compaction methods used in placing fill material;
- treatment of any shallow workings shafts or adits;
- location and historic value (if any) of surviving buildings.

Where no records exist, investigations will be required on site in order to determine all or some of the above information.

4.5 Demolition works

The demolition works should be carried out by a specialist demolition contractor with previous experience of the features normally encountered on coal and steel sites. Problematic features include: plant and equipment, large steel structures, massive foundations, chimneys and cooling towers. The contractor should also be familiar with the procedures to be adopted when handling toxic or contaminated materials.

Where necessary, works should be carried out to decontaminate plant and buildings prior to demolition, in order to minimise the spread of contamination during actual demolition works. Steps should in any case be taken during demolition works to minimise contamination of otherwise uncontaminated areas.

Any areas to be protected should be fenced-off at any early stage. Demolition works should be carefully planned prior to actual works commencing on site. Following an investigation of available records and site conditions as outlined in Box 4.6, a detailed method statement for the demolition procedure should be prepared including calculations, drawings and the proposed sequence of operations. Provision should be made for avoidance by the contractor of any areas which have been fenced off for protection.

Box 4.6: Detailed information to be obtained during planning of demolition and site clearance works:[258]

- historical uses of different parts of the site and subsequent changes of use;
- locations of mineshafts, adits and shallow workings;
- uses of pipework and process plant;
- assessment of hazardous materials which may be present;
- available records and drawings;
- survey of structural conditions;
- survey of cladding materials and materials used in construction e.g. asbestos;
- location and condition of public utility services and buried site services;
- details of buried foundations;
- details of underground basements, tanks, culverts, drains, flues;
- likely effects of demolition works on any adjacent properties and structures;
- evaluation of any restrictions on any adjacent properties and structures;
- details of any pre-stressed concrete units or post-tensioned concrete beams;
- interpretation of method and sequence of original construction;
- investigation of existing structure by exploratory probing, coring, drilling, cutting and testing;
- examination of records of any archaeological/conservation designation or listing.

An assessment should be made of any operations involving materials which may be hazardous to health, and of the possible exposure risk to site operatives, for example:[70]

- removal of asbestos;
- work with lead, including flame cutting of painted steel structures;
- other metals such as zinc which give off fumes if flame cut;
- synthetic mineral fibres;
- chemical residues from previous site usage;
- contaminated soils resulting from the spillage and leakage of oils, solvents and other chemicals.

Before demolition commences the contractor should ensure that all power is cut off, water and gas mains disconnected and protection given to existing utility services, as necessary. The demolition methods employed will depend on individual site circumstances but will range from traditional impact methods to the controlled use of explosives. Typical methods used are as follows:[260, 227]

- crane and ball;
- bulldozers;
- hydraulic excavators with grabs, buckets, breakers, magnets, shears, pulverisers and poles;
- thermal cutting;
- diamond sawing/drilling;
- concrete crushing/bursting;
- thermic lance;
- explosives.

The demolition contractor will be expected to comply with current EC Directives or equivalent regional legislation with regard to working practices:[174]

- health and safety requirements;
- control of hazardous substances;
- use of equipment, tools and machines;
- use of personal protection equipment;
- construction sites regulations;
- protection of workers exposed to carcinogens;
- safety signs regulations;
- any other relevant regulations.

The requirement in European Community Council Directives for Risk Assessments to be carried out means that clients and their professional advisers are also involved in the demolition process through their checking of method statements and safety plans. A prior assessment of the risk incurred during demolition and site clearance operations will not only help to reduce risk *e.g.* to workers' health and safety, of pollution and liability, but will also aid the avoidance of unforseen wastage, the loss of valuable materials and the cross contamination of materials.

Where off-site disposal of waste materials is to be carried out, these materials should be analysed and recorded and disposed of safely in accordance with legal requirements (see Section 11.2.5 regarding off-site disposal).

Arrangements should be made for supervision by qualified archaeologists/industrial archaeologists during demolition works so that any features of value which are exposed can be recorded, especially if the site is known to contain features of historic interest.

4.6 Recovery and recycling of materials

The cost of the demolition contract may be all or partly offset by the value of waste materials recovered during the demolition works.

In some cases equipment and processing plant may be sold off intact for re-use elsewhere. Specialist engineers are normally called in to dismantle the plant for transport and re-erection at the new site. In other cases the majority of equipment from a steelworks undergoing closure has been sold and shipped to other countries outside Europe, where it is reassembled to commence a new productive life.

Most of the machinery from disused coal and steel sites is unsuitable for re-use and will therefore be demolished and broken up for its scrap value. Specialist demolition techniques are often required for this process because of the large size of many of the items which are to be broken up.

Some buildings such as steel framed structures may be suitable for dismantling and re-erection elsewhere. Where buildings are to be demolished, building components such as steel beams, columns and trusses may be sold off for re-use, or broken up for scrap. Other building materials such as timber, bricks, stone, tiles, slates and steel cladding may be reclaimed for re-use elsewhere. Concrete and building rubble which is uncontaminated may be crushed and graded for use as fill material. Building rubble containing significant quantities of timber will not be suitable for use as fill beneath building structures. Building rubble containing, for example, gypsum plaster, could be subject to expansion and heave and therefore may also be unsuitable for use beneath buildings. It is therefore very important to adequately test materials prior to re-use.

Steelworks slag may be suitable for crushing, grading and re-use as general fill material, or as aggregates for road construction (see Chapter 9). However, some steel slags may be subject to expansion and could be unsuitable for use beneath structures (see Section 9.2). This problem requires special attention since serious problems have been experienced when buildings have been adversely affected by expansive slags in the ground beneath. Laboratory tests have been developed which will identify expansive slags (see Section 9.2.4).

In some cases the recovery of iron and steel fragments from waste tips on steel sites may be viable using magnetic separation techniques.

Burnt colliery spoil is suitable for re-use as fill material. However high sulphate contents may sometimes be present which could lead to sulphate attack of concrete. Pyritic shales may also be subject to heave making them unsuitable for use beneath structures (see Section 5.6).

Coal recovery from colliery spoil heaps and lagoons should be always be considered appropriate when the coal content is high enough to make the work commercially viable (see Chapter 8).

Both steelworks and collieries may contain shallow reserves of coal suitable for opencast working, which can also provide some of the finance for site clearance and reclamation *e.g.* Section 17.2.2.

4.7 Industrial archaeology

An understanding of industrial archaeology is important not only because of the historic and architectural importance of the subject but also for practical reasons such as the understanding of the development of construction techniques and the physical constraints on old structures in terms of stability and general safety (see Box 4.7).

Any audit of a coal or steel site either in preparation of abandonment or after closure should consider the history of the site, location of mineshafts, ancient structures and interesting features in its development (see Section 2.7).

Box 4.7: Industrial heritage

The first reaction among local people on closure of a coal mine or steel works is usually to clear it all away and make a fresh start. If this is not done, it is common to find that attitudes change, surprisingly quickly, to a preference for conservation of the industrial heritage. This is part of an underlying trend, which has been gaining momentum in some countries for several decades, towards greater awareness of the significance of old industrial sites and processes.

The industrial archaeological interest of an old industrial site involves individual buildings and plant, of course, but it is increasingly recognised that the overall pattern is at least as important. Preservation of a single building in isolation, can be fairly meaningless. The significance of a whole site to the development, way of life and culture of a community can, by contrast, be extremely significant. As part of the preparation for reclamation of the Glengarnock iron and steelworks in Scotland, sociological studies were carried out into the history of the steelworkers, as well as an industrial archaeological investigation which was carried out on site.[47]

The case study on Duisburg-Nord Country Park (see Section 17.8.2) is an example of a site where steelworks plant, which was in production in recent decades, is to be conserved for its historical and educational value.

Any feature identified as being of interest should be recorded photographically and by measured drawings, and if considered necessary an archaeological or structural examination should be undertaken by responsible persons. Copies of results should be lodged with the appropriate bodies.

Contracts for the demolition and regrading of a site should provide for:

- the archaeologist to carry out supervision during the work;
- the reporting and, if necessary, examination of any artifacts or features of historical interest;
- the adequate fencing off of any features on site which are to be retained.

It should be made clear to the contractor that the persons carrying out any archaeological or structural examination on site will cooperate in every way possible to avoid delaying work. Funds should be made available to ensure that any rescue dig or removal of features from site for preservation can be undertaken expediently. The contractor or developer may well improve his standing by providing assistance to the archaeologist and by holding site 'open days' for other interested individuals to visit the site of any feature exposed or under excavation.

Some coal mines, and many steel sites contain several generations of plant and processes alongside each other, which makes them exceedingly important from the historic point of view. The coal and steel industries are waste producers on a large scale so that many items of interest can be found below present ground levels or under the waste tips themselves (see Box 4.8).

4.8 Re-use of existing buildings and structures

Existing buildings should be retained and adapted for new use where appropriate. Buildings and features of architectural interest or of industrial archaeological importance should be retained wherever possible (see Section 4.7).

When existing buildings are to be retained, architects and structural engineers should first be consulted to carry out a feasibility study to confirm that the buildings are structurally sound and can be refurbished and adapted to new use at acceptable cost.

In some cases, where existing structures are to be demolished, subject to investigation it may be appropriate to re-use the existing foundations or floor slabs to support new structures.

Features such as existing roads, bridges, retaining walls, embankments, storage tanks, and drainage may also be suitable for incorporation into the new development.

Box 4.8: Features of industrial archaeological interest

The following are of particular interest on iron and steel sites:

- ore mining techniques, shafts, engine house chimneys, adits, tramways;
- ore preparation, kilns, roasting and calcining areas;
- smelting, bloomeries (furnace, hearth, forge), water power arrangements, charcoal sheds, pig beds, blast furnaces, engine houses, steam engines;
- forges, hammers, anvils, finery, reverberatory furnaces, slags;
- foundries, casting pits, air furnaces, kilns, moulds;
- boring mills, earthworks, water wheel pits;
- secondary trades - forges, rolling mills, grinding mills;
- steel works, Bessemer plant, cementation furnaces, crucible furnaces, finery, open-hearth plant, puddling furnaces, slags.

The following are of particular interest on colliery sites:

- mining techniques, adits, shafts, early surface mining sites;
- power, steam engine houses, boiler houses, chimneys, generator and compressor houses, winding equipment, haulage apparatus, water wheel pits, horse gin circles;
- pumping, engine houses, shafts, lodges, reservoirs;
- ventilation, fan houses, drifts, methane drainage plant;
- heapstead, including headframes, coal tipplers, screens;
- coal preparation plant, waste disposal arrangements;
- general-powder magazines, offices, stables, workshops, saw mills, pithead baths, on-site housing;
- transport, weighbridge, sidings, canal wharves, railways;
- secondary processes - briquetting plant, coke works, tar works;
- ancillary plant, brickworks, timber yard.

There are many examples where all colliery buildings have been re-used as an industrial estate (*e.g.* Madeley Wood Colliery, Shropshire, UK) or as a business estate (*e.g.* Ledston Luck Colliery, Yorkshire, UK). At Ledston the winding house is a listed building but is used as a store, the pit canteen is now a flourishing cafe, and the site now employs almost as many people as when it was a mine. Similar examples exist elsewhere in Europe *e.g.* Riesa steelworks in Germany (see Section 17.7), Maximilian colliery in Hamm-Werries (Germany) and Hottinguer Colliery in Epinac (France), now a paint factory.

Some mines can be converted to tourist centres, as has been done in a number of countries, with or without an underground feature. Individual mine buildings have also been converted to housing, and older colliery sites have been converted to a variety of uses such as farms and recreational centres *e.g.* Grange Colliery, Shropshire, UK.

5 COLLIERY SPOIL HEAP CHARACTERISTICS

Chapter contents

5 COLLIERY SPOIL HEAP CHARACTERISTICS

5.1 Introduction

5.1.1 Deep mining

The deep mining of coal generates large quantities of spoil materials. The volume of such spoil is dependent on site-specific factors of which the most important are:

- the amount of mine development work required;
- the geology of the area;
- the coal:spoil ratio;
- the mining methods employed.

Since the industrial revolution, which initiated the use of coal as the major energy source throughout the industrialised countries, spoil has been deposited in tips, heaps and in tailing ponds, and has been used to fill voids.

Dereliction resulting from deep coal mining has increased in significance as mining technology has become more mechanised. In the past shafts were relatively shallow and manual sorting of non-combustible material and 'small coal' was carried out below ground and the material with no saleable value was back stowed.

Waste was tipped from tramway tubs and railway wagons producing gently graded spoil heaps which consumed large areas of land as they fanned out from the original dumping ground.

Mechanisation of coal mining, involving the use of mining machines to extract the coal, led to more rapid and less labour intensive mining but resulted in an increase in the proportion of waste brought to the surface.

Modern coal cutting resulted in spoil being extracted along with the coal and transported to the surface by conveyor for separation. The increased scale of roadway construction also led to larger quantities of spoil being produced. The ratio of coal to spoil production has often been as low as 1:1, although for many pits the ratio is higher. The introduction of computer controlled coal cutting operations has resulted in the ratios being improved.

The introduction of mechanised tipping by overhead ropeways and buckets and conveyors led to greater volumes of spoil being deposited per unit surface area by allowing spoil heaps to rise above 50m in height. The mechanically produced conical or ridge tips became common features of coalfield landscapes (see Photograph 5.1). The lack of compaction of the spoil in the tipping process left large air-filled voids which in combination with their high coal content made them prone to spontaneous combustion and instability (see Chapter 7).

Photograph 5.1: Colliery spoil heaps at Loos en Gohelle, Nord-Pas de Calais, France (source: EPF)

Modern tips throughout Europe are often created by tipping from rubber tyred dump trucks which result in spoil heaps of a plateau-like structure, frequently no more than 15m in height. Tipping in this manner enables the waste material to be easily compacted by the dump trucks as they run over the spoil heap. This compaction and the shape of the waste heap makes future reclamation somewhat easier and increases tip stability.

At some mines the coal is washed to improve its quality. The washery process can involve the use of chemical flocculants and produces a fine-grained waste that in the past was deposited in lagoons as a slurry but in more modern operations may be mixed with run-of-mine waste after being dewatered in a filter press (see Section 8.3.4).

5.1.2 Open pit mining

In open pit mining extraction of coal follows the removal and storage of overburden. This overburden is generally stored so as to retain the different types of soil material removed from the profile and thereby facilitate restoration once the extraction of coal has been completed. Restoration is generally carried out as part of the mining operation and where this occurs open pit mining does not produce colliery spoil heaps. However early extraction by opencast methods was often not accompanied by restoration and overburden and spoil materials from between coal seams may have been deposited in many small heaps or only roughly levelled. As a result some dereliction has resulted from opencast mining. This takes the form of undulating land comprised of a mixture of spoil and overburden. The characteristics of the materials are similar to those of deep-mined spoils although chemical and physical characteristics will not be as extreme where the proportion of subsoil mixed with spoil is high.

5.2 Impact of spoil heaps

The physical and environmental impact of spoil heaps is dependent on their location, size, age and the nature of the tipped materials.

Significant impacts include:

- visual (see Figure 13.1);
- air pollution from combustion and dust (see Section 7.3);
- water pollution arising from weathering and erosion of spoil (see Section 12.2).

Colliery spoil heaps may also be considered local landmarks and therefore worthy of retention rather than being reshaped. Large scale reclamation of derelict sites can endanger the preservation of industrial monuments, and past reclamation projects intended for amenity provision have, in some cases, incorporated sites of industrial archaeological value (see Sections 4.7 and 17.8.2). Similarly some spoil heaps are of ecological value and may be preserved for this reason.

5.3 Spoil characteristics

5.3.1 Heterogeneity of spoil material

Colliery spoil consists of material from the sedimentary strata adjacent to the coal seams, waste produced from the sinking of shafts and other works, dirt and fragments of coal. If washery waste has been deposited the residues of the chemicals used in the washing process may also be found. Materials from demolished buildings, railway sidings and other wastes will also be mixed with spoil materials (see Photograph 5.2).

Spontaneous combustion of spoil can occur, particularly in loose tipped, highly aerated heaps with a high coal content. Combustion in a tip can lead to the fusion of large blocks of spoil, collapse of the surface of tips, production of noxious gases and a fused red 'shale' (see Table 5.1 and Section 7.3). The combination of these factors, the mining of more than one seam at one site and hence deposition of spoil with different characteristics, the use of more than one tipping method at a site and importation of spoil from other mines result in considerable heterogeneity of physical and chemical characteristics of most colliery spoil heaps.

5.3.2 Chemical characteristics of spoil

Mineralogically colliery spoil is made up of components of two important and distinct origins:[187]

- the detrital minerals; those which were incorporated into the sediments of the coal basin as a result of weathering and erosion of the surrounding area such as quartz, feldspar, illite, kaolinite, montmorillonite and muscovite. These minerals, already weathered, will weather further only slowly;
- the diagenetic minerals; those which were formed during sedimentation and later geological events in the rocks themselves. The two commonest types are carbonates, such as ankerite and siderite and the carbonates of calcium, magnesium, iron and manganese, and sulphides, of which the most frequently occurring and important is iron pyrites.

The diagenetic minerals weather rapidly on exposure. An indication of the relative proportions of the various minerals in colliery spoil is given in Table 5.2. The data in Table 5.2 indicate that colliery spoil is variable and that spoil from different sites can have very different compositions.

Freshly exposed colliery spoil often has a pH of 7 or more and may have a high electrical conductivity level. The high conductivity levels are often associated with high concentrations of sodium, calcium and magnesium and indicate elevated concentrations of dissolved salts. The most common salts of these elements are carbonates, sulphates and chlorides of which the carbonates are the most easily leached.

Table 5.1: The sequence of spontaneous combustion in colliery spoil heaps (after Goodman and Chadwick, 1978[104])

Temperature, °C	Effect
20	Over 90% of the oxygen content absorbed. Where gases escape they contain (v/v dry basis): < 0.005% O_2; 0-10% CH_4; 2-10% CO_2; 2-10% N_2.
100	As burning increases, trace amounts of CO are released.
200	Depending on the oxidation state and flow rate of the gas stream, varying amounts of CH_4, CO, CO_2, C_2H_6, C_4H_8, H_2S and SO_2 are released.
300	Carbonisation occurs and coals become fluid.
450	All interstitial water is eliminated.
500	Fumes of $(NH_4)SO_4$ and NH_4Cl are produced; NH_3, tar, SO_2, CO, SO_2, H_2S and N_2 released.
600	Micaceous material begins to break down.
800	Clay content is eliminated and brick produced.

Photograph 5.2: Demolition waste deposited in a colliery spoil heap in South Yorkshire, UK (source: Richards, Moorehead and Laing Ltd)

Table 5.2: Proportion and composition of minerals found in colliery spoil

Mineral	Components	Quantity (%)	
		Bouroz (1964)[38]	Bradshaw and Chadwick (1980)[40]
Quartz	SiO_2	32.5	90 (Quartz and clay minerals combined)
Clay minerals	alumino-silicates	5.6	
Ankerite	Fe, Ca, Mg, Mn; CO_3	-	5 (Ankerite, siderite and iron pyrite combined)
Siderite	Fe; CO_3	2	
Iron pyrite	FeS_2	-	
Gypsum	$CaSO_4$	-	1-2 (Gypsum and jarosite combined)
Jarosite	K, Fe; SO_4	-	
Feldspar	K, Al; SiO_2	2	-
Chlorite	Mg, Fe, Al; SiO_2, OH	2	-
Calcite	Ca; CO_3	trace	-
Amorphous material	SiO_2, $Al(OH)_3$, $Fe(OH)_3$	-	1-2

Conductivity levels often drop as these salts are leached out of the surface layers. A strong link between net percolation of water and salinity levels on colliery spoil has been noted.[75]

Despite the short-term nature of high levels of salinity in colliery spoil because of leaching, the consideration of salinity is important during reclamation because movement of spoil will expose unweathered saline materials.

The variation in spoil chemical characteristics that can occur over time and with depth from the surface is illustrated in Table 5.3 and Figure 5.1. Colliery spoil may contain significant concentrations of the trace elements found in coal (see Table 5.4). Surface concentrations of some trace elements may be high because of additional aerial deposition from coal combustion in the vicinity (see Section 10.3).

One of the more important consequences of weathering on colliery spoil is the fall in pH caused by the oxidation of iron pyrites. Shales from iron ore mining in Lorraine are also pyritic (see Section 9.1.2). Pyrite is stable until exposed to air and water, when it is oxidised to produce, amongst other products, sulphuric acid. The reactions involved in pyrite oxidation are summarised in Box 5.1. Above pH 4, iron-oxidising bacteria are relatively inactive and the oxidation of pyrite is slow and regulated by the supply of oxygen, but below pH 4, oxidation is much faster due to the activity of iron-oxidising bacteria,[195] the overall rate being determined by that of Equation 5.5 or 5.7 (Box 5.1). Pyrite oxidation may be by chemical oxidation alone, or by a combination of chemical and bacterial oxidation. The latter is considerably faster than the former: iron-oxidising microorganisms, such as *Thiobacillus ferrooxidans*, can significantly increase the rate of pyrite oxidation by mediating the oxidation of Fe^{2+} (Equation 5.5, Box 5.1). The rate of Fe^{2+} oxidation may be increased by a million-fold and the overall rate of acid generation increased by up to twenty-fold. These bacteria tend to be most active between pH 2 and 4.

Table 5.3: Mean electrical conductivity and associated cation levels in freshly exposed spoil in 1975 and in subsequent years at two sites in the United Kingdom

	Thorne (S Yorkshire)			Abertysswg (Mid-Glamorgan)		
	1975	1978	1982	1975	1978	1980
Conductivity (Ms/cm)	6.7	0.36	0.45	0.56	0.15	0.26
K (mg/kg)	35	15.6	20.0	41	17.3	17.2
Na (mg/kg)	821	19.6	13.4	8.7	13.0	8.5
Mg (mg/kg)	184	12.8	16.0	22.5	3.7	4.7
Ca (mg/kg)	368	31.5	31.9	36.2	5.2	9.0

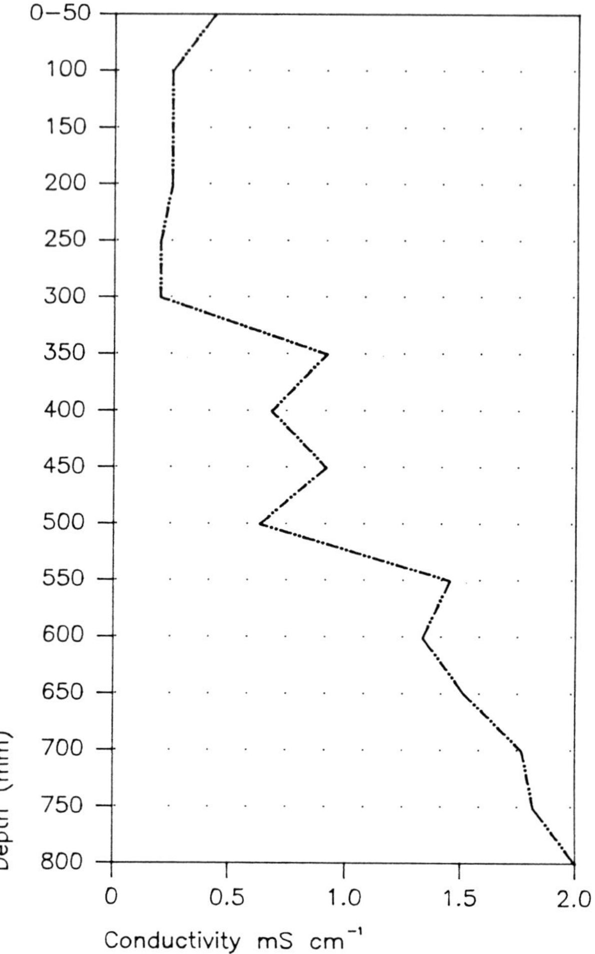

Figure 5.1: Electrical conductivity down the spoil profile six years after spoil exposure at Thorne in South Yorkshire, UK

Table 5.4: Some records of the elemental content of coal and coal ash (after Bouska, 1981[39])

Element	Coal	Coal Ash	Element	Coal	Coal Ash
Aluminium	0.4-6.8%	6-16%	Molybdenum	3-10	up to 0.6%
Antimony	<200	0.3%	Nickel	-	0.001-8%
Arsenic	<1-60*	80-10,000	Niobium	-	10
Barium	90-27,000	0.01-4.7%	Nitrogen	0.2-3%	-
Beryllium	0.8-2.8	<1-200	Phosphorus	21-240	up to 1594
Bismuth	<1-10	200	Platinum	-	0.5
Boron	1-300	18-14,000	Potassium	-	up to 3.2%
Bromine	4.1-41.6	-	Radium	<0.001	<0.001
Cadmium	-	3-30	Rhenium	0.08-0.3	-
Calcium	very high	1.3%	Rubidium	30-250	-
Caesium		27-11	Scandium	5	24-400
Chlorine	0.43-0.77%	-	Selenium	10-30	-
Chromium	up to 0.1%	0.02-1.3%	Silicon	-	14-19%
Cobalt	92	0-2000	Silver	<10	<10
Copper	-	150-4000	Strontium	-	290-2380
Fluorine	0-175	-	Sulphur	0.2-15%	-
Gallium	<1-500	<10-4000	Thallium	0.3-2.3	-
Germanium	1-300	<0.01-7.5%	Tin	20	3-6000
Gold	-	<1mg/kg	Titanium	up to 2.1%	0.2-14.6%
Indium	-	2mg/kg	Tungsten	up to 0.2%	up to 5%
Iron	up to		Uranium	up to 1%	up to 1%
Lead	up to 3000	up to 3000	Vanadium	<100-1000	up to 3.5%
Lithium	30-250	27-500	Yttrium	-	<100-800
Magnesium	-	1.7%	Zinc	10-7800	100-5000
Manganese	-	0.01-2.2%	Zirconium	-	0.5%
Mercury	<0.1-1.3	-			

Concentrations are in mg/kg unless otherwise stated.

Box 5.1: Pyrite oxidation[25, 92, 137, 195, 259]

The overall reaction can be written:

$$FeS_2 + {}^{15}/_4\,O_2 + {}^{7}/_2\,H_2O \rightleftharpoons Fe(OH)_3 + 2\,SO_4^{2-} + 4\,H^+ \quad (5.1)$$

if the pH is greater than 2.3, or

$$FeS_2 + {}^{15}/_8O_2 + {}^{13}/_2Fe^{3+} + {}^{17}/_4H_2O \rightleftharpoons {}^{15}/_2Fe^{2+} + 2\,SO_4^{2-} + {}^{17}/_2H^+ \quad (5.2)$$

if the pH is less than 2.3.

Intermediate reactions 5.3 to 5.7 have been identified.

$$FeS_{2(s)} + {}^{1}/_2\,O_2 + 2\,H^+ \rightleftharpoons Fe^{2+} + 2S^{o}_{(s)} + H_2O \quad (5.3)$$

$$2S^{o}_{(s)} + 3O_2 + 2H_2O \rightleftharpoons 2SO_4^{2-} + 4H^+ \quad (5.4)$$

$$Fe^{2+} + {}^{1}/_4\,O_2 + H^+ \rightleftharpoons Fe^{3+} + {}^{1}/_2\,H_2O \quad (5.5)$$

$$Fe^{3+} + 3H_2O \rightleftharpoons Fe(OH)_{3(s)} + 3\,H^+ \quad (5.6)$$

$$FeS_{2(s)} + 14\,Fe^{3+} + 8\,H_2O \rightleftharpoons 15\,Fe^{2+} + 2\,SO_4^{2-} + 16\,H^+ \quad (5.7)$$

Notes:

$_{(s)}$ Solid

(5.3) Pyrite is oxidised by oxygen. This is a slow reaction, the rate limited by the diffusion of oxygen releasing ferrous ions and elemental sulphur.

(5.4) Elemental sulphur is oxidised to sulphate ions and acidity, as H^+.

(5.5) Under aerobic conditions, ferrous ions are oxidised to ferric ions. This reaction is catalysed by iron-oxidising microorganisms such as *Thiobacillus ferrooxidans.*

(5.6) Above pH 3.5, ferric ions are not stable in water. Ferric hydroxide is formed which precipitates, and pH is lowered further. Ferric hydroxide is very insoluble, so few Fe^{3+} ions are left in solution.

(5.7) Any Fe^{3+} ions remaining in solution are free to oxidise pyrite. Further acidity is generated by this reaction. Since ferric ions are stable below pH 3.5 and are not removed by precipitation, this step is significant in the production of very acid spoil or drainage.

Key features of the reactions shown in Box 5.1 are as follows:

- pyrite can be oxidised by two oxidising agents: oxygen and Fe^{3+}. Oxygen gas from outside the deposit is the ultimate oxidising agent;
- oxidation generates acidity;
- the oxidation of Fe^{2+} (Equation 5.5) is the slow step;
- once Fe^{3+} ions have been produced, these rapidly oxidise pyrite further unless removed by precipitation;
- iron-oxidising bacteria, such as *Thiobacillus ferrooxidans*, can significantly increase the rate of pyrite oxidation by catalysing the oxidation of Fe^{2+} (Equation 5.5);
- above pH 4, iron-oxidising bacteria are relatively inactive and the oxidation of pyrite is slow and regulated by the supply of oxygen, but below pH 4, oxidation is much faster due to the activity of iron-oxidising bacteria,[195] the rate being determined by the rate of Equation 5.5 or 5.7 (see Box 5.1).

In an acidic environment, a cycle of acid generation develops in which ferrous ions released from the pyrite are oxidised by *T. ferrooxidans* to ferric ions, which can then oxidise pyrite and generate large quantities of acid. This acid generation can have various effects:

- it may provide a very acidic substrate for plant growth, preventing vegetation from becoming established;
- revegetation of reclaimed spoil heaps may revert as a result of acid generation, killing off any established vegetation;
- acid generation can result in acid mine drainage which in turn can acidify ground and surface waters (see Section 12.2.3) and result in visual pollution by the deposition of metal oxides, most notably iron oxide, on stream beds;
- high sulphate concentrations formed can cause deterioration of concrete (see also Section 10.5.2).

There are several factors that can influence the kinetics of pyrite oxidation by bacteria,[207] including:

- pyrite reactivity, which can vary considerably;
- amount of pyrite and its surface area;
- bacterial contact with the pyrite;
- oxygen availability;
- pH;
- temperature;
- the presence of inhibitors.

This oxidation step can continue, if conditions are suitable, until all the pyrite has been oxidised. This can take thousands of years.

The rate of fall in pH of a colliery spoil site is not a simple function of pyrite reactivity. The presence of carbonate minerals such as siderite and ankerite in spoil can neutralise the acidity produced and thereby buffer the pH.[51, 187] In neutralising the acidity secondary minerals such as gypsum ($CaSO_4$) and jarosite ($KF_3(OH)_6(SO_4)_2$) are produced. A considerable range of pHs, pyrite contents and acid neutralising capacities have been found in colliery spoils (see Figure 5.2).

The pH profile through an extremely pyritic spoil is shown in Figure 5.3. Acidification of spoil also occurs by other means including leaching of bases, secretion of humic acids by plants, nitrogen fixation by legumes and addition of fertilisers. The result is that the pH of all colliery spoils will fall over time except for those where the rocks from which they are derived are calcareous (see Figure 5.4).

The consequence of unneutralised acidity on colliery spoil can be detrimental to plant growth. Cation exchange sites will become dominated by hydrogen ions with the resultant loss of bases previously occupying the sites. At low pH iron, aluminium, manganese, copper and zinc will be solubilised creating toxic conditions. Aluminium and manganese can be very toxic in low concentrations in acid conditions.

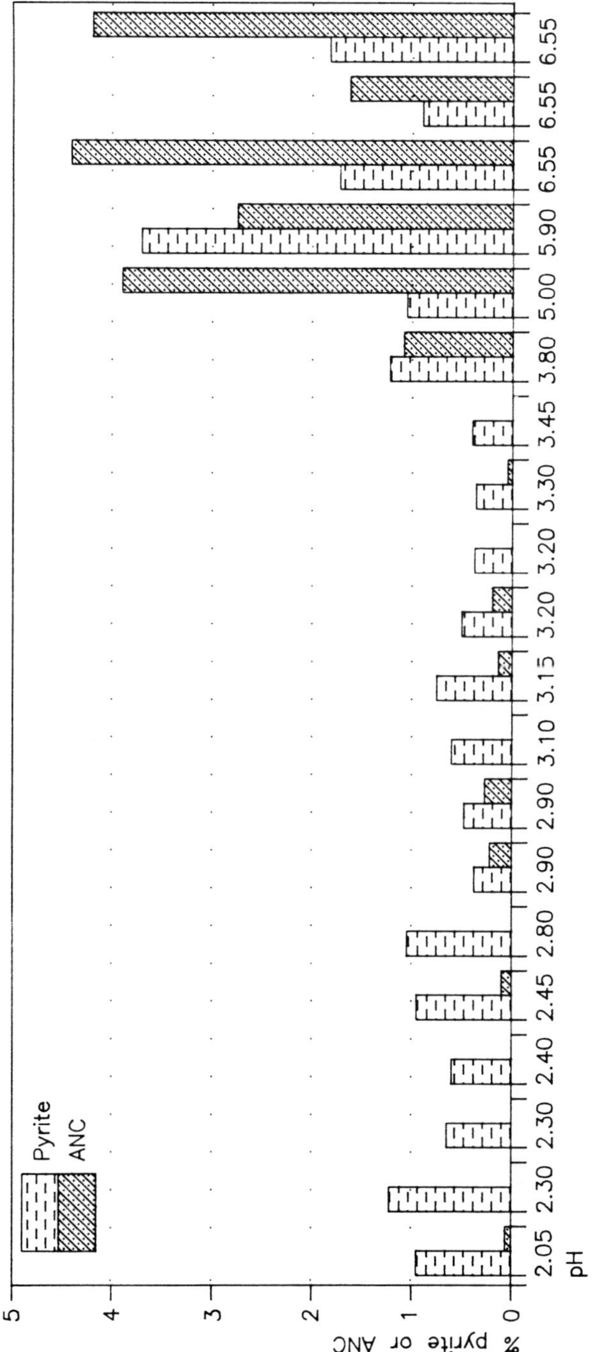

Figure 5.2: Percentage pyrite content and acid neutralising capacity (ANC) ranked against pH for twenty spoils from various parts of England and Wales (after Costigan, Bradshaw and Gemmell, 1981[58])

Low pHs can also affect the availability of the major nutrients needed for plant growth: nitrogen, phosphate and potassium. Phosphate can be fixed and made unavailable to plants at low pH and potassium may be made unavailable by the formation of jarosite and lost from the clay minerals through the destruction of clay lattices. Even moderately low pHs can cause aluminium and manganese toxicity and inhibition of microbial activity resulting in a build-up of undecomposed organic matter and inhibition of nitrogen mineralisation.

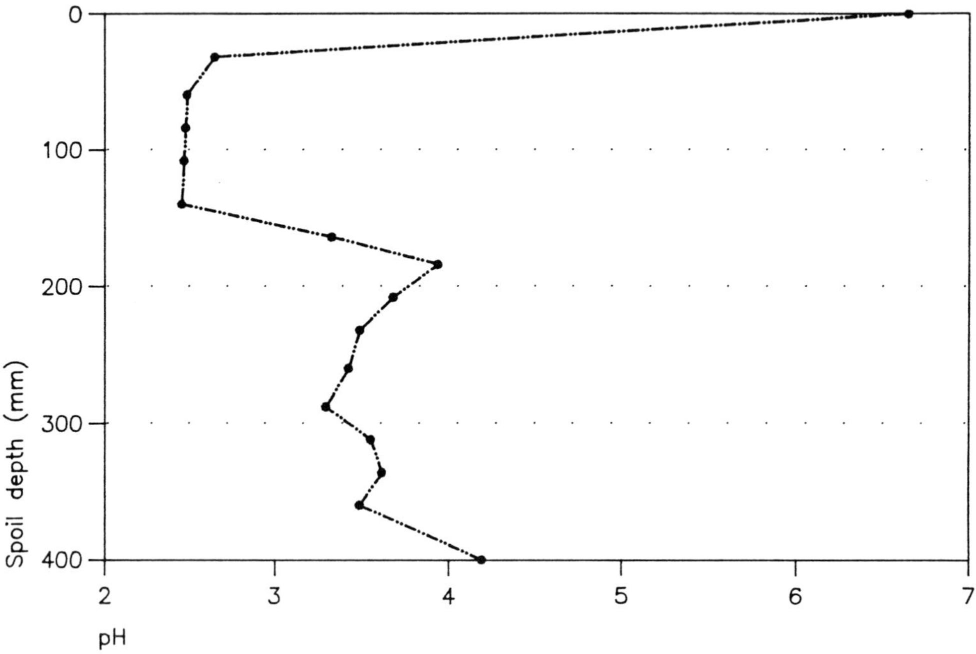

Figure 5.3: pH variation with depth in a highly pyritic spoil. The spoil had been limed at the surface resulting in a high pH and had been cultivated to a depth of about 350mm resulting in greater penetration of air and water and the resultant oxidation of pyrite to produce very low pHs

Plant-available nitrogen and phosphorous concentrations are very low on unameliorated spoil although colliery spoil contains substantial amounts of fossil nitrogen. Spoils vary in their ability to sorb or 'fix' phosphate, and thus make it unavailable to plants. There is a major distinction in phosphate fixation capacity between spoils which have been burnt at a temperature above 450°C and unburnt spoil, those having been burnt at a high temperature having a lower phosphate fixing capacity because the fused shale has a lower surface area.[76] Potassium is not usually deficient except where the pH is low.

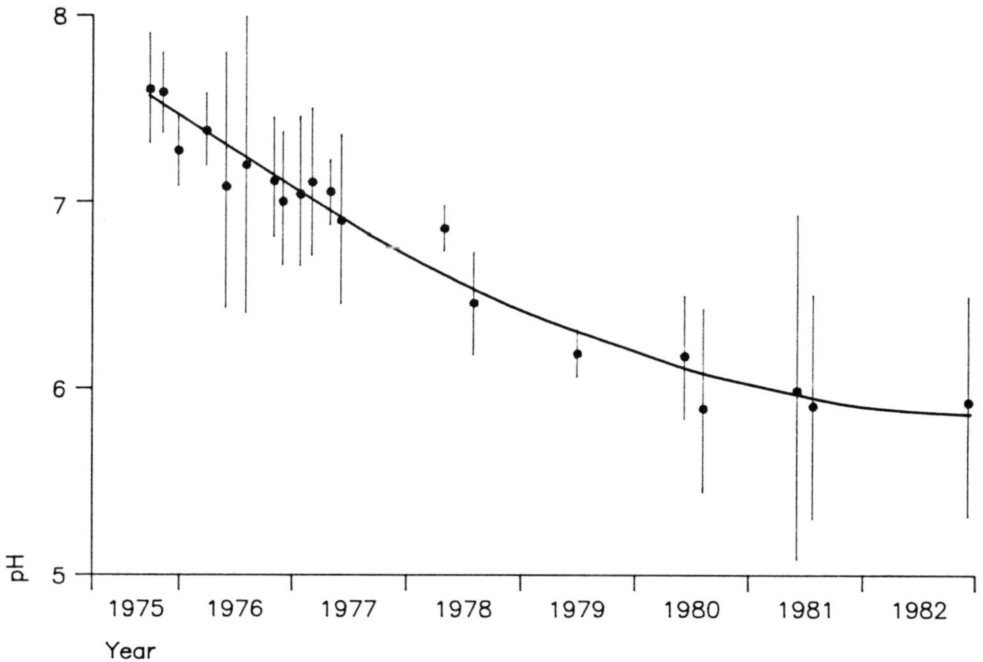

Figure 5.4: pH trend over time in a vegetated non-pyritic spoil (after Palmer *et al.*, 1986[186])

5.3.3 Physical characteristics of spoil

Freshly exposed colliery spoil can weather rapidly; large boulders may be reduced to silt and clay-sized particles in a short time (Table 5.5). The rate of weathering depends in part on the extent to which sediments were compressed during their formation. Sediments which have been subject to pressure, such as those associated with anthracite deposits, do not weather as readily as some of those associated with bituminous coals.

The proportion of fine-grained particles is important in the water relations of the spoil. A high proportion of fines results in greater moisture holding capacity and better plant establishment. Colliery spoil, because of its particle size distribution and lack of organic matter, is also susceptible to surface waterlogging in the winter and drought in the summer.[205]

Poor structure in spoils is caused by the paucity of finer grained particles, or where these do occur, to their lack of aggregation which gives a dense, massive structure rather than an open, crumbly structure. Poor aggregation is due mainly to the scarcity of plant roots, microorganisms and organic matter which can bind the fine-grained particles together into crumbs and thus open up the matrix.

Table 5.5: Weathering of spoil materials (after Skarzynska, 1987[219])

Sample	Fraction %			
	Boulder	Gravel	Sand	Silt and clay
Original	38	58	3	1
After 22 months *	6	61	20	13

* 40kg samples were placed in boxes and compacted to 0.1m and exposed to the atmosphere for 22 months.

The lack of aggregation combined with pressure and compaction results in the spoil particles packing closer together. This close packing reduces the pore space between particles, and the lack of macropores gives the spoil a low infiltration rate, since it prevents excess rainfall draining quickly through the spoil under the influence of gravity. The resulting surface waterlogging can lead to puddling which blocks the smaller pores and aggravates the drainage problems further. Poor drainage can lead to the presence of standing water in depressions on flat areas of spoil during wet periods, and erosion effects brought about by the surface run-off of water on slopes (see Photograph 5.3). Lack of infiltration leads to low spoil moisture availability and drying. Following drying of the surface during summer periods, water stress to plants is increased, and the lack of organic matter, which acts as a moisture conserving material, adds further to the problem. Water stress is intensified by high spoil surface temperatures.

Photograph 5.3: Erosion of washery waste in a poorly designed reclamation scheme (source: Richards, Moorehead and Laing Ltd)

Surface physical properties are worsened by the practice of compacting spoils to reduce voids and to reduce the likelihood of spontaneous combustion, subsidence and landslip.

Table 5.6 shows the influence of compaction of colliery spoil by different types of machinery and Table 5.7 the effect of compaction on infiltration rates.

Reorientation of fine-grained particles during and after rainfall may produce a low porosity and low permeability layer at the colliery spoil surface. Anaerobic conditions may develop under the surface pan and the low porosity of the pan will mean that it is easily saturated by rain, so infiltration will be low and runoff high. Upon drying the surface can form a hard crust up to many millimetres thick. The extent of panning will be dependent on the size distribution of the spoil and the degree of weathering. A smooth surface, whether caused by vehicle trafficking or surface panning, is a great impediment to plant establishment because of difficulty in root penetration and risk of desiccation.

5.3.4 Calorific value

The calorific value of spoil is variable and dependent on:

- the coal content of the spoil;
- the carbon content of the sediments adjacent to the coal seam and forming the bulk of the spoil materials;
- the amount of other combustible materials such as wood, oily materials and general rubbish.

The coal content of older tips is usually higher than modern tips because early methods of coal processing were less efficient than modern ones (see Section 8.1).

The calorific value of the spoil together with other factors such as degree of compaction and pyrite content will determine the likelihood that the

Table 5.6: The influence of modern earth moving machinery on spoil compaction as indicated by penetrometer readings. The higher the reading the greater the compaction (from Ayerst, 1978[15])

Traffic	Penetrometer reading (kg/cm^2)
None - newly tipped	19.9
Levelled by bulldozer	51.0
Levelled, plus bulldozer track (one pass)	98.4
Levelled, plus bulldozer track (four passes)	189.2
Lorry track (laden)	219.1

Table 5.7: Influence of compaction on infiltration rates

	Bulk density (g/cm^3)	Infiltration rates (cm/min), after compaction at: (kg/cm^2)		
		0	0.5	5.0
Unvegetated	1.48	0.31	0.26	0.05
Vegetated	1.50	0.67	0.25	0.05

spoil will combust. This is discussed in detail in Section 7.2. The calorific values of some colliery wastes is presented in Table 7.1.

5.3.5 Surface temperature

Most colliery spoils are black in colour and therefore absorb significant quantities of heat and light generated by solar radiation. Burnt colliery waste is red in colour and as a result reflects some of the incoming solar radiation.

The major site influences on surface temperature are slope, aspect and the nature of the spoil material. The importance of other factors varies with the season. In the northern hemisphere, south-facing slopes are the first to warm in the spring but can produce conditions that become limiting to plant growth in the summer months. The nature of the spoil influences the amount of heat energy absorbed and its later dissipation by conduction and re-radiation. The more mineral matter, or conversely less organic matter and water content, the greater the thermal conductivity. Thus, the spoil conditions most likely to produce temperatures damaging to vegetation are very black spoils on sites with little organic matter and a low moisture regime. Temperatures of >50°C have been found on British spoil heaps in the summer months. In Nord-Pas de Calais surface temperatures of spoil heaps were found to vary dependent on whether there was a vegetation cover, and were much affected by combustion within a tip (see Table 5.8). On well-vegetated grey spoils temperatures are lower than on unvegetated black spoils but temperature between the surface and 100mm below can still differ by as much as 5°C (see Figure 5.5).

Table 5.8: Spoil temperatures (°C) at different depths in spoil heaps in Nord-Pas de Calais (after Petit, 1980[191]).

Depth (mm)	Spoil heap 85, Hénin Beaumont		Spoil heap 119, Ostricourt
	Bare spoil	Vegetated	(Burning)
20	46	28	-
50	32	26	36
100	29	23	49
150	-	-	53

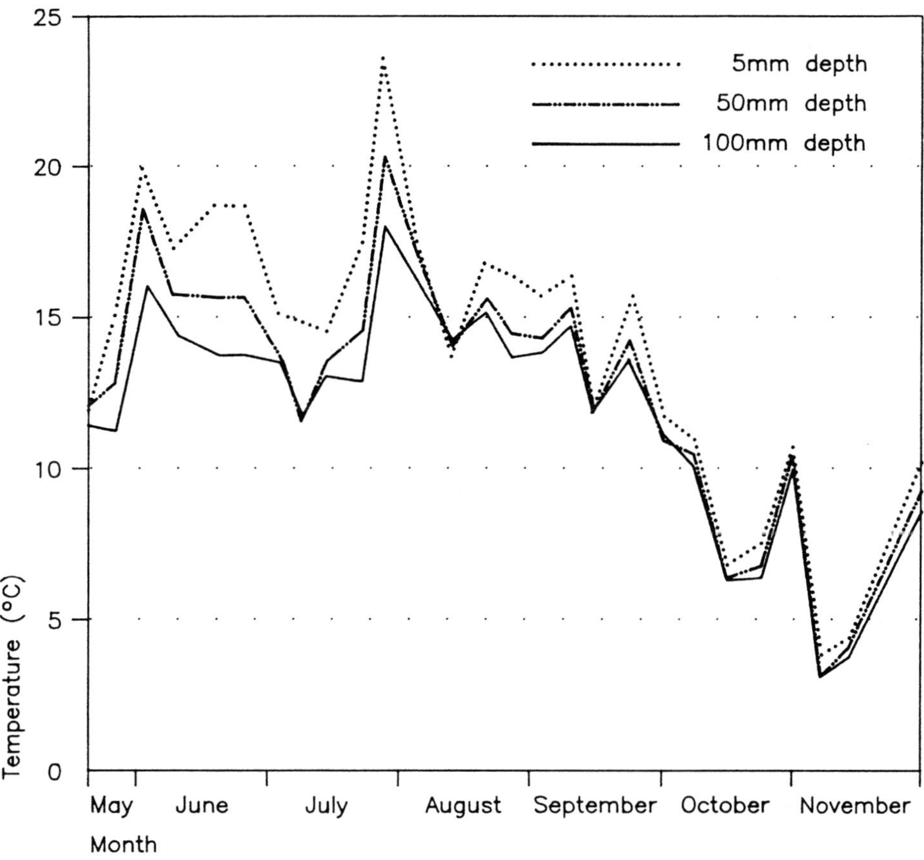

Figure 5.5: Temperature differences in vegetated grey colliery spoil, at different depths during the growing season at a South Yorkshire spoil heap, in the UK.

5.4 Colliery spoil as a substrate for plant growth

5.4.1 Introduction

Compared with other types of derelict land colliery spoil has characteristics which make it one of the least favourable for plant growth (Figure 5.6). The chemical and physical characteristics of colliery spoil have been described in previous sections and their influence on plant growth is summarised in Box 5.2.

Of the characteristics in Box 5.2 the most important are acidity (low pH) and nitrogen status.

5.4.2 Acidity

The greatest levels of acidity found in colliery spoil are caused by pyritic oxidation (see Box 5.1). Usually where pyritic oxidation is ongoing and the natural neutralising capacity of the spoil has been consumed, the pH is so low (<3.5) that no plants will establish or survive and the spoil is devoid of other life except some microorganisms which can withstand the low pH (see Photograph 5.4). At such low pH values very high soluble concentrations of elements such as aluminium and zinc may be found which make the spoil extremely phytotoxic. In some cases pHs of approximately 3.5 may be found in old well-vegetated spoil heaps. Here the toxic elements associated with this acidity will have been leached out of the surface layers over the years and any pyrite present will have been oxidised. It is likely that organic matter will have built up in surface layers. Under these conditions plants tolerant of low pH can establish and survive.

5.4.3 Nitrogen

A major factor limiting plant growth on despoiled land is lack of nitrogen. For vegetation to establish and grow the plants need a source of nitrogen *e.g.* ammonium or nitrate, to form amino acids, protein and other compounds associated with cell growth.

Waste material / Constraint	Stability	Combustion	Slope angle	Flooding stress	Toxicity	Compaction	Temperature	Wind erosion	Nutrients	Stoniness	Uneven surface	Erosion	Soil fauna and microbes
Colliery spoil	○	○	○	○	○	○	○	○	○	○	○	○	○
Smelter slag	○	○	○	○	○		○	○	○	○	○	○	○
Slate and shale	○		○	○			○	○	○	○	○	○	○
Metal wastes	○		○		○	○	○	○	○	○	○	○	○
Quarry pits			○	○		○	○	○	○	○	○	○	○
Brick pits			○	○		○	○	○	○	○	○	○	○
Peatland		○		○	○		○	○	○			○	○
China clay	○		○	○			○	○	○			○	○
Ironstone	○		○	○		○	○	○	○		○	○	○
Chemical waste	○				○	○	○	○	○			○	○
PFA				○	○	○	○	○	○			○	○
Sand and gravel			○	○		○	○	○	○		○	○	○
Domestic refuse		○		○	○	○		○			○		

Figure 5.6: Constraints on vegetation development on different wastes (after Kent, 1982[135])

In fertile soils nitrogen is supplied from the breakdown of organic matter in the soil and supplemented by nitrogen in rainfall. The organic matter in the soil is replenished by the decomposition of plant and animal remains by microbes. This is the basis of nitrogen cycling and a typical nitrogen cycle is shown in Figure 5.7. In nitrogen deficient systems,

Box 5.2: Physical and chemical characteristics of colliery spoil and their impact on plant growth

Characteristics	Implications for plant growth.
pH	<4 very few species survive 4.0-5.5 many grasses and clovers grow successfully >8 limited species survival.
Electrical conductivity	High conductivity (>2.5mS/cm) indicates high levels of dissolved salts which will affect seedling establishment and plant growth. High conductivity is usually associated with high and low pHs. High conductivity is often only a problem with freshly exposed spoils.
Nitrogen status	Plant-available nitrogen is very low and restricts colonisation on many spoils.
Phosphate status	Plant-available phosphate is very low. Added phosphate, e.g. in rainfall, rapidly becomes unavailable to plants through sorption by clay minerals. Low phosphate concentrations restrict root growth and nitrogen fixing plants such as legumes.
Potassium status	Potassium supply is adequate for natural colonisation and is replenished from the breakdown of clays. Potassium may be limiting in productive swards.
Trace elements	Colliery spoil has an adequate supply of trace elements for plant growth. Some elements may reach toxic proportions, e.g. aluminium, manganese, zinc and copper, particularly at low pH.
Organic matter	Readily decomposable organic matter is virtually non-existent on freshly exposed colliery spoil even though traditional methods of soil analysis for organic nitrogen, organic carbon and organic matter will indicate that there are substantial quantities of these determinants. This is because these methods are recording 'fossil' organic compounds which do not readily decompose. Low organic matter content results in poor nutrient cycling, poor moisture retention and poor soil structure.
Physical	Compaction, surface panning, poor moisture retention, poor soil structure, high surface temperatures all impede plant establishment and productivity.

Photograph 5.4: Vegetation, which has developed on a thin layer of limed spoil, peeling away from the more acid underlying spoil (source: Richards, Moorehead and Laing Ltd)

cycling is at a low level because inputs of nitrogen are low and is also often impeded because nitrogen is bound up in compounds of high carbon to nitrogen ratio or immobilised in microbial biomass (see Box 5.3). In such situations the carbon to nitrogen ratio in organic matter means that microbial decomposition is limited by nitrogen supply and any available nitrogen arising through mineralisation is quickly taken up and immobilised by microbes. This interruption of cycling is often assisted by factors other than microbial action, such as low pH. Mineralisation and nitrification are the most common areas of interruption of the nitrogen cycle on reclaimed land (see Figure 5.7). Figure 5.8 illustrates how microbial substrate availability in terms of ammonium and fluctuations in pH (as affected by lime additions) influence nitrification, a microbially controlled process. The message is clear: for nitrogen to be supplied to vegetation there has to be a steady supply of ammonium to be nitrified and the pH has to be high enough for that nitrification to take place.

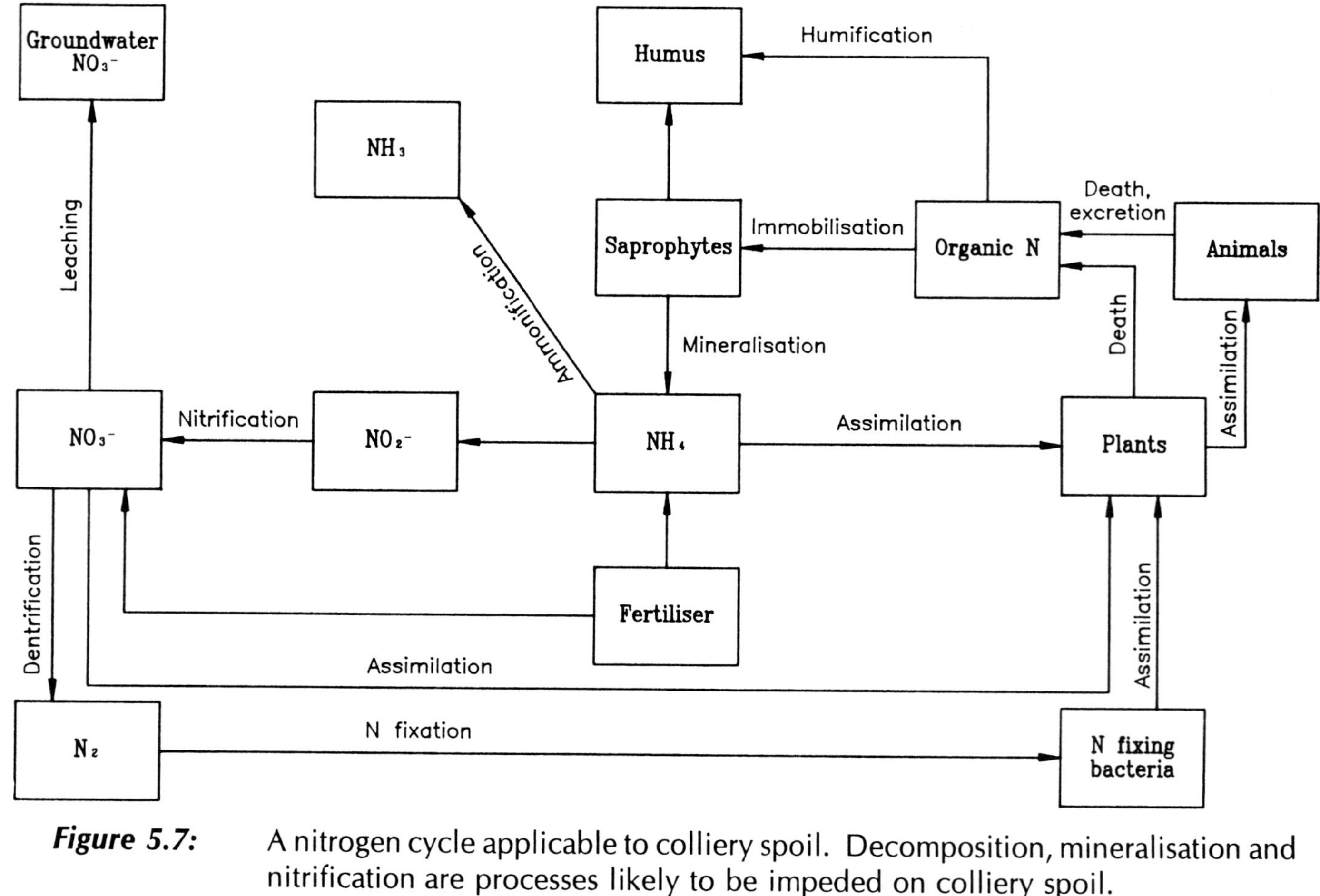

Figure 5.7: A nitrogen cycle applicable to colliery spoil. Decomposition, mineralisation and nitrification are processes likely to be impeded on colliery spoil.

Box 5.3: Nitrogen cycling

In order for nitrogen to be cycled it has to be in compounds where the carbon to nitrogen ratio is low enough for microbes to be able to break them down. The critical carbon to nitrogen ratio is between 25-30:1 and above this level *i.e.* where there is a greater proportion of carbon, microbial breakdown of organic matter may not occur.

Many grass, herb and tree roots and the woody and dead parts of plants have a carbon to nitrogen ration higher than 30:1 and in order to break these materials down microbes need a supply of nitrogen from elsewhere. In fertile soils this nitrogen is available so materials of a high carbon to nitrogen ratio are broken down and the nitrogen becomes available to be taken up by plants again. In infertile soils there is no freely available nitrogen and nitrogen can become bound up in compounds of high carbon to nitrogen ratio or microbial biomass.

Other mechanisms may cause resistance to microbial breakdown of organic matter *e.g.* (after Stevenson 1982[231]):

1. Stabilisation of proteinaceous constituents (*e.g.* amino acids, peptides, proteins) through their reaction with other organic soil constituents (*e.g.* lignins, tannins, quinones, reducing sugars).

2. The formation of biologically resistant complexes by the chemical reaction of NH_3 or NO_2 with lignins or humic substances.

3. Protection of organic N compounds from decomposition by their adsorption on to clay minerals.

4. Stabilisation of organic N compounds by the formation of complexes with polyvalent cations.

5. The siting of organic N in pores or voids physically inaccessible to microorganisms.

The significance of these mechanisms will be greater in infertile soils.

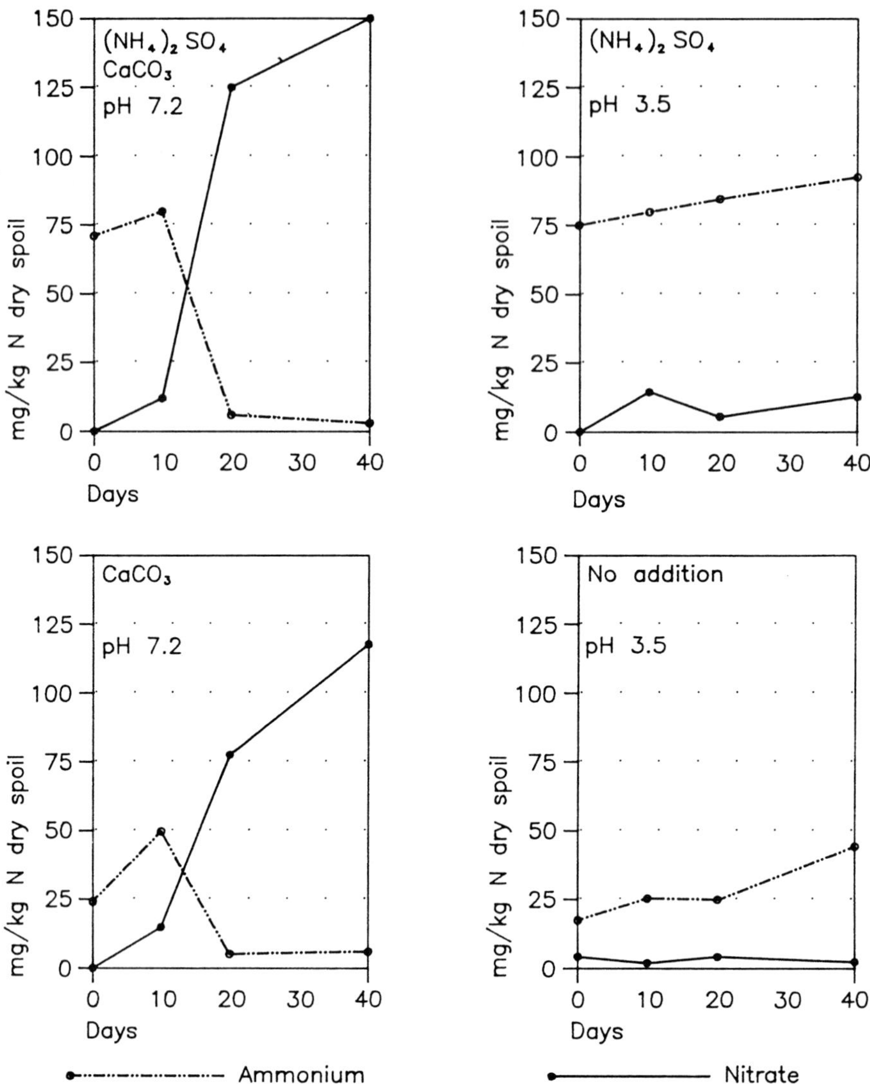

Colliery spoil incubated at 25°C and 10% moisture for 40 days with additions of ammonium sulphate, calcium carbonate or without additions. The conversion of ammonium to nitrate is microbially controlled and inhibited by low pH. Additions of calcium carbonate raise the pH and facilitate conversion of ammonium to nitrate which is then available to plants.

Figure 5.8: Nitrogen transformations on reclaimed colliery spoil (after Williams, 1975[261]).

5.5 Natural vegetation on colliery spoil heaps

The natural vegetation of colliery spoil heaps has been much studied. Some studies have related the vegetation found to classical succession patterns *i.e.* vegetation development and species composition goes through a recognised sequence of stages to reach a 'climax' vegetation type. Other studies have found no correlation between vegetation type and time since tipping but a correlation with spoil factors such as pH (see Figure 5.9). Both findings are probably correct:

- where pH is low there will be no succession until the pH is ameliorated, and colonisation will be by those plants that can tolerate low pH only;
- where pH is not limiting colonisation will follow a successional pattern with species composition dominated by local species.

Although successional theory would indicate that trees are not primary colonisers, at some sites colonisation by pioneer trees such as birch occurs within a few years of tipping ceasing (see Photograph 5.5). The rate of colonisation will depend on the extent to which spoil characteristics inhibit plant establishment and growth. In addition to pH, both nitrogen accumulation and salinity level have been shown to influence the rate of ingress of new species.[183] Examples of successional sequences which have been suggested are shown in Figures 5.10 and 5.11. Some workers have also classified species found on spoil heaps on the basis of life-form (see Table 5.9). Comparison of data on life-form of vegetation on colliery spoil with that from Mediterranean and temperate Europe (non-spoil heap) allows the following conclusions to be drawn:

- the high proportion of phanerophytes (trees and shrubs) indicates their success at colonisation of spoil compared to herbaceous species;
- the low level of geophytes (bulbs and corms) indicates that ecosystem development is at an early stage.

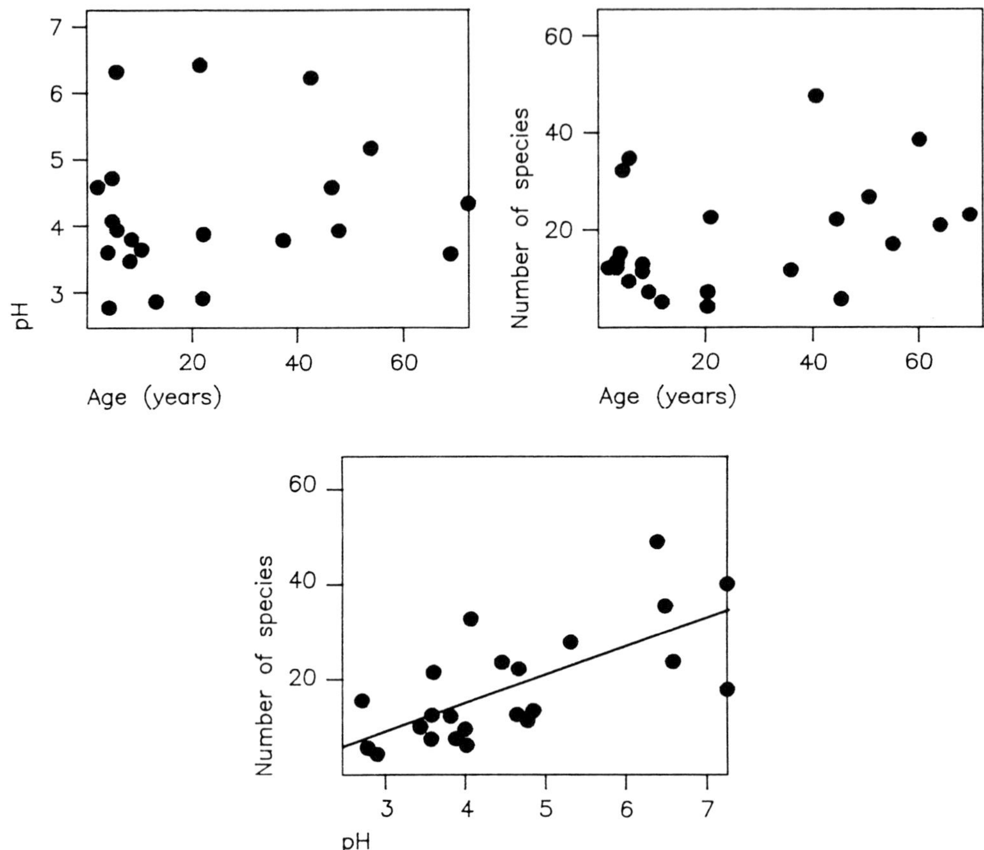

Figure 5.9: The number of species on a spoil heap may correlate more with pH than age of tip although there may be no correlation between age of tip and pH (after Bradshaw and Chadwick, 1980[40])

Photograph 5.5: Birch trees colonising acid spoil with little ground cover in the Ruhr, Germany (source: Richards, Moorehead and Laing Ltd)

Table 5.9: Classification of colliery spoil heap species on the basis of life-form in France and the United Kingdom (data derived from Petit, 1980[191] and Hall, 1957[110]).

	Nord-Pas de Calais	UK	Denmark†	Italy†
Phanerophytes	17	18	8	12
Chamaeophtes	3	3	3	6
Theophytes	32	19	20	42
Hemicryptophytes	45	58	56	29
Geophytes	3	3	12	11

† Distribution established by Raunkiaer for temperate Europe and Mediterranean Europe (non-spoil heap)

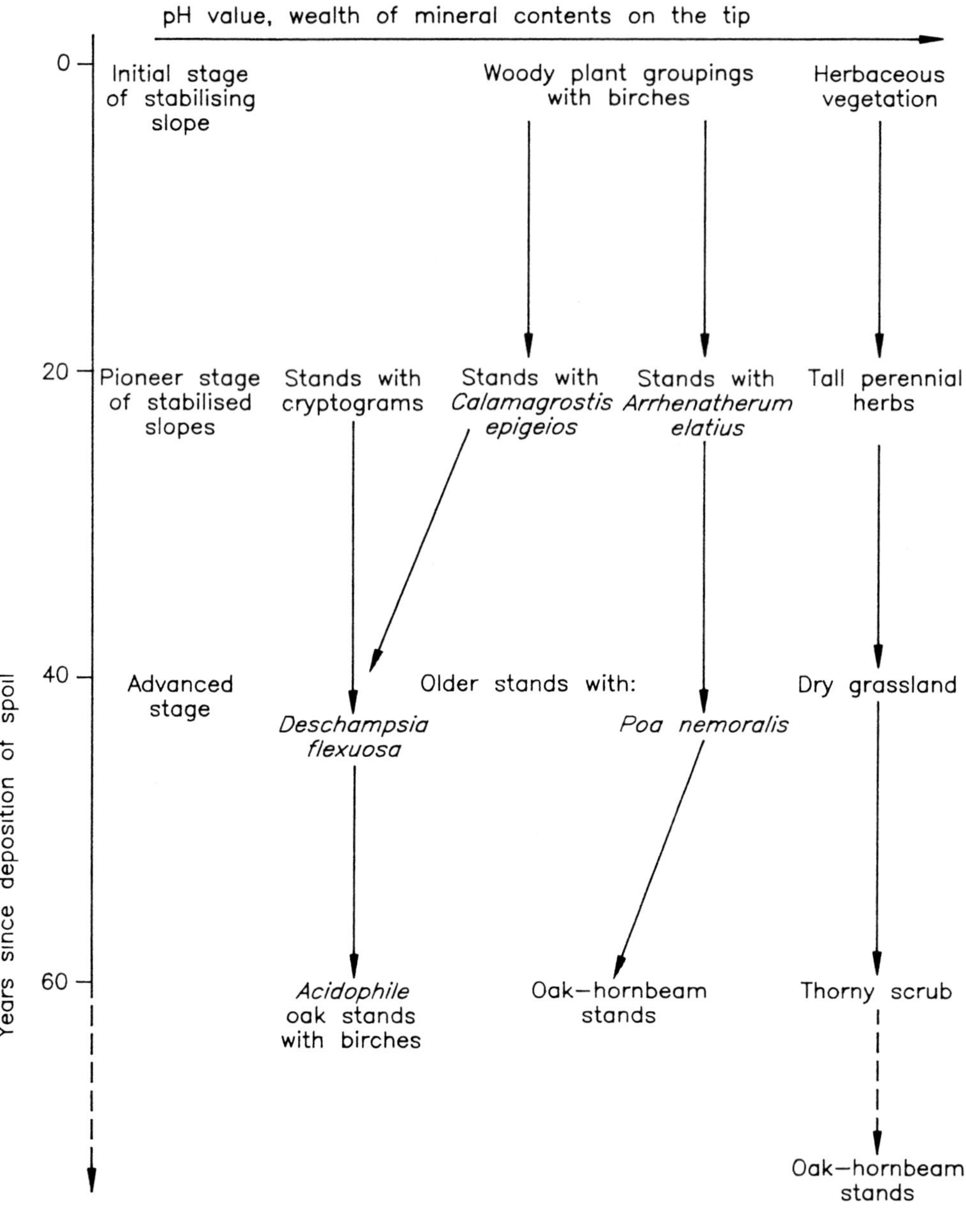

Figure 5.10: Suggested successional sequences on the spoil heaps of Nord-Pas de Calais (after Petit, 1982[192])

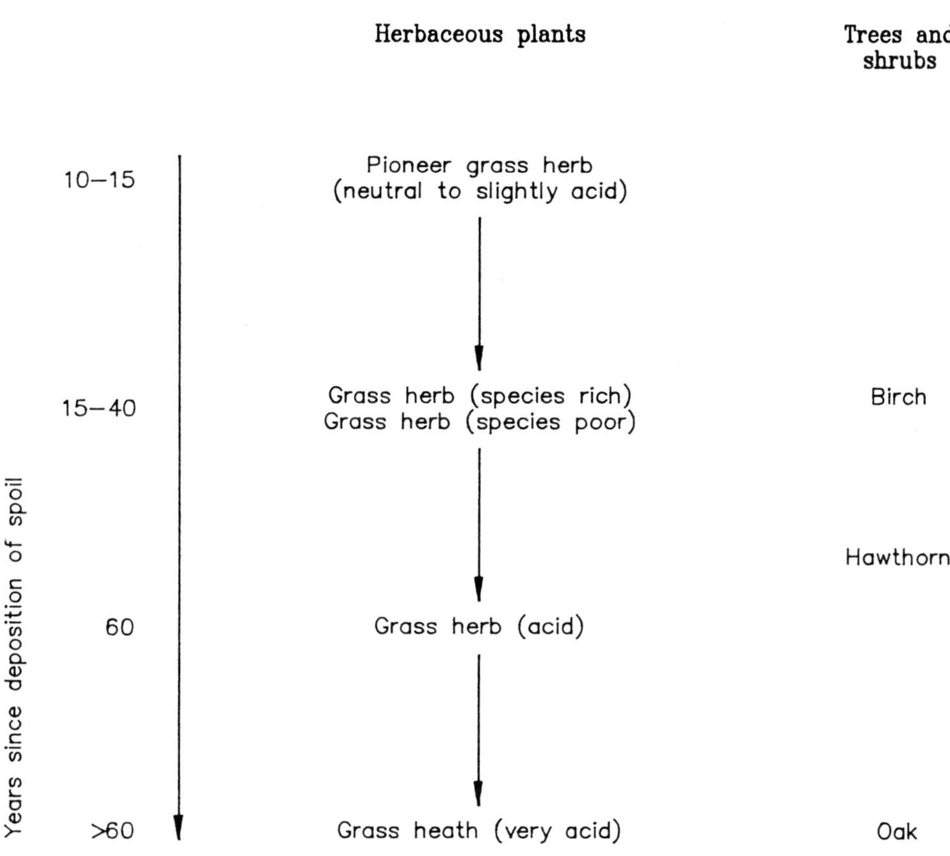

Figure 5.11: Successional sequence of vegetation postulated by Hall (1957)[110] using data from a number of spoil heaps in the UK

Mycorrhizal fungi which enable plants to increase their uptake of nutrients have been found to thrive on colliery spoil and undoubtedly assist early colonisers such as birch (see Box 5.4 and Photograph 5.5). On some acid spoils birch trees supported by mycorrhizal fungi but with no herbaceous layer may be found (see Photograph 5.6).

Because colliery spoil heaps take on the vegetation characteristics of the surrounding land a very large number of plant species have been found growing on them. Colliery spoil heaps are not well known for supporting rare species and are more important as tracts of semi-natural vegetation and refugia of species becoming uncommon in the locality because of industrialisation, urbanisation or agricultural use. Such spoil heaps may also be important foraging areas for birds, insects and small mammals and their predators. In some former coal mining areas naturally vegetated heaps have become more important because other heaps have been reclaimed to vegetation which is not ecologically rich.

5.6 Use of colliery spoil

Colliery spoil is non-toxic and has the potential for use as fill material. Burnt colliery spoil is favoured over unburnt spoil because it is unlikely to combust and its engineering properties are preferable. The use of colliery spoil has been reviewed for the UK,[222] Spain,[47] and for Germany.[147] The general conclusion is that some colliery spoils will meet specifications for engineering use and that the use of other materials, such as sandy material from washery waste, need further research.

Testing should be carried out on spoil materials to assess their suitability prior to incorporation into engineering works (see Box 5.5). The use of colliery wastes in the construction industry has not been widespread because of the ready availability of other aggregates. However, the trend toward encouraging the use of secondary aggregates, because of environmental objections to primary aggregate production, may lead to an increase in the use of colliery spoil for secondary aggregates.

Box 5.4: Mycorrhizal fungi

Mycorrhizal fungi are one of the most extensively occurring groups of beneficial soil microorganisms. Few plants lack them. Mycorrhizal fungi form an intimate mutualistic association with plant roots (mycorrhizae) that extends the absorptive area of the roots (sometimes thousands of times) and contributes greatly to mineral nutrition, water absorption, and root system stabilisation of the host plant. There are many types of mycorrhizae. The two principal groups are the endomycorrhiza and the ectomycorrhizia. In the ectomycorrhizae the fungus penetrates root cell walls and then radiates hyphae (thin filaments) into the soil. In the ectomychorrhiza the hyphae form a sheath around the roots. Both trees and herbaceous plants can be infected and in both types the absorptive area of the roots is increased by the fungal hyphae. Many of the more successful colonisers on colliery spoil have been found to be infected by mycorrhizal fungi.

Photograph 5.6: Fruiting bodies of mycorrhizal fungi in association with colonising birch trees (source: Richards, Moorehead and Laing Ltd)

Box 5.5: Testing of colliery spoil for construction purposes

In order to determine the suitability of colliery spoil for construction purposes testing should include:

Laboratory testing

Geotechnical
particle size distribution
density
relative density
moisture content
particle strength
shear strength

Chemical
pH
sulphate
pyrite content
coal content

Colliery spoil may have been contaminated by-products of coal carbonisation, demolition materials such as asbestos or other non-coal related wastes. Analysis may need to be carried out for these substances also (see Chapters 2, 9 and 10).

***In-situ* testing**

penetration tests
density
moisture content
load tests to measure settlement characteristics

Increasing emphasis on sustainable development may also encourage such use of colliery spoil. Colliery spoil has been used extensively in engineered fill for road embankments and also for housing and industrial developments. Local use as fill material is likely to continue provided any potential combustibility can be dealt with (see Section 7.7). The extent of settlement of freshly placed material will be dependent on the depth of the material, method of placement and degree of compaction. Material which has been in place for many years may already have undergone some settlement and be less likely to be subject to further settlement. Colliery spoil has also been used successfully as cover material in reclamation schemes for other wastes (see Section 14.4.4).

Prior to construction on colliery spoil it may be necessary to cover the spoil with a layer of inert material as a precaution against combustion (see Section 7.7).

Pyritic spoil can heave due to formation of gypsum (calcium sulphate), when sulphuric acid formed during pyrite oxidation reacts with lime containing materials. Such heave has been most often documented in relation to the use of pyritic shales as fill or where construction has taken place on pyritic shales *in situ*.[113, 197, 176]

6 COLLIERY SPOIL HEAP STABILITY

Chapter contents

6 COLLIERY SPOIL HEAP STABILITY

6.1 Introduction

The characteristics of existing colliery spoil heaps vary widely depending on the nature of the spoil and the method of deposition (see Chapter 5). Old tips were generally formed in an uncontrolled manner by loose tipping, often by mechanical means. The tips so produced were conical or irregular in shape with a high profile in relation to the local topography.

Little attention was paid to the subject of tip stability until tip failures began to occur. Most tip failures resulted from errors in design, construction and operation or from poor maintenance. Often failures resulted from action being taken without recognition of the dangers involved. High moisture contents or water seepage are associated with most failures in spoil heaps and lagoon banks.

The problem of tip stability was brought into focus by the disaster at Aberfan, South Wales, UK in 1966 when a serious flowslide occurred and thousands of tonnes of liquified spoil swept down the mountainside into the village below killing 144 people[5, 158] (see Photograph 1.1). The events at Aberfan led to the introduction of the Mines and Quarries (Tips) Act, 1969, and the more detailed Mines and Quarries (Tips) Regulations, 1971, which now govern the design, construction and management of tips in the United Kingdom. A considerable amount of work was carried out by the United Kingdom National Coal Board following the Aberfan disaster and a technical handbook was produced[169] which is the source of some of the information in this chapter.

After Aberfan the stability of many tips was investigated and where necessary precautionary or remedial works were carried out. These measures usually involved regrading slopes and improving drainage.

With new tips, where modern engineering and management practices have been introduced, the likelihood of these tips becoming unstable has been virtually eliminated.

Where sites are being reclaimed or redeveloped, it is therefore important to investigate any tips which are to remain and establish whether or not an adequate factor of safety exists against failure, and whether any remedial measures are necessary to ensure long-term stability.

The main responsibility for the design, construction and management of tips lies with engineers, and many fields of expertise in addition to civil engineering are called upon, such as soil mechanics, geology, hydrogeology and hydrology. In addition, since the long-term behaviour of earth structures can be influenced by the vegetation which they support, disciplines other than engineering, such as soil science, ecology and horticulture, need to be involved. However it is essential that practical experience is also taken into account. The design and construction of large scale earth structures is not an exact science and the interpretation of local conditions, an understanding of which is vitally important to the successful completion of any project, must be given full regard.

6.2 Spoil and tip characteristics

The effect of the mechanisation of mining and spoil deposition methods on spoil heap characteristics is described in Section 5.1.

The various methods of tip construction and the characteristics of these methods are discussed in Box 6.1. Typical tip configurations for older tips are shown in Figure 13.1.

Box 6.1: Common tipping methods

Dry materials end tipped over high faces.
Where tips are formed of dry material tipped over high faces, the slopes generally form at the angle of shearing resistance of the loose material *i.e.* at the angle of repose (see Figures 6.1(a) and 6.1(c)). The water table in the tip is generally low or absent, except where the tips have been built over springs or where surface water is allowed to flow into the tip. Many tips in this category are partially or completely burnt.

Wet material tipped over high faces.
Where heaped tips are formed of wet materials such as washery discards, the slopes formed are generally irregular with the upper parts of the slopes steeper than the lower parts. Periodic slumping may occur leaving steep slopes near the crest of the heap and shallower slopes considerably less than the angle of shearing resistance near the base. The wet material in the shallower slopes may be subject to prolonged creep and bulging. Perched water tables are often found within these tips with localised seepages emerging from the lower slopes. Many tips of this type are partially-burnt, but generally to a lesser extent than dry tips.

Thick layer tips.
Where a tip has been formed of dry materials placed in thick layers, each layer will have similar characteristics to those described above for end tipped dry material. With wet material slumping may occur. In general the wetter and more clayey the material and the thicker the layer, the greater the probability that slumping will have occurred. Stratification usually takes place, and perched water tables often occur at the junctions between the layers (see Figure 6.1(b)). In addition, some such tips may have been subject to combustion.

Thin layer tips.
More recent construction practice is to place material in layers approximately 300mm thick and compact the material either randomly using construction plant or systematically using rollers. The side slopes are usually formed to less than the angle of shearing resistance of the tip material. Such tips are less permeable than tips of the same material built by other methods, and the water table is slower to respond to changing weather conditions. Tips of this type are not liable to spontaneous combustion.

Lagoon banks.
Most old lagoon banks were built by thick layer tipping with steep outside slopes approaching the angle of shearing resistance of the tip material. Heightening of the banks was often carried out 'fir tree fashion' by an additional bank of spoil placed partly on the existing bank and partly on the lagoon deposit (see Figure 6.1(d)). Perched water tables often occur at each of the layers, and seepage may emerge on the outside slope of the bank. Recently built lagoon banks are generally of coarse discard placed in thin compacted layers with outside slope gradients of less than 1 in 2. Drainage provisions within or below the banks prevent seepage water from the lagoon deposits emerging on the outside slopes.

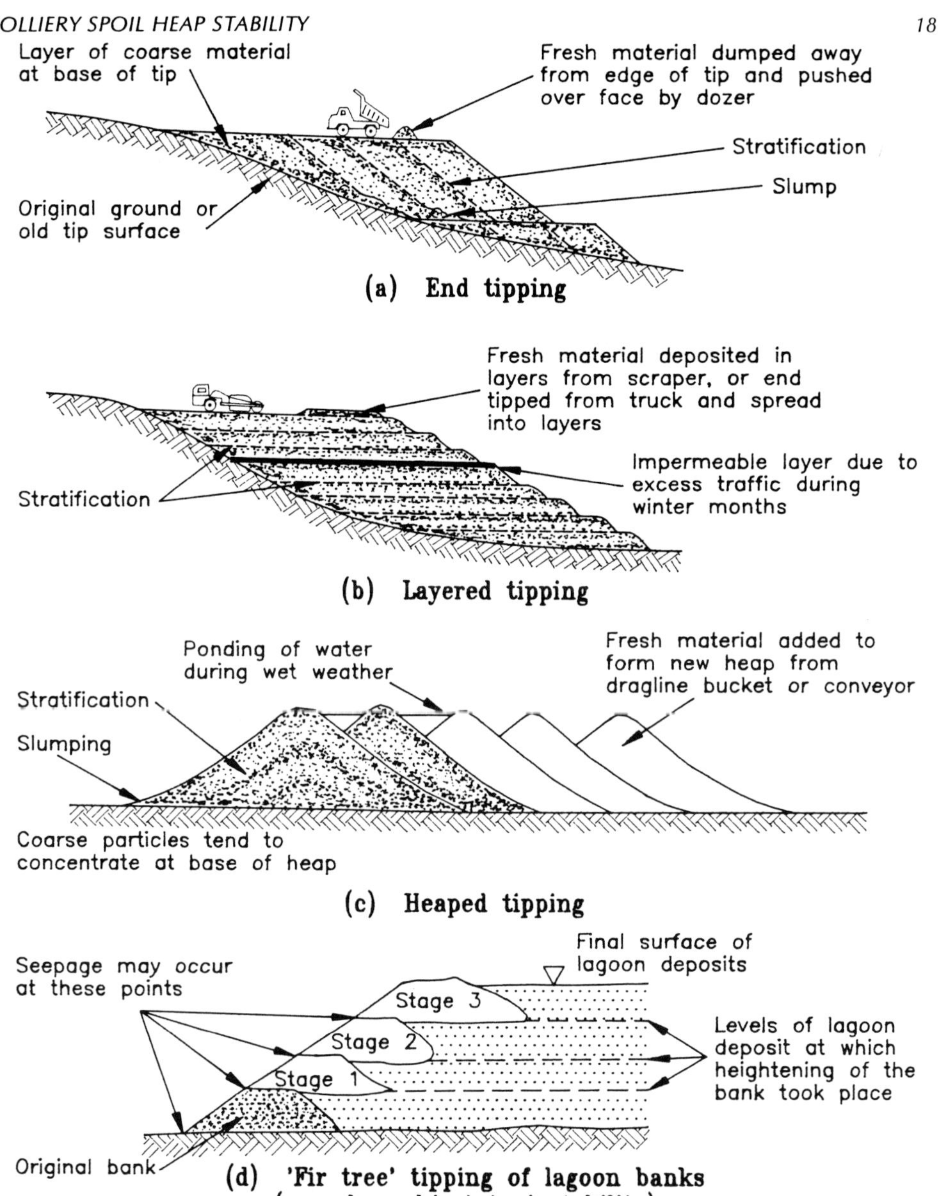

Figure 6.1: Some examples of construction methods for tips and lagoons (after Geoffrey Walton Practice, 1991[100] redrawn with the permission of the Controller of HMSO)

6.3 Factors affecting stability

Although the tip may appear stable this should not be taken to indicate that the tip has an adequate factor of safety against failure and will therefore remain stable. The factor of safety is the ratio of the disturbing forces over the resisting forces. Existing tips originally formed by tipping over high faces but which stand with slopes less than about 20° (1 in 2.75) are likely to have been unstable at some time in the past, and may still be unstable.

A number of factors may serve to reduce the stability of an existing tip by increasing the disturbing forces or reducing the resisting forces present in the tip or its foundations (see Box 6.2). Water in the tip can adversely affect stability in a number of ways, as follows:

- by reducing the strength of the material in the tip or the material on which the tip is founded;
- by increasing the weight of tip material thereby providing additional disturbing forces;
- by generating water pressures in the tip thereby reducing the effective shear strength of the tip material;
- by generating seepage pressures leading to piping (internal erosion; see Box 6.2).

6.4 Modes of tip failure

6.4.1 Introduction

The main types of slope failure in spoil tips and lagoon banks are shown in Figure 6.2 and are discussed in the following sections along with other more localised failure modes.[100]

Box 6.2: Factors which may cause a reduction in tip stability

Additional loading on the top edge of a slope by further tipping, vehicle loading or water seepage into the tip;

Steepening of the slope due to excavation, or erosion due to uncontrolled drainage. Mining subsidence beneath the toe of the slope could also lead to marginal slopes becoming unstable;

Removal of support at the toe of the slope by excavation or water erosion. Wave action in a lagoon may cause undercutting of a bank;

Increase in water level (pore pressure) within the tip or its foundation leading to a reduction of effective shearing resistance along potential failure surfaces. This may occur due to surface water seepage, lack of drainage provisions for springs, seepage from lagoon deposits, blocked culverts, or due to the effect of mining subsidence. Freezing of the surface of a slope may also lead to a build up of seepage water;

Disturbance of the tip or its foundations which may reduce effective stresses between particles with a consequent rise in pore pressure. Sudden disturbances may be caused by: vibration e.g. from blasting or pile driving, mining subsidence, impact loading from tipping, or the effect of slippage in an adjacent part of the tip;

Piping, which is internal erosion within the tip due to the passage of water, thereby forming voids and reducing stability;

Softening or swelling due to the effects of water seepage or ponding may lead to a localised reduction in stability, and any resulting slippage could affect the overall stability of any adjacent tip or its foundations;

Rapid drawdown by the removal of wet deposits or water retained within a lagoon may lead to a rotational slippage of the lagoon bank;

Spoil heap combustion may cause the formation of voids which could result in a localised collapse, although sometimes burning may improve stability by an increase in the shear strength or by the fusing of spoil material.

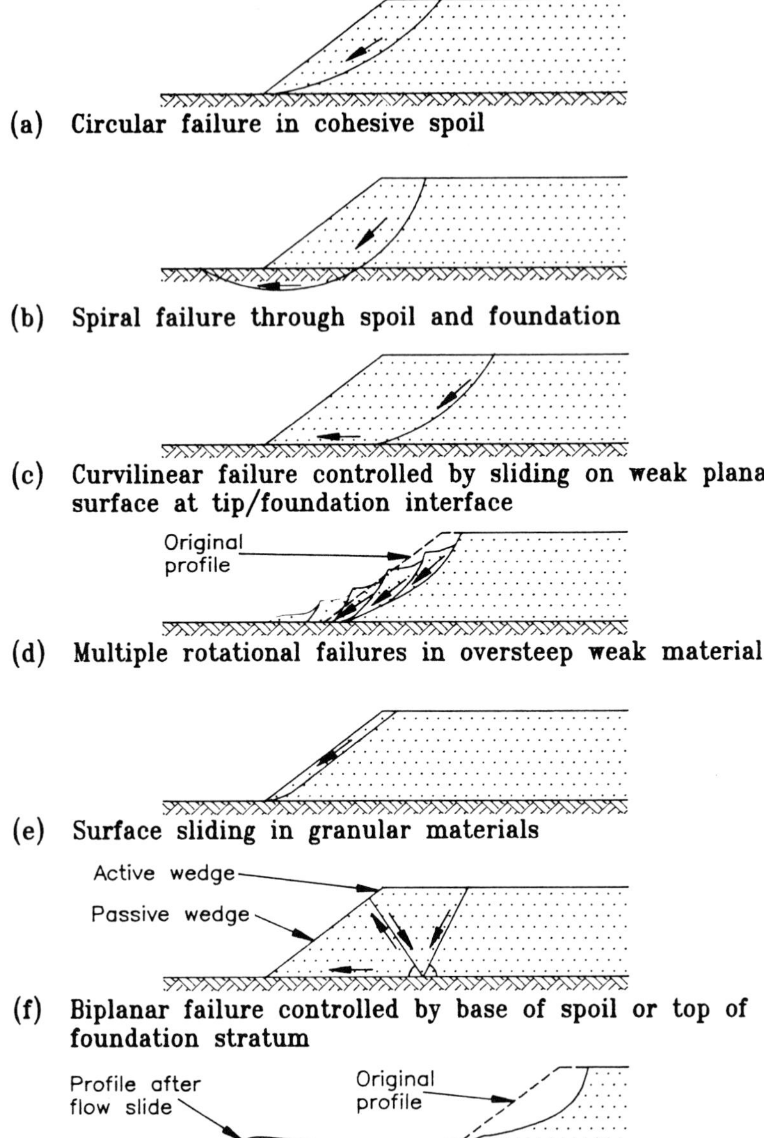

Figure 6.2: Modes of tip failure (after Geoffrey Walton Practice (1991),[100] redrawn with the permission of the Controller of HMSO)

6.4.2 Rotational slips

In rotational slips, movement takes the form of rotation about an axis which is generally outside the slope itself. The shape of the failure surface varies depending on the characteristics of the material in the tip (see Figures 6.2(a) to 6.2(d)). The first sign of slippage is usually a tension crack in the spoil above the failure surface, sometimes accompanied by slumping of the spoil in front of the crack (see Photograph 6.1). The rate of failure is generally relatively slow *i.e.* developing over a period of hours to days.

6.4.3 Surface slips

For surface slips the failure surface is parallel to the front face of the tip (see Figure 6.2(e)). Such slides generally take place in dry, cohesionless, granular material tipped at or above the angle of repose, and may occur as the surface layer dries out and any temporary cohesion is lost. Surface

Photograph 6.1: Example of a rotational slip failure (source: I.G. Brown)

slips may also occur where the strength of the surface layer has been reduced by weathering or water action, or following the removal of the toe of the tip by excavation or erosion.

6.4.4 Biplanar slips

Biplanar slips involve shearing along two planes of differing orientation, with the formation of an upper active wedge of material and a lower passive wedge (see Figure 6.2(f)). The upper wedge displaces the lower wedge and shearing occurs at the tip foundation. Such failures are common in tips where the foundation material is soft and weak.

6.4.5 Flow slides

Flow slides occur when the soil mass in transformed into a liquified state as a result of disturbance following saturation. Disturbance may occur due to rotational failure, mining subsidence or vibration from earthquakes, blasting or heavy plant, Collapse of the soil structure takes place, but closer packing of the grains of spoil material is prevented by the viscosity of the water, and liquefaction occurs allowing the spoil to flow downhill as a slurry. Flow slides typically occur in poorly compacted or saturated spoil heaps consisting of sand or silt sized material, and are a common failure mode in lagoon materials following breaching of the lagoon bank. Flow slides occur rapidly and material can travel significant distances.

6.4.6 Piping failure

Piping is a localised failure caused by internal erosion within the tip as soil particles are washed out by the passage of water. Collapse due to piping may sometimes trigger other forms of failure.

6.4.7 Cavitation collapse

This is a localised collapse of underground voids resulting from events such as piping, collapsed culverts or underground combustion. General tip stability is not usually affected, except sometimes for lagoon embankments, although sudden collapse may be a source of danger to persons at the surface.

6.4.8 Mud runs

A mud run is a localised failure caused by a rapidly moving flow of water-borne soil having the consistency of mud. This is brought about following heavy rainfall by flows or issues of water eroding gulleys in the side slopes and forming mud runs at the base of the tip (see Photograph 6.2).

Photograph 6.2: Example of water erosion forming surface gulleys (source: Welsh Development Agency)

6.4.9 Settlement and heave

Settlement and heave may occur to varying degrees within different parts of a tip, sometimes giving rise to differential movement.

Settlement occurs as a result of loading due to self weight or imposed loads. Collapse settlement may sometimes occur as a result of a reduction in strength of a material following saturation.

Heave may take place for a variety of reasons, such as:

- rotational failure;
- shear failure of weak material in the tip foundations;
- upward seepage pressures;
- following removal of surcharge loading *e.g.* when a tip is removed;
- chemical changes of the tip materials;
- the action of frost;
- rehydration of desiccated clay *e.g.* following the removal of trees.

6.5 Investigation and stability analysis

6.5.1 Introduction

Stability analysis plays an important part in the process of reviewing tip stability, particularly in situations where conditions affecting the tip are to be altered, or have already been altered. The investigation and stability analysis should be carried out by specialist engineering or geotechnical consultants with experience of slope stability problems.

Investigation and assessment of slope stability is discussed in the following sections.

Investigations will be required on an existing tip if:

- the appearance of the tip suggests that failure might occur, thus endangering persons or property;
- an existing tip, although appearing stable, by its size or situation constitutes a significant threat to the safety of persons or property;
- works are to be carried out on or in close proximity to an existing tip which may have an adverse effect on the stability;
- there is serious public concern that a tip may constitute a threat to persons or property.

The extent of the investigations carried out should also take into account the possible secondary consequences of a failure, such as:

- possible blockage of an adjacent watercourse leading to a sudden failure and release of water;
- possible displacement of lagoon deposits leading to overtopping (*i.e.* overflowing) and erosion of the lagoon bank with consequent failure;
- possible disruption of nearby water mains leading to a release of water which could accentuate the failure;
- possible impact of material from a rotational slip in the upper parts of a tip slope leading to liquefaction and flowslide failure in the lower part of the tip.

6.5.2 Initial assessment

An initial assessment should be carried out based on a walk-over survey and desk study (see Section 2.3) to establish the geological structure, geomorphology and any historical events influencing stability. The following information should be sought:

- the nature of the site before the tip was built, particularly with regard to original watercourses, drains, culverts and any history of previous tipping or opencast workings;
- the nature of the materials tipped and their location within the tip, particularly fine-grained materials such as tailings. Fine-grained residues present in buried lagoons or as areas of dumped materials could lead to the formation of a zone of weakness;
- the stages of construction of the tip and its varying shape and size in order to allow the identification of weathered surfaces, surfaces of low permeability, boundaries between different tip materials and areas where segregation may have occurred;
- the methods of deposition of material within the tip which may provide guidance as to the density, relative permeability and homogeneity of the tip materials;
- the presence of any zones or surfaces of weakness within the tip or its foundations deduced from knowledge or records of past problems or failures on the tip;
- the positions of all water mains and their control valves.

6.5.3 Other relevant information

Other important information to be established and taken into consideration in assessing tip stability is outlined in the following sub-sections.

Topographical details

Survey plans and sections of the tip will be required to show the site and its surroundings: before tipping was started, during the development of the tip, and at the time of the site investigation. These plans and sections will be used to complement studies of the geology, hydrogeology, rainfall and hydrology, and in compiling a history of tipping. The plans and sections will also provide information to be used in the stability analysis.

Subsidence plans

Plans of all past and proposed mining which might adversely affect tip stability should be prepared to show the cumulative effects of past and future extraction.

Geology

A study should be made of the geology and geomorphology of the tipping site. This should include details of the nature and distribution of all superficial deposits as well as details of dip and characteristics of the solid strata below the site.

Hydrogeology

An assessment should be made of the hydrogeological conditions, particularly with regard to the possible formation of springs or artesian pressures which might influence tip stability. The effects of mining subsidence on the hydrogeology should also be considered.

Ground movements

Consideration should be given to evidence of any ground movements in the natural ground beneath the tip due to rotational slips, creep, rock slides, cracks and fissures, or erosion by flowing water.

Rainfall and hydrology

Data on the rainfall and hydrological conditions should be collected. This will be necessary for the design of drainage systems. Information may also be required for the stability analysis with regard to fluctuating water levels in the tip or its foundations.

Site investigations

Subsurface exploration should be carried out using trial pits, trenches or boreholes (see Section 2.5). Boreholes should extend through all strata likely to have an influence on the stability of the tip, and will normally be carried down to bedrock. At least three boreholes, but preferably more, will be required to provide sufficient information for the stability analysis; one borehole at the toe of the slope, one mid way up, and one close to the crest of the slope. All boreholes should be sealed by grouting after sampling.

If groundwater levels and pore pressures need to be monitored over a period of time at various levels within the tip and its foundations, piezometers should be installed.

An assessment should be made of the shear strength of the tip materials based on *in situ* and laboratory testing of representative samples.

6.5.4 Stability analysis

Stability analysis must be undertaken systematically, and will involve the following activities:

- establishing the geometry of the structure to be analysed;
- collecting relevant input data for the analysis;
- assessing the likely mode or modes of failure;
- estimating security using slope stability analysis to calculate the factor of safety against failure.

Geometry

Most common methods of slope stability analysis are two dimensional and are based on the slope cross section. Critical cross sections should therefore be established from the topographical survey information.

Input data

The relevant input data should be based on the information discussed in Section 6.5.3. The shear strengths of the materials in or beneath the tip should be obtained from:

- laboratory testing of representative samples;
- *in situ* testing;
- empirical strength relationships;
- experience based on previous test results and published values;
- past analyses of previous failures.

Mode of failure

The most critical modes of failure should be considered. Modes of failure are discussed in Section 6.4.

Slope stability analysis

Slope stability analysis is normally carried out by computer program on an iterative basis using one of a number of design methods which are available.[100] Factors of safety can be calculated for various situations, and the most critical factor of safety should be established by varying the following parameters as appropriate depending on the circumstances:

- mode of failure;
- range of failure surface geometries;
- geometry of the tip *i.e.* height, slope angle, foundation gradient;
- water table;
- shear strengths of the tip materials and tip foundations.

Minimum acceptable factors of safety are shown in Box 6.3.

6.6 Remedial measures on existing tips

6.6.1 Introduction

Remedial measures on existing tips can be categorised under the following headings:[169]

- precautionary work;
- remedial work;
- maintenance.

6.6.2 Precautionary work

Precautionary work is required where stability analysis indicates that a tip may become unstable, or where features observed on site suggest a failure may occur. Precautionary measures may also be required in the following circumstances:

- to counteract the adverse effect on a tip due to external influences *e.g.* regrading a slope, or constructing a berm at the toe of a tip to counteract the effect of subsidence from proposed mine workings;
- to reduce risks where it is suspected that a failure may occur *e.g.* by closing roads in the vicinity, or by evacuating nearby property;
- to reduce risks where failure is not expected but where a tip or lagoon presents a risk to persons or property *e.g.* regrading a large steeply sloping tip located near houses.

Box 6.3: Minimum factors of safety suggested for tip design (after Geoffrey Walton Practice, 1991[100])

Design criteria	Factors of safety	
	Case 1	**Case 2**
Design based on peak shear strength parameters	1.5	1.3
Design based on residual shear strength parameters	1.3	1.2
Tip designed to appropriate strength parameters and subject to seismic or other ground accelerations	1.2	1.15

$$\textbf{Factor of safety} = \frac{\text{Resisting forces}}{\text{Disturbing forces}}$$

Case 1: applies where persons or property may be at risk

Case 2: applies where persons or property would not be endangered by failure.

Note: These factors of safety are given for guidance only. Higher or lower factors of safety may be appropriate depending on the adequacy and reliability of the input data and design assumptions.

6.6.3 Remedial works

Remedial work is required where slippage has occurred or where movement of the tip indicates that it is unstable. Depending on the circumstances, the following precautionary or remedial works may be carried out to increase or restore stability:

- construction of a berm one third or half way up the slope to increase the resisting forces (see Figure 6.3(a));

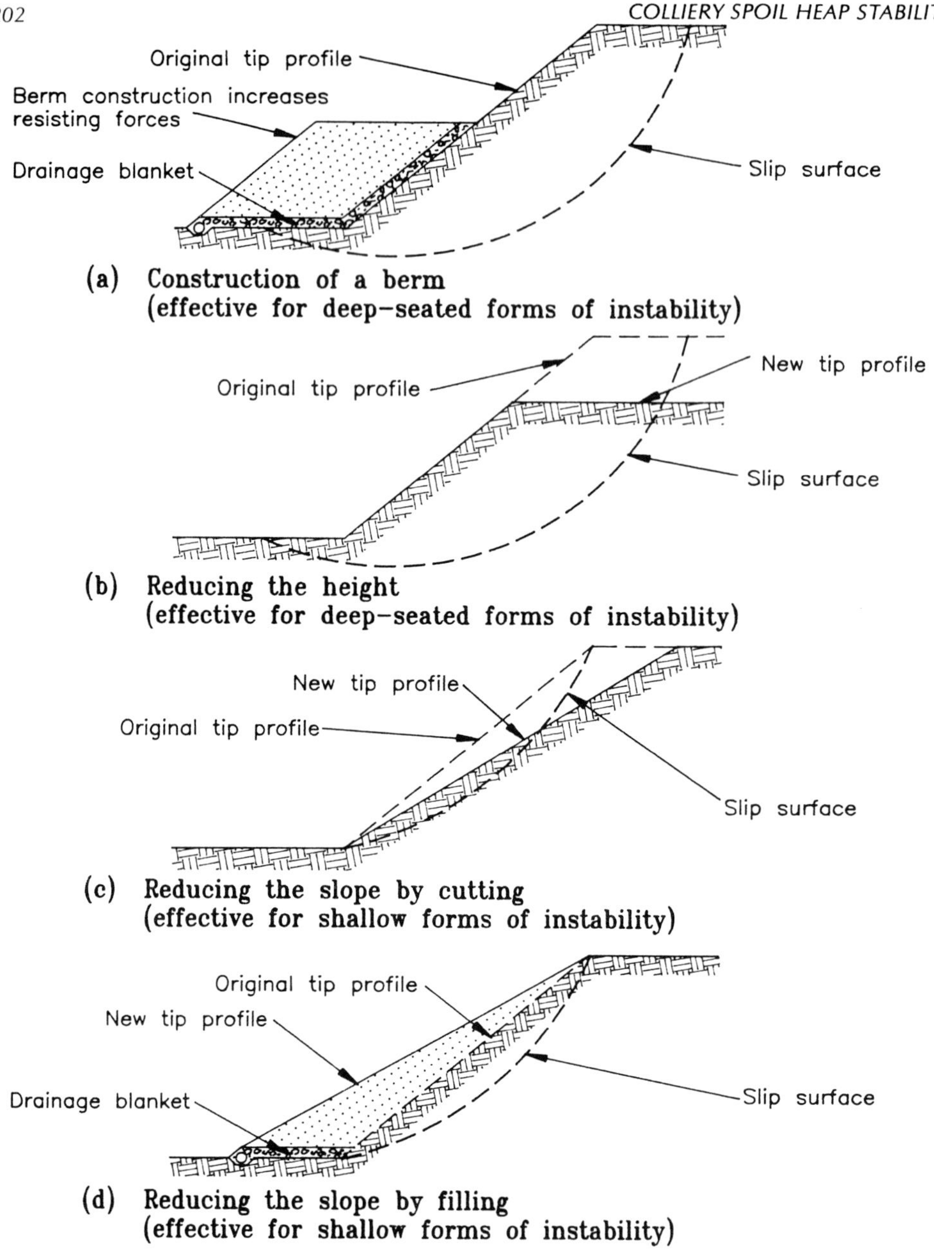
Original tip profile
Berm construction increases resisting forces
Drainage blanket
Slip surface
(a) Construction of a berm (effective for deep-seated forms of instability)
Original tip profile
New tip profile
Slip surface
(b) Reducing the height (effective for deep-seated forms of instability)
New tip profile
Original tip profile
Slip surface
(c) Reducing the slope by cutting (effective for shallow forms of instability)
Original tip profile
New tip profile
Drainage blanket
Slip surface
(d) Reducing the slope by filling (effective for shallow forms of instability)

Figure 6.3: Methods of improving the stability of spoil heaps (after National Coal Board, 1970[169])

- reduction of the height of the tip (see Figure 6.3(b));
- flattening of the slope to reduce the disturbing forces (see Figure 6.3(c));
- reducing the gradient of a slope by introducing new material (see Figure 6.3(d));
- lowering the water table in the tip by improving the existing drainage system or by installing a deep drainage system using relief wells or bored filter drains (see Figure 6.4);
- construction of a retaining wall, or the installation of sheet piling at the toe of the tip. For sheet piling, consideration should be given to the possible adverse effects of pile driving on tip stability;
- preventing surface erosion through the use of vegetation (see Section 13.6.4).

The remedial work should be designed so that during construction there is a progressive increase in stability. Any work involving changes to existing conditions should be carefully designed to avoid adverse secondary effects occurring. The following aspects should be considered:

- erosion which might follow the disturbance of existing vegetation or a mature weathered surface;
- instability which might arise from the exposure of buried lagoons or other weak zones during regrading;
- reduction in stability which might arise from an inappropriate sequence of earthmoving operations, from incorrect siting of drainage works, or from an incorrect choice of earthmoving plant or procedures;
- the starting of a fire or the encouragement of burning;
- the proper collection of running or standing water and its continuous removal prior to earthmoving operations;
- the long-term effectiveness of any previous remedial work such as emergency drainage works;
- further instability caused by the deterioration and collapse of steep scarps remaining after a slip.

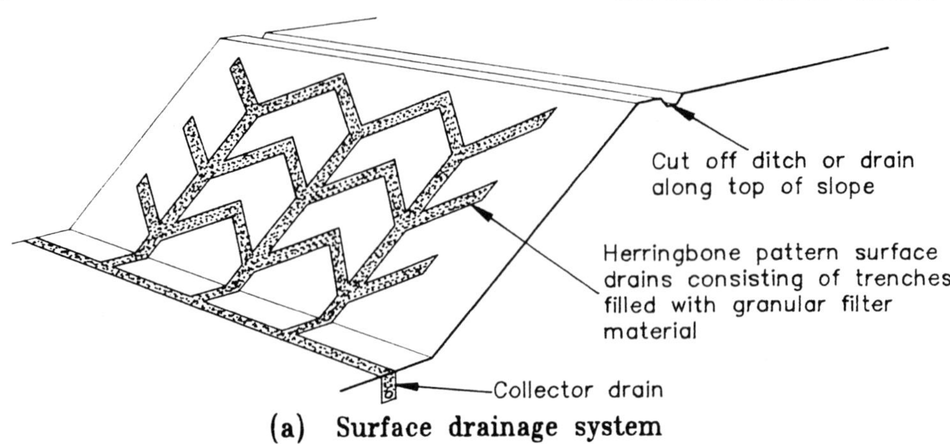

(a) Surface drainage system

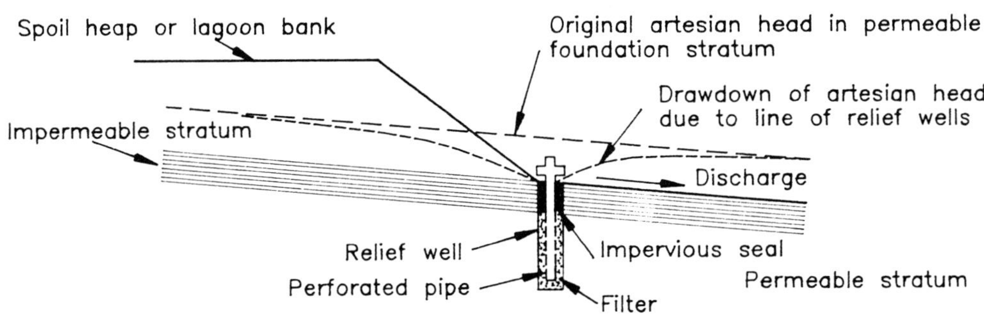

(b) Deep drainage using relief wells

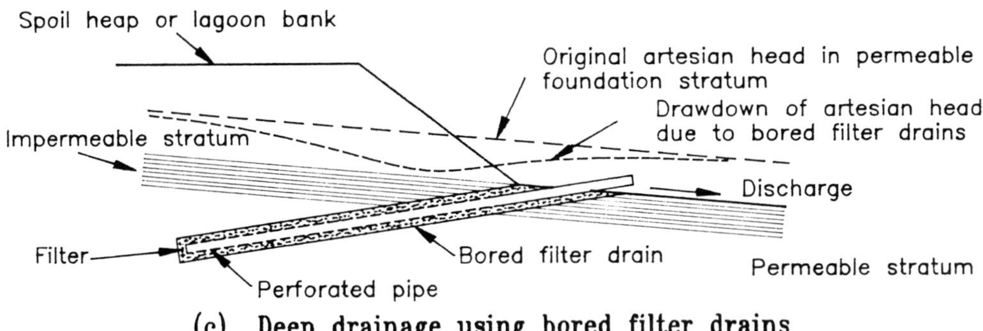

(c) Deep drainage using bored filter drains

Figure 6.4: Improving slope stability by shallow and deep drainage (after National Coal Board, 1970[169])

6.6.4 Maintenance

The need for maintenance

Maintenance work is required on all tips whether they are active or disused. Any features of concern noted during inspection or maintenance should be reported in writing to the responsible authority so that a re-appraisal can be made of the stability of the tip. The following subsections discuss items which require maintenance.

Drainage system

Maintenance of the drainage systems should be carried out at periodic intervals, and particularly after heavy rain, to prevent the unnecessary entry of water into the tip. Such work will include:

- the clearance of vegetation, spoil and refuse from screens, settlement ponds and ditches;
- rodding pipes to remove sediment;
- cleaning silt traps;
- cleaning of drainage inlets and outlets;
- diverting any flow or accumulation of water into the permanent drainage system;
- maintenance of filters and settlement ponds particularly where river pollution may be a problem.

Lagoons

Maintaining lagoons can involve the adjustment of inlet and draw-off or overflow arrangements, the clearance of any blockages, and the rectification of any undercutting due to wave action. Where lagoons have become redundant but have not been filled, the draw-off or overflow arrangements should be maintained so as to control accumulation of water which may occur in the future. A specific programme should be prepared for the infilling/abandonment of any disused lagoons and other features.

Haul roads

Haul roads remaining in use may require periodic regrading and repair. Drains and culverts should also be inspected and maintained in good order to avoid flooding.

Vegetation

Existing vegetation should be maintained in a healthy condition. Grass should be controlled by cutting or grazing, which also acts as a useful precaution against damage caused by burning.

6.7 Construction methods for new tips and the reshaping of existing tips

6.7.1 Design requirements

Design requirements for reshaping existing tips are similar to those for new spoil heaps and lagoon banks and are likely to include consideration of the following items:[100]

- site investigation (including a desk study of available data);
- stability analysis;
- calculation of volumes;
- design of surface and subsurface drainage measures;
- selection of earthworks equipment;
- specification of tipping rules and details of construction.

Additional requirements for lagoons and attenuation ponds should include:

- determination of required lagoon area and capacity;
- design of an efficient surface drainage network for the site;
- specification of inflow, outflow and emergency outflow arrangements.

6.7.2 Method of placing spoil

Spoil should be placed in heaps on lagoon banks in a controlled manner in accordance with good earthworks practice such as that set out in Box 6.4.

6.7.3 Tip geometry

Safe slope gradients for spoil materials and for terracing of slopes should be determined following stability analysis (see Section 6.5.4)

6.7.4 Lagoon banks

Special care is required in the design and siting of lagoon banks, particularly where these are to be built on or adjacent to existing tips or lagoon deposits. Generally, lagoon banks are designed either as a drainage bank to allow maximum drainage from the lagoon deposits, or as an impermeable bank which restricts drainage from the deposits.

6.7.5 Drainage measures

Drainage provisions should be made to reduce pore pressures within the tip and to collect and discharge surface water and groundwater and prevent saturation or ponding. Drainage measures may include the provisions in the following paragraphs.[169]

Tip underdrainage

This may take the form of a drainage blanket, radial or herringbone drains, or relief wells depending on the site conditions (see Figures 6.4(b) and 6.5). Filter materials for drainage purposes should be selected to avoid long-term deterioration. Where suitable filter granular materials are available only in limited quantities the use of geotextiles may be appropriate.

Internal drainage

Additional drainage blankets may be required within the body of the tip to reduce pore pressures and prevent the development of perched water tables.

Box 6.4: Methods of placing spoil[169]

Method A - in layers not greater than 300mm thick (uncompacted or tracked in with a bulldozer);

Method B - in layers not greater than 300mm thick and compacted with a minimum of four passes of a towed smooth-wheeled roller having a weight not less than 5 tonnes per metre width, or equivalent roller;

Method C - in layers not greater than 5m thick; within the range 300mm to 5m the thickness of layers must be the minimum commensurate with stability requirements and practical considerations. If placed in a lagoon bank the maximum thickness should be 1.5m.

Notes:

- where shear strength (*i.e.* stability) is the main requirement the spoil should be placed in thin layers by Method A or B;
- where permeability for drainage purposes is the main requirement the spoil should be placed in uncompacted layers by method A or C;
- lagoon banks and spoil liable to combustion should be placed in thin layers in accordance with Method A or B.

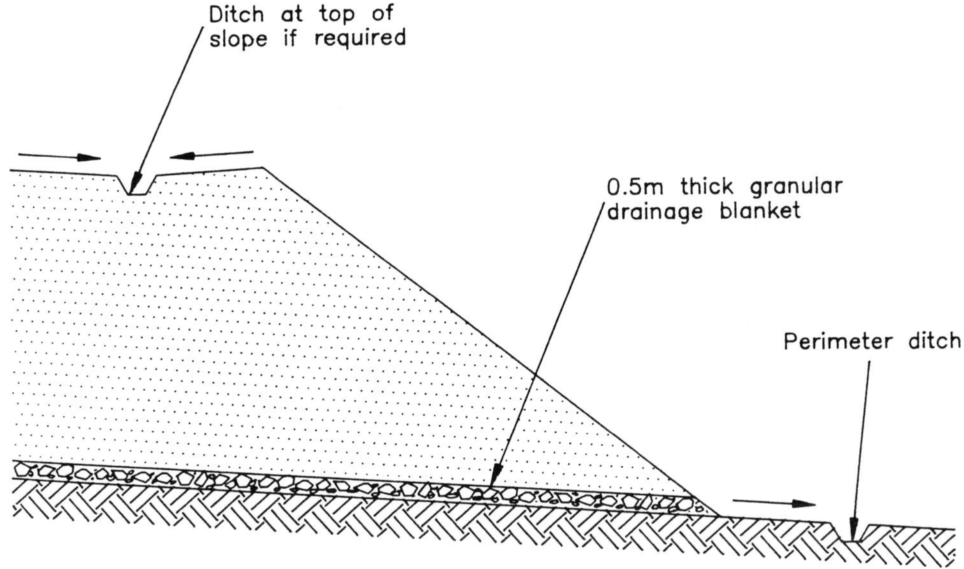

(a) Provision of drainage blanket

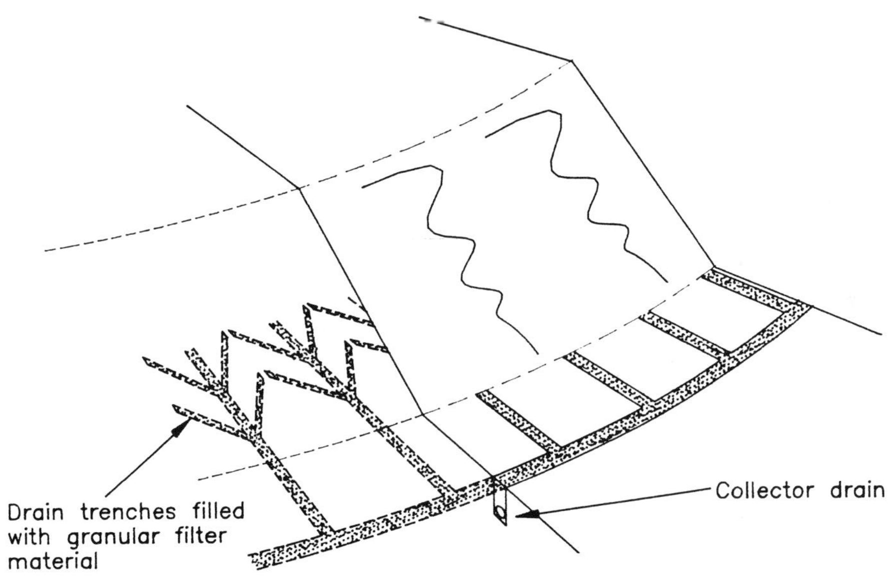

(b) Provision of herringbone drainage system

Figure 6.5: Provision of drainage beneath new tips

Surface drainage

A main drainage ditch should be provided around the bottom edge of the tip to collect surface water and water from internal drainage. The ditch should be located at least 3m from the toe of the heap, and may need to be lined depending on the ground conditions. A ditch may also be required along the top edge of the tip in order to intercept surface water reaching the tip from higher up the slope (see Figure 6.5(a)).

6.8 Inspections

At periodic intervals inspections should be carried out by a suitably qualified engineer to check the stability of active or disused tips in accordance with relevant legal requirements. Items likely to be included in the inspection are shown in Box 6.5.

Box 6.5: Periodic inspection of tips

Defects highlighted by poor performance must be attended to, and a system of inspection, reporting and implementation of recommendations should be established within the responsible organisation.

The items to be included in the inspection are as follows:[170]

Changes in situation and shape
Any changes with regard to development on adjacent land, clear space at the toe of the tip, dimensions and volumes of the tip, and widths at the bottom and top of slopes will require inspection. Assessment of the probable consequences of a failure of each section of the tip and of overtopping of each lagoon should be made.

Drainage
Inspection of surface drainage with regard to the condition, adequacy and freedom from obstruction of watercourses. Inspection of the efficacy of internal drainage of spoil heaps, lagoons and foundations is also necessary as is the inspection of surface drainage arrangements of lagoons and any erosion at inlets or outlets.

Tipping operations
For active tips, inspection should be made with regard to: the characteristics of the spoil, adherence to specification and tipping rules, method of placing and compaction, location and condition of lagoons, any excavation works or emptying of lagoons, any precautionary, remedial or improvement works, and any other work.

Cracks or movement of the tip and foundation
Inspection of the following is necessary: the positions and sizes of any cracks or movement and comments on causes, any land slips, and records of movements. Special attention should be given to the locations of buried lagoons and areas which may be affected by mining subsidence.

Fire
Inspection of the position and extent of any fire, fumes given off, and any control measures carried out.

Safety
Inspections (where appropriate) of personnel, public, trespassers, notices, lighting, fences, dangerous practices, adequacy of manhole covers and screens, walkways to lagoon controls and communications are useful and necessary.

Vegetation
Inspection of the extent and type of vegetation is useful as a guide to the successful management of the tip, its integrity and moisture characteristics, and its ecological value and balance.

7 COLLIERY SPOIL HEAP COMBUSTION

Chapter contents

7 COLLIERY SPOIL HEAP COMBUSTION

7.1 Introduction

Colliery spoil heap combustion is a common occurrence, particularly on older spoil heaps formed by loose tipping over the edge of the heap. At some sites underground fires are known to have been burning continuously for several years.

Little information is available on the current extent of spoil combustion, although in the main coal mining areas of Europe several incidents are reported each year. Historically the problem was much worse. It was estimated in 1967, for example, that of the 2,000 loose tipped spoil heaps then owned by the United Kingdom National Coal Board 15% were classified as burnt out and more than half as burning.[235] Box 7.1 provides some examples of underground combustion, and Box 7.2 gives examples of explosion and landslide.

In recent years civil engineering design has required that spoil is compacted during tip building and this has considerably reduced the risk of combustion.

Spoil composition and consistency varies widely, but loose tipped heaps containing a high proportion of carbonaceous materials are most likely to combust.

Spoil heaps may be ignited accidentally by the lighting of fires or the tipping of hot ashes. Self-heating may also occur, which can lead to spontaneous ignition.

When the supply of oxygen is limited, slow combustion may occur by smouldering, but without the spoil material breaking into flame. Smouldering may proceed underground for long periods with little visible evidence at the surface.

Box 7.1: Examples of underground combustion

1. Lanarkshire Steelworks, Scotland

During redevelopment of a former steelworks site an underground fire was eradicated which was known to have been burning for 45 years[134]. The fire within colliery waste was causing problems with noxious fumes, subsidence of buildings, hardstandings and railway sidings, and disruption of utility services. Early limited attempts to contain the fire were largely unsuccessful, and the fire continued to spread until 4.7ha of the site had at some stage been affected. During the reclamation works the underground fire was treated by excavating the burning material and spreading it out to cool prior to recompacting it in layers.

2. Ramsey Bing, Loanhead, Scotland

Treatment of this burning colliery spoil tip was hampered by the fact that the heap is located in the middle of the town and is surrounded by houses and industrial buildings.[61] A pilot phase of the works was carried out which involved treating part of the tip by excavation, cooling and recompaction. Dust levels were strictly controlled by damping down with water, and dust monitoring was carried out to avoid disruption to a nearby precision engineering works. The pilot phase proved to be successful, and the main phase of the works was then carried out including capping and landscaping of the 11.9 hectare site.

3. Oakthorpe, Leicestershire, England

Combustion can also occur within underground workings. Burning within abandoned shallow workings beneath Oakthorpe occurred over a number of years as a result of spontaneous combustion.[27] This resulted in elevated temperatures beneath buildings, and subsidence damage to buildings and roads. Treatment was carried out by drilling and grouting using pulverised fuel ash and cement grout to extinguish the burning and stabilise the workings.

Box 7.2: Examples of explosion and landslide[79]

The following incidents took place in Pas-de-Calais, France.

1975, Fosse 6 d'Auchel, Calonne-Ricouart.
An accident occurred involving an explosion and avalanche of burning material.

The spoil heap was started in 1913 and tipping continued until 1961. The tip was semi-circular in shape with a radius at the base of 150m. The height was 96m and the slope angle was 32 to 36°. Two villages were located close to the bottom of the spoil heap.

Removal of material from the spoil heap began in 1971, and at the time of the accident material was being worked on four staggered levels.

The incident took place after excessively heavy rain when an explosion occurred followed by a landslide which caused burning material with a temperature of approximately 800°C to flow like larva on to the outskirts of the village below.

The cause of the accident is not known precisely, but at the time was thought to have been triggered by mining subsidence or by a pocket of gas disturbed by the excavation of the heap. It is not known whether the heap had been subject to underground combustion, but this seems likely to have been a contributory factor.

1930, Fosse 5 d'Auchel (conical spoil heap).
An explosion and landslide occurred leading to one person being killed and four seriously burned.

1930, Fosse 5 d'Auchel (conical spoil heap).
A further landslide of hot material occurred causing damage to property.

1962, Bruay-en-Artois (flat spoil heap).
One person dead following a gas explosion.

1968, Blignières (flat spoil heap).
Seven injured in an explosion and landslide of hot material.

1973, Fosse 3 d'Auchel (conical spoil heap).
Explosion and landslide causing two injuries by burning.

1974, Fosse 3 d'Auchel (conical spoil heap).
Major landslide of hot material halted by a dike 100m from housing.

The ignition of materials within a spoil heap, whilst not always being obvious at the surface of the tip, is associated with particular safety and public health risks.

The main hazards from burning spoil heaps are concerned with:

- release of noxious gases;
- subsidence damage;
- damage to buried services.

7.2 Causes of combustion

7.2.1 Factors involved

The following factors determine whether or not a spoil heap is susceptible to combustion:

- the composition of the spoil;
- the grading of the spoil;
- the compaction of the spoil;
- the method by which the spoil heap has been formed;
- the size of the spoil heap;
- whether the spoil heap has steep faces exposed to the wind;
- whether the spoil heap has been capped with dense, non-combustible material.

Spoil can be divided into three main categories on the basis of combustion properties:

- subject to spontaneous ignition;
- combustible but not subject to spontaneous ignition;
- non-combustible.

7.2.2 Spontaneous ignition

Introduction

Colliery spoil materials can spontaneously ignite due to self heating caused by the slow chemical oxidation of carbonaceous material in the spoil. The following sections describe factors which affect spontaneous combustion.[169]

Temperature

The oxidation of carbonaceous material is exothermic, and if the rate at which heat is given out exceeds the rate at which it is dissipated then the effect becomes cumulative and the temperature increases. The temperature rises until smouldering becomes possible if the supply of oxygen is adequate. Spontaneous combustion is thus facilitated by the direct oxidation of coal residues.

Pyrite

Pyrite (FeS_2) is often to be found in colliery spoil (see Section 5.3.2). The oxidation of pyrite is strongly exothermic, and the presence of pyrite in conjunction with the carbonaceous material increases the tendency towards spontaneous ignition.

Coal rank

Lower rank coals are more reactive and more susceptible to self heating than higher rank coals. Low rank coals have a lower carbon content and higher oxygen content than high rank coals.

Moisture

The presence of free moisture is essential for the oxidation of pyrite and can accelerate the heating process. The high moisture content in low rank coals is indicative of a tendency to self heating, and thus to spontaneous ignition.

Void ratio and particle size

The void ratio of the material in the spoil heap affects the ease with which air can pass through the heap and thus supply oxygen to stimulate exothermic reactions. Where heaps consist of large, single-size material, movement of air through the heap can be sufficient to dissipate a build up of heat through a relatively large volume of material. In heaps consisting of well graded or fine-grained material there is likely to be little air movement and the heat generated will tend to be retained. Heating will however stop once available oxygen is consumed.

With heaps of intermediate gradings, conditions for spontaneous heating and ignition are ideal and areas of high temperature may form within the spoil heap which can eventually break into flame. The rate of oxidation and build up of heat increases when there is a high proportion of the smaller particle sizes because of their large surface area to volume ratio.

7.2.3 Accidental ignition

Accidental ignition generally occurs due to a sustained application of heat from the surface. Such heating can be caused by a number of factors including:

- a bonfire;
- a grass fire;
- tipping of hot ashes;
- underground electric cables;
- a boiler house or furnace.

Combustion may continue below ground even after the surface fire has been extinguished.

7.3 Hazards

7.3.1 Noxious gases

Gases produced by burning spoil heaps include the following:

- carbon monoxide;
- carbon dioxide;
- sulphur dioxide;
- hydrogen sulphide.

These gases tend to accumulate in low lying areas where ventilation is poor. This accumulation presents a risk to operatives working nearby, particularly vehicle operators working in poorly ventilated cabs. Such gases are hazardous to health either by direct toxicity or by asphyxiation (see Box 7.3).

The gases produced can travel considerable distances through the ground and may enter and accumulate within nearby buildings thereby putting residents or users of the buildings at risk.

7.3.2 Subsidence

The combustion of spoil heaps leads eventually to localised subsidence of the burnt ground. Hidden cavities are thus formed which may subsequently be subject to sudden collapse. Access by earthmoving equipment and personnel is therefore potentially hazardous.

Underground fires spreading beneath buildings present a threat to the structural stability of the building and may lead to eventual collapse. Incidents have occurred where it has not been possible to bring underground fires under control, and buildings have had to be abandoned and eventually demolished.

Box 7.3: Physiological effects of gases produced by burning spoil heaps

Carbon monoxide

Carbon monoxide is the most dangerous of the noxious gases as it cannot be detected by smell, taste or irritation, even when present in potentially lethal concentrations. Low exposure causes headaches, weakness and nausea. Increased exposure leads to shortness of breath, palpitations and vomiting followed by unconsciousness and in severe cases, death.

Carbon dioxide

Carbon dioxide is an asphyxiant. Low exposure causes increased lung ventilation. Increased exposure causes laboured breathing and headaches with eventual loss of consciousness.

Hydrogen sulphide

Hydrogen sulphide is readily detectable by smell and taste. Low exposure causes irritation to the eyes and throat and eventually headaches. Increased exposure leads to unconsciousness. The ability to detect the odour tends to disappear during prolonged exposure.

Sulphur dioxide

Sulphur dioxide is detected by taste at lower levels and by smell at higher levels. Low exposure causes irritation to the nose and throat and eventual irritation to the eyes. Increased exposure can lead to death.

Combustion does not lead to a loss of stability in all instances, and burning may sometimes lead to an increase in shear strength. At high temperatures fusing of materials within the tip may occur (see Table 5.1).

7.3.3 Combustion damage to underground structures and utility services

Any underground structures and services such as cables and pipework on sites subject to combustion may be affected by heat damage. Underground service ducts or drains may sometimes act as paths along which combustion or combustion products such as gases may spread.

7.3.4 Risk of explosion

When attempting to treat underground fires, the use of excessive water could result in an explosion. The use of water jets on a burning heap will lead to the formation of steam. Hot carbon in contact with steam can form 'water-gas' which is a mixture of carbon monoxide and hydrogen. A mixture of water-gas and air, when ignited, will explode with great violence.

A coal dust and air mixture is also explosive, therefore where there is a significant coal content in the heap there is a risk of explosion if excessive dust is produced (see Section 7.6 regarding safety precautions).

7.3.5 Dust

Dust levels will be affected by the weather conditions. Rainy conditions will help to reduce the spread of dust whereas windy conditions will assist the spread of dust. The emission and spreading of dust from a burning spoil heap can cause nuisance particularly in periods of dry weather. This could affect nearby residents or sensitive local industries, but is unlikely to be a health risk unless dust levels are excessive (see Box 7.4).

7.4 Assessment

7.4.1 Introduction

There are many factors which affect whether or not a spoil material is combustible. Laboratory test results provide a method of assessing potential combustibility, but the assessment of test results is a matter of careful judgement particularly with regard to the risk of spontaneous ignition. The following parameters are of particular relevance to spoil combustibility.

7.4.2 Calorific value (CV)

This is the test most commonly used for assessing combustibility and involves measuring the quantity of heat released from a sample on combustion. It is generally agreed that samples with a CV of 10 MJ/kg or greater are almost certainly combustible, whilst samples with a CV of 2 MJ/kg or less are unlikely to be combustible.[120] Samples with a CV of 7 MJ/kg or greater are considered to be at risk from smouldering, although under laboratory conditions smouldering may occur in samples with a CV of less than 7 MJ/kg.

The calorific value of some spoils in Europe are shown in Table 7.1. Colliery spoils deposited in the nineteenth century and early twentieth century are likely to have higher calorific values than those in Table 7.1 because of their higher coal content.

7.4.3 Loss on ignition

This test is sometimes used for assessing combustibility, and involves measuring the loss in weight of a sample after ignition. Samples with a loss on ignition of 25% or greater are generally considered to be combustible.

Table 7.1: The calorific value of some colliery spoils (after Cañibano and Leininger 1987[47])

Country	Calorific value (MJ/kg)
Germany	2.0 - 6.0
Spain	2.2 - 5.5
UK	5.0 - 8.0
France	1.7 - 5.0

7.4.4 Combustion potential test

This is a test developed in the UK to assess the combustibility of spoil material under conditions similar to those on site. The test involves heating the material in a special container with a controlled air flow and monitoring the rate of temperature increase.[46]

7.4.5 Change in site conditions

Experience has shown that materials which are at risk of spontaneous ignition through self heating will ignite in weeks rather than years.[145] Therefore, providing there is no change in external conditions, older spoil heaps which have not previously been subject to spontaneous combustion are considered unlikely to become so in the future.

7.4.6 Temperature monitoring

The extent and depth of material which is undergoing combustion may be assessed by *in situ* temperature monitoring using steel pipes driven into the heap in a grid pattern. Jointed sealed pipes of 50mm diameter or less can be driven into the heap to a distance of several metres using a pneumatic hammer. Temperature readings may then be taken at varying depths using a thermocouple linked to a digital thermometer. Where deep fires are involved, boreholing may be necessary with the steel pipes being grouted into the boreholes using a pulverised fuel ash/cement grout. Portable infra red temperature indicators may be of use for an initial survey of surface temperatures.

Aerial surveys using thermographic equipment are an appropriate technique for monitoring large spoil heap fires.

Temperatures as high as 1200°C have been recorded within combusting spoil heaps. Ignition is likely to occur at temperatures of 200°C or greater. Temperatures below 200°C but significantly above background temperatures indicate a risk of self ignition.

7.4.7 Surface indications

Indications at the surface that underground combustion is occurring are as follows:

- areas of charred and blackened ground;
- the emission of smoke and noxious gases;
- the emission of steam following rain;
- unfrozen ground during cold weather when adjacent ground is frozen;
- areas of dead or dying vegetation (although initially slightly elevated temperatures could make the vegetation more lush);
- elevated ground temperatures;
- areas of localised subsidence.

7.5 Methods of treatment of spoil heap combustion

7.5.1 Factors determining treatment method

The method of treatment will depend on the following:

- the size and nature of the heap;
- the location of the heap;
- the extent and rate of progress of the fire;
- the proximity of the site to areas of habitation;
- the availability of suitable extinguishing materials.

Expert advice should be sought from experienced consultants as to the best method of treatment to be used in specific circumstances.

Methods to extinguish or control underground fires include:

- excavation;
- trenching;
- blanketing;
- grout injection.

7.5.2 Excavation

Digging out is the most common method of successfully dealing with spoil heap fires. The technique involves excavating the burning material and spreading it out to cool (see Photograph 7.1). After cooling the material can be removed to a suitable tipping area, compacted and covered with a layer of inert material. Two or more cooling areas may be required to allow sequential treatment of the burning material.

Machinery used for digging out usually consists of heavy earthmoving equipment such as large tracked excavators, draglines or drag scrapers. The excavation of burning materials can pose risks to the safety of operatives and plant.

Photograph 7.1: Treatment of a spoil heap fire by excavation (source: Richards, Moorehead and Laing Ltd)

One disadvantage of digging out is that once the heap is disturbed, fresh surfaces are exposed to the air. Further combustion may result and increase the production of fumes and dust. The careful use of water sprays may be employed to damp down excavated material and control dust and fumes.

7.5.3 Trenching

Trenching may be used to isolate and limit the spread of the fire. The trenches should be at least 2m wide and 2m deep, or deeper if the depth of burning material is greater. Trenches may be left open for inspection or, if it is important to contain the fire the trenches should be filled with a slurry of limestone dust and water, or other inert material such as clay or pulverised fuel ash.

7.5.4 Blanketing

Blanketing involves covering the burning area with a layer of inert material such as limestone dust, clay, fine sand, or ground shale. The blanketing layer must be sufficiently thick to prevent air access to the affected area and adjoining slopes, thus starving the combusting material of oxygen. Periodic inspection and maintenance is required to ensure a continuous airtight seal. Blanketing may sometimes be used in conjunction with trenching in order to contain and prevent further spread of the fire.

7.5.5 Grouting

Grouting has been used successfully to control spoil heap fires. The technique involves injecting a slurry of clay, shale or limestone dust directly into the burning material in a controlled manner. Limestone dust is the preferred grouting material because carbon dioxide is produced as the limestone is heated, which starves the fire of oxygen.

Grouting may be more appropriate to limit the spread of fires than to actually extinguish them.

Particular care is required where high temperatures are involved, as there may be a risk of explosion. In these cases grouting may not be a suitable solution to the problem.

7.5.6 Compaction

For shallow underground fires, compaction is a suitable method of treatment in order to limit the access of oxygen to the fire. Treatment has been successfully carried out using dynamic compaction which involves repeatedly dropping a heavy weight on to the material using a crane.[16]

7.5.7 No action

In isolated locations or where the fire is small and the potential spread of the fire is limited the fire may be left to burn out of its own accord.

7.6 Safety precautions

Work on burning spoil heaps is hazardous and has resulted in serious accidents, some of which have been fatal. Recommended safety precautions, in relation to some of the hazardous parameters discussed are summarised in Box 7.4.[169]

7.7 Methods of prevention of combustion

The construction of modern spoil heaps by compacting material in layers, has significantly reduced the risk of combustion, and no spontaneous ignition is thought to have occurred in recently constructed tips.

The recompaction of older spoil heaps to reduce the risk of combustion would be expensive, unless the work to regrade and recompact the material is to be carried out as part of general land reclamation or ground improvement works.

In some cases it may be appropriate to cap existing spoil heaps which are considered to be at risk. A 1m thick layer of inert material is normally considered to be sufficient to prevent accidental ignition of the combustible spoil by an intense fire at the surface. A lesser thickness of capping layer *e.g.* 0.6m, may suffice provided the site will not subsequently be disturbed, although for housing and gardens a depth greater than 1m may be necessary to avoid combustible material becoming exposed during future activities on the site.[120]

Box 7.4: Recommended safety precautions

Noxious gases

- Before commencing treatment consultations should be held with all interested parties such as those dealing with environmental health, health and safety, planning, utilities, and also with local industries, local residents, and adjoining property owners.
- A programme of gas testing should be agreed, including personal gas monitors where appropriate.
- Site staff should be made aware of the symptoms of carbon monoxide poisoning and instructed to leave the area immediately if these or other adverse symptoms develop.
- Safety staff should be present on site, including first aid staff trained in the treatment of gas poisoning.
- Breathing apparatus should be available at all times.
- In the event of equipment breakdowns, operators should be instructed to abandon their machine and leave the area immediately.
- Telephone communication should be provided for use in emergencies.
- The cabs of site machinery should be well ventilated.
- Personnel should be kept under close supervision to ensure operatives do not get into difficulties.

Dust

- Dust levels should be monitored to ensure that they are kept within acceptable limits.
- Dust control measures may vary depending on weather conditions.
- Work programmes may need to be adjusted to take account of prevailing wind conditions.
- Dust levels may be controlled by the careful use of water sprays, including the wetting of haul roads.

Box 7.4 continued

Explosion

- Where the spoil is at red heat the use of excessive water should be avoided as this could lead to the production of water-gas with the subsequent risk of explosion. Grouting should only be carried out in the cooler parts of the heap.
- Where there is a high coal content the production of excessive dust should be avoided as this also presents the risk of explosion.

Collapse of cavities

- Plant and personnel should use access routes which have been investigated and are known to be safe.
- Areas of suspected cavities should be tested by probes or by a crane with a dropping weight.
- Any badly fissured areas should be avoided.
- Personnel on foot should be provided with harnesses and lifelines.

Use of earthmoving equipment

- Draglines should be used rather than conventional excavators in areas where underground voids are suspected.
- In the case of large fires the use of automatic drag scrapers may be appropriate to avoid the need for operatives to come into close contact with the burning material.

Other safety precautions

- Work areas should be cordoned off and warning signs provided in order to prevent accidental access by unauthorised personnel.
- Site operations should be overseen by the safety officer.
- Safety procedures should be agreed in advance with site personnel.

Where spoil is considered to be at risk from spontaneous ignition, capping alone may not be sufficient to prevent the possibility of combustion. Additional measures may therefore be required to compact the material adequately, both to prevent it drying out, and to reduce the ingress of oxygen. Sites at risk from spontaneous ignition are not normally considered suitable for development.

For development sites the capping layer should be designed to provide adequate protection from accidental ignition of the underlying spoil throughout the life of the development. Buildings and hardstandings will also provide protection, apart from boiler houses, plant rooms or electricity sub stations where underground heating could take place.

All underground services placed in colliery spoil should be laid in inert material. This method applies particularly to electrical cables and heating pipes and these may need to be insulated in high risk situations. Public utilities should be consulted regarding any special requirements they may have.

Special design considerations may be necessary in order to allow buildings to be erected on sites containing combustible spoil. For example, at one site strip footings were adopted which were supported on pulverised fuel ash in trenches taken down through the combustible material to natural ground, thus forming an effective barrier against migrating underground fires.[145]

Where sites are known to contain combustible material, the lighting of fires or the tipping of hot ashes or other forms of heating which could lead to ignition of the spoil should be prohibited.

8 COLLIERY SPOIL WASHING

Chapter contents

8 COLLIERY SPOIL WASHING

8.1 Introduction

In many of the spoil heaps and tailings lagoons associated with coal mining there exist appreciable quantities of coal, much of which is amenable to recovery. The amount of coal varies widely but patterns can usually be discerned and related to the history of operations. In the UK, for instance, it is rare to find large quantities of coal in the very old or very modern tips, reflecting the total lack of coal preparation or highly efficient removal respectively.

The motivation for coal recovery is primarily financial, but in the context of the reclamation of derelict land it is not essential to operate at an overall profit provided the revenue from coal sales can offset the total cost of an integrated land reclamation scheme.[241] Other motivating factors are those of valuable resource recovery, reduced risk of subsequent combustion, increased stability of the final spoil heap and provision of local employment during the washing operation.

Coal recovery from tips has been practised in the UK, particularly in South Wales, for some thirty years,[194] and in 1992 accounted for around 1 million tonnes of product annually, mainly for power generation. Several schemes have also operated in Belgium, but interest in other parts of the European Community has been lower. Interest in tip recovery is strong in the USA and is increasing in places such as the former USSR, in Poland and in the Czech and Slovak republics.

The coal content of colliery spoil varies from 2-3% for spoil from modern coal washing facilities to, in rare circumstances, greater than 30% coal.

With the development of technology, especially for recovery of fine-grained particles of coal and tailings dewatering, it is now possible to consider coal recovery from large tips containing only about 7% coal, and this means that some of the earlier inefficient operations on relatively rich tips are being assessed for reworking.

Specifications for the coal and discard produced by recovery operations on colliery spoil heaps vary widely depending on the market for the former and end-use of the latter. Coal ash contents from about 15 to 30% might be acceptable for conventional power stations, and possibly double these values acceptable for fluidised bed power generation. It is desirable, on economic grounds, to limit the coal content in the discard, but there may also be other reasons why the level should not be too high. These may include susceptibility to combustion where coal contents are high, and the increased stability of spoil heaps which contain low coal contents. A reasonable target would be a maximum of 2% w/w of material floating at specific gravity (SG) 1.6 in the >250 micron size fraction of the discard, provided that the tip does not contain appreciable quantities of low density non-combustible material such as clinker.

8.2 Characterisation of colliery spoil for feasibility of coal recovery

8.2.1 Product quality

In order to assess the feasibility of coal recovery from colliery spoil, it is necessary to determine not only the amount but the quality of product which will be produced. The quality will typically be quantified in terms of parameters such as ash, moisture and volatile content (Proximate Analysis - see, for example, BSI 1016[42]), sulphur content, size analysis, calorific value, abrasive and caking properties. The appearance of the product can also be an important consideration for some markets because of the presence of very small amounts of aesthetically undesirable contaminants such as clinker, wood and other low density substances commonly found in tips but not in run-of-mine material.

The investigation of tips and of lagoons will be considered in the following sections.

8.2.2 Tip material

Samples for laboratory analyses should be taken during routine site investigations. These can be from trial pits or boreholes. Trial pits provide samples which are probably more typical of what would be fed to a coal preparation plant but suffer from limitations on depth of sampling (see Section 2.5). Shell and auger techniques are better than rotary drilling for borehole investigations, but even this method is limited to a depth of about 60m in the type of material concerned.

Sample size and spacing are important,[4, 131] but even though guidelines exist for conventional site investigations (see Section 2.6), these are not always appropriate to colliery spoil heap examinations. It would be necessary to sample inordinately large amounts of material to perform accurate determinations of the coal content in the coarser size fractions. However, significant quantities of coal are rarely found to exist in the >100mm size fraction, and are uncommon in the >50mm size. For reasons of sample size and statistical accuracy it is convenient to sample approximately 30kg from each location using lined, 150mm diameter shell and auger boreholes at approximately 50m spacings. Sampling frequencies at <50m spacings tend not to significantly enhance the accuracy of investigation. On a large tip the sample would typically be a composite of the material encountered in a depth band of approximately 5m.

In addition to the traditional drilling methods geophysical techniques exist for the initial assessment of coal contents in colliery spoils. Resistivity methods have been used at several locations, but whilst being useful in locating zones of very low coal content, such as burnt material, they are still not capable of accurate quantitative determination of coal yields. Further developments in this promising field could lead to reduced prospecting costs in the future.

Because coal properties vary both within and between coalfields there is no universally accepted method for determining the data required for the performance of feasibility calculations or plant design. The procedure outlined in Figure 8.1 has been used successfully for several projects. All samples are subjected to ash determinations and a selection of samples are subjected to rigorous size and sink-and-float analyses at a range of specific gravities (SG). Tables 8.1(a) and 8.1(b) show a set of data from one such washability analysis.

It is clearly seen that the amount of low density material increases as particle size decreases, and that the amount of material of intermediate density (middlings, see Box 8.1) also decreases as a proportion of the size fraction with decreasing particle size. This is a very common trend in colliery spoil heaps. For this particular sample it is apparent that a yield of 7.7% of the tip would be obtained with an ash content of 3.5% if efficient separation could be performed at SG 1.4. This yield increases to 12.4% if separation is made at SG 1.6, but the quality of product is diminished as reflected by the increased ash content of 23.6%. This sensitivity of ash content to SG of separation is due to the relatively large amount of middlings in this tip.

In most cases it would be prohibitively expensive to perform such rigorous washability studies on all samples. However, a useful indication of the likely yield of coal is obtained by correlating the yield at a specified SG of separation (commonly 1.6) with the ash content of the raw samples. Linear correlations are usually obtained unless the tip is very heterogeneous or contains large proportions of clinker *i.e.* burnt

Box 8.1: Colliery spoil middlings

Middlings represent coal-shale particles with a density somewhere between that of clean coal and shale. Typical densities of middlings may thus be in the order of 1.8 to 2.0t/m^3. The material tends to have a high ash content.

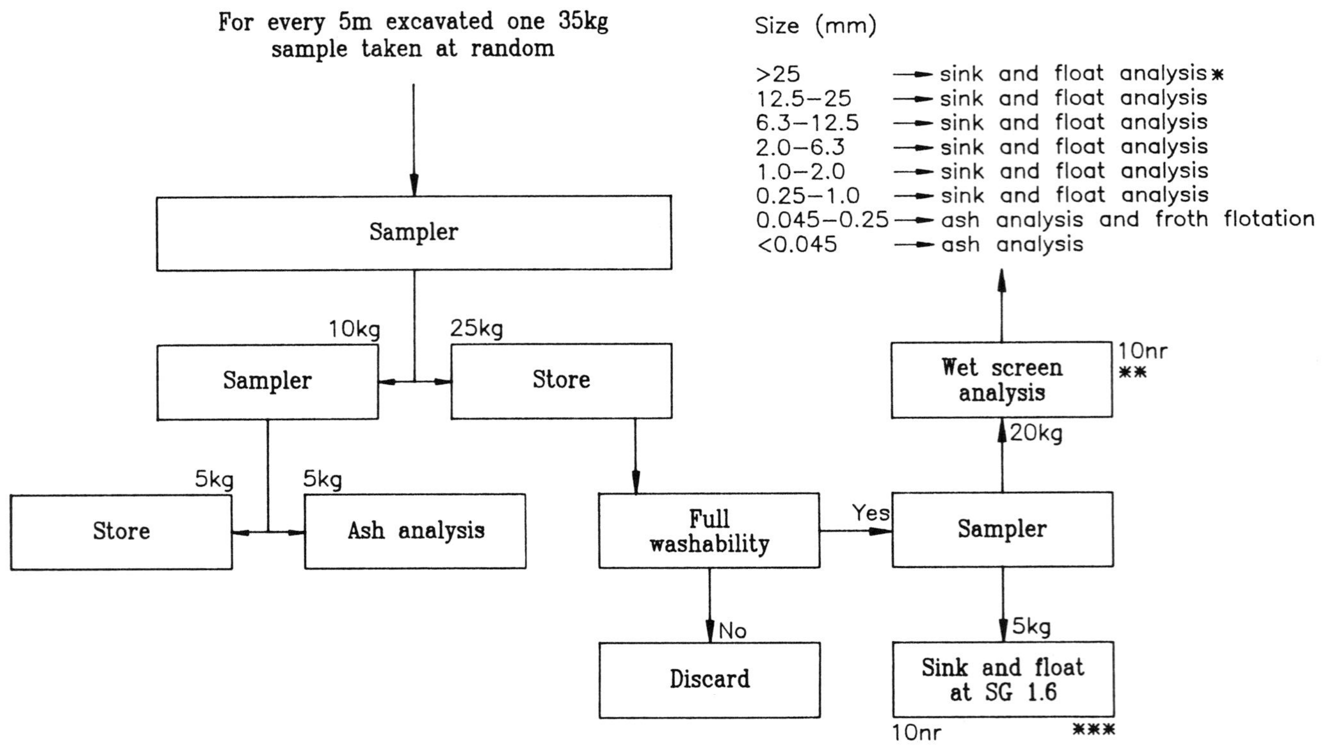

* SGs for sink and float study:– 1.4, 1.5, 1.6, 1.7, 1.8, 1.9.
** Samples to be chosen by the engineer after completion of ash analyses.
*** Full proximate analyses, sulphur and calorific value to be determinedon these float products.

Figure 8.1: Typical flowchart for laboratory analyses of samples for assessment of potential for coal recovery

Table 8.1: Washability data for typical colliery spoil material

(a) % WT of sample in given SG range

size mm	Specific gravity ranges <1.4	1.4 -1.5	1.5 -1.6	1.6 -1.7	1.7 -1.8	1.8 -1.9	>1.9
>25	0.0	0.0	0.1	0.0	0.0	0.2	7.1
12.5-25	0.2	0.1	0.2	0.1	0.1	0.3	8.6
6.3-12.5	0.5	0.3	0.5	0.3	0.3	0.4	10.1
2.0-6.3	1.7	0.8	0.8	1.0	0.6	0.9	10.6
1.0-2.0	1.5	0.4	0.3	0.4	0.3	0.4	8.0
0.25-1.0	3.8	0.7	0.5	0.4	0.3	0.5	11.6

(b) Ash % in given SG range

size mm	Specific gravity ranges <1.4	1.4 -1.5	1.5 -1.6	1.6 -1.7	1.7 -1.8	1.8 -1.9	>1.9
>25	0.0	0.0	30.9	0.0	0.0	49.8	88.9
12.5-25	6.1	16.3	24.8	38.8	44.1	53.1	87.9
6.3-12.5	6.2	16.3	24.0	35.3	44.7	52.2	86.4
2.0-6.3	4.3	15.1	23.8	35.9	44.6	52.2	88.8
1.0-2.0	3.6	14.2	23.3	34.6	44.6	52.0	88.1
0.25-1.0	2.7	11.2	21.2	31.9	40.5	49.0	89.2

solid residues. Figure 8.2 gives an example of such a correlation. After determination of the ash distribution in the raw samples, it is possible to obtain a weighted average value for the yield of coal. The process can be repeated at other separating densities to relate yield and quality.

Results of laboratory analyses must be treated with care for two reasons. Firstly, the sampling process tends to discriminate against the larger particles; those over 100mm rarely reaching the laboratory. The yields must therefore, be reduced to account for the amount of coarse material in the tips. This is usually estimated by subjecting large samples to screen analysis at coarse sizes. If suitable continuous plant is available then this task is less laborious and more accurate. Secondly, laboratory results should not be used to predict the coal recovery achievable with a full scale operation without consideration of the inefficiencies of the equipment used on the industrial scale.

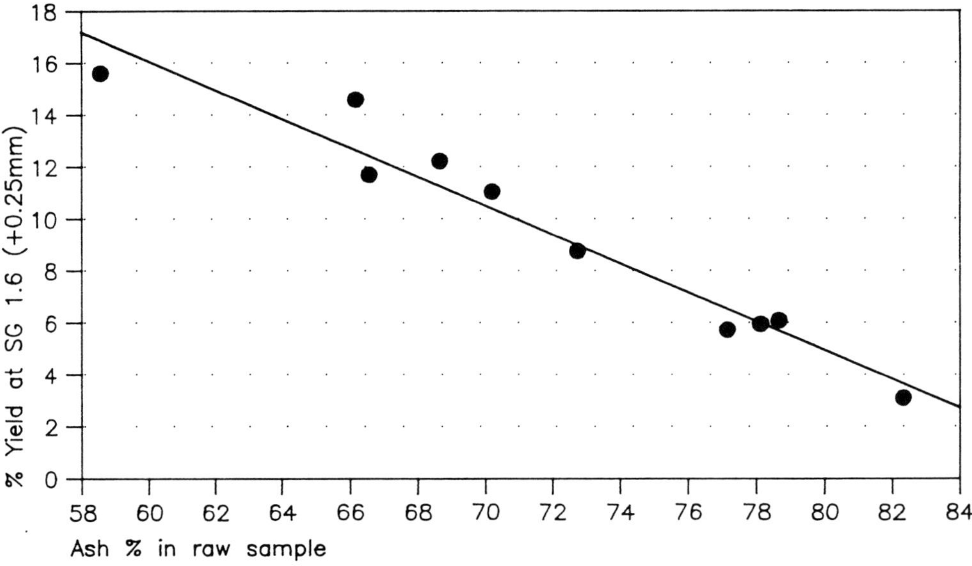

Figure 8.2: Correlation between yield at specific gravity (SG) 1.6 and ash content of the raw sample (site in South Wales, UK)

As an illustration of the trends in coal content with particle size commonly encountered in colliery spoil, washability data for ten UK tips have been combined to produce the data plotted in Figure 8.3.[180] The coal content almost invariably increases with decreasing particle size[47] and the low-density but high-ash clinker material predominates in the coarser size fraction. Coal recovery is rarely practised at particle sizes of less than 0.25mm, but the potential for extra yield from this fraction is high and is therefore receiving attention from plant suppliers and operators. The percentage of material less than 0.25mm in UK tips has been found to vary from 1 to 50%.[198]

8.2.3 Tailings lagoons

Characteristics of the materials deposited in lagoons vary widely. It is however rare to find more than about 5% of material of a size >2mm.[169] It is also to be expected that the coal content is highest in the coarsest size fractions, as illustrated by the data of Table 8.2.[260] Coal recovery may simply involve size classification in simple operations using, for example, small hydrocyclones. However, the economics of coal reclamation from such lagoons are critically dependent on the cost of dewatering the resulting slurries.

In certain circumstances the coal content may be high enough to allow blending of the material, as excavated, with clean coal from other sources for use in power stations. Modern fluidised bed combustors may also be able to handle the material directly.

8.3 Coal recovery technology

8.3.1 Introduction

The unit operations used to recover coal from tips are largely similar to those used for run-of-mine applications (see Section 12.2.1) but adapted to cope with the much higher proportions of high-density materials. The

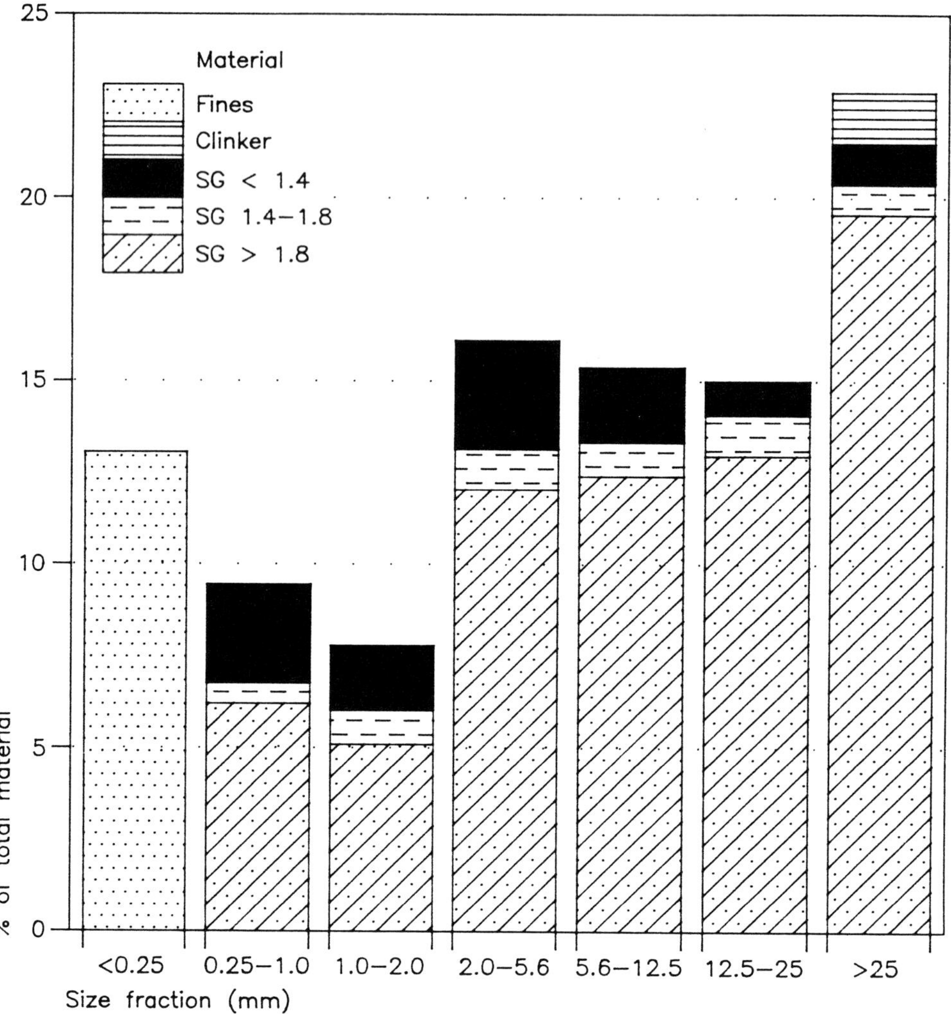

Figure 8.3: Distribution of tip material according to size and specific gravity (SG), data for ten tips in South Wales, UK

Table 8.2: Typical size and ash analyses for a tailings lagoon

Size micron	% wt	% ash	Cumulative % wt	Cumulative % ash
>1000	4.4	3.3	4.4	3.3
500-1000	7.0	5.9	11.4	4.9
106-500	17.6	26.3	29.0	17.9
<106	71.0	66.7	100.0	52.5

processes will involve size reduction, screening and classification, gravity concentration, froth flotation, solid-liquid separation, and drying equipment, which are well described in standard texts.[182]

Figure 8.4 is a schematic flowsheet showing the sequence of operations commonly used for tip processing. The individual operations will be discussed in turn.

8.3.2 Feed preparation

It is essential to provide the coal recovery sections of the facility with a constant flow of material of the desired particle size. This is achieved by suitable combinations of stockpile blending, hoppers, feeder, conveyors and screens wherein the coarse, coal-lean fraction is rejected at a size varying from about 30 to 100mm, depending on tip properties. Avoidance of compaction of material, causing blockages in hoppers and chutes, is important at this stage.

For tips with high contents of fine-grained particles which tend to be cohesive, it is often necessary to incorporate devices such as barrel scrubbers into the feed preparation section in order to separate the individual particles. Care must, however, be exercised to avoid production of extra fine-grained particles, particularly of coal. This can occur through excessive abrasion during the washing process, especially where softer shale materials are being reprocessed. If the washing plant

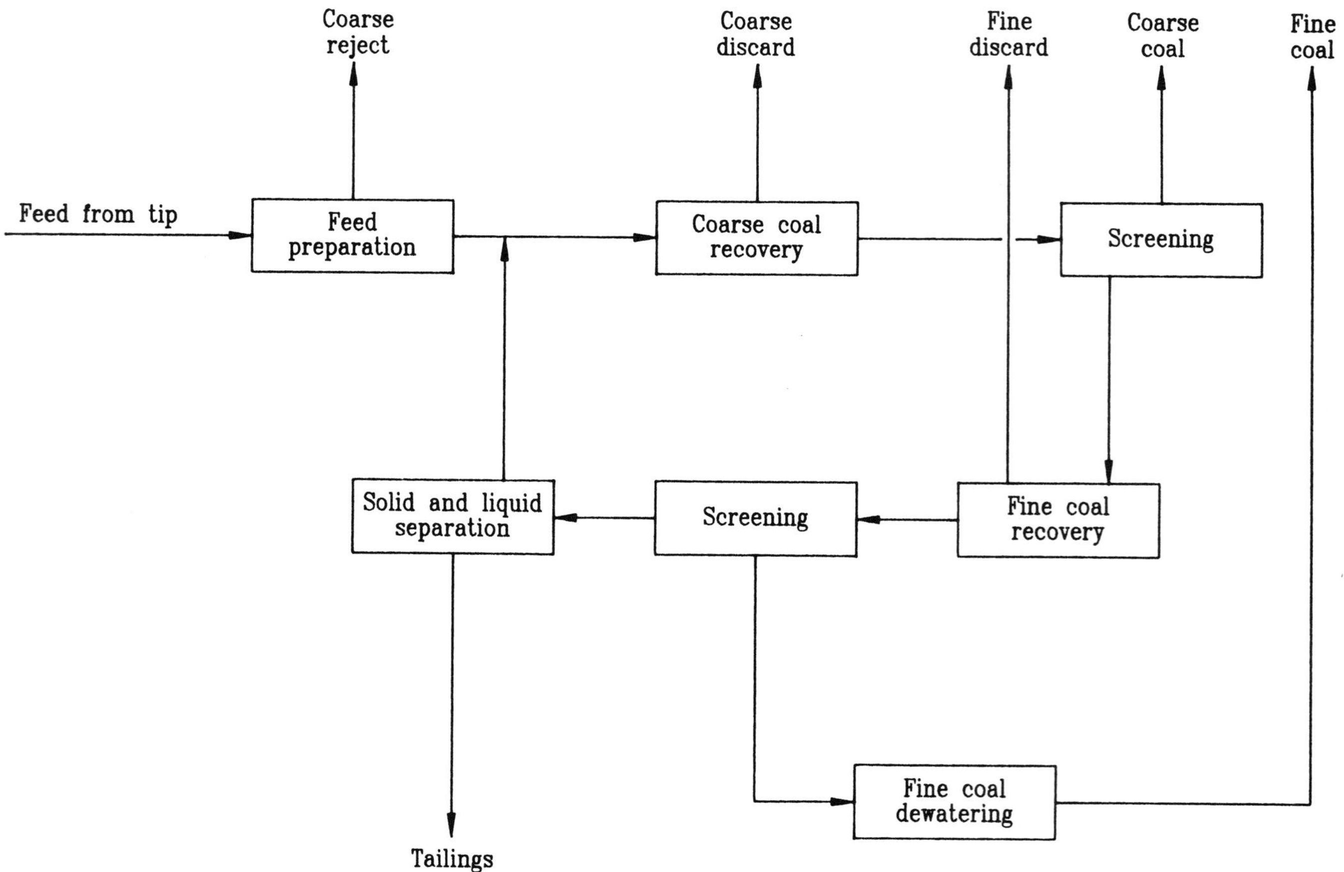

Figure 8.4: Simplified flowchart for coal recovery from colliery spoil

is overloaded with fine material this can severely affect the feasibility of a project.

8.3.3 Coarse-grained coal recovery

The size of interest for coarse-grained coal recovery is from about 100mm to 1mm. Coal recovery from material >100mm is considered impractical, and jigs, barrel washers and dense medium methods will only operate effectively above 1mm. No single machine is capable of operating on material of particle sizes from 0-100mm. The unit operations employed are dependent on the quantity and type of materials in the tips, and also on the market for the recovered coal. If the tip contains appreciable amounts of middlings (see Box 8.1) and if it is essential to meet strict limitations on ash content in the coal, dense medium separation, normally in the form of cyclones, might be necessary. Dense medium separation relies on the use of a liquid medium with an SG between that of the coal and discard. This separating medium often consists of the slurry formed by the fine-grained particles in the colliery spoil and water, although slurries of magnetite (a type of iron ore) form lower viscosity suspensions which can be carefully regulated by means of magnetic separators. However, because of the high capital and running costs it is rare to see plant with magnetite suspensions as media, shale mediums being generally preferred, particularly on the smaller sites.

Jigs are cheaper than dense medium systems and have been successfully employed on a large number of colliery reclamation schemes. Modifications to the discard handling mechanisms are required and careful control of discard removal must be exercised to minimise coal losses. Jigging involves the use of pulsating water flows, rather than dense media, and the effective separation density increases as the particle size decreases. This is an important consideration in view of the distribution of coal with particle size discussed earlier.

Barrel washing, normally in combination with shale-medium cyclones for the fraction below 5mm, is probably the most widely used coal recovery system for colliery spoils. It is a low cost option because of its simplicity and ease of operation, and barrels have therefore been found to have application on tips containing as little as 250,000t of spoil. Figure 8.5 illustrates how a barrel washer uses the flow of water to transport coal, leaving the heavier discard to be conveyed against the slope by the scrolls. Provided that the barrel is fed at a constant flow with reasonably consistent material, and that the slurry density is carefully controlled, it can be operated at an efficiency approaching that of a jig.[260] A typical partition curve for a barrel washer is shown in Figure 8.6. The partition curve is the classical method of reporting the efficiency of coal separation achieved by a particular piece of equipment. From it the Epm value (see Box 8.2) and the separating SG, *i.e.* the SG at which 50% of the coal with that SG is recovered in the product, can be calculated.

The Epm value for the barrel washer of Figure 8.6 is slightly higher than would be associated with jigs on similar duties but significantly higher than values found in dense medium processes. Dense medium process Epm values can be as low as 0.06. The separating SG of about 1.5 was found to be insensitive to particle size in the application shown in Figure 8.6, possibly indicating some shielding of fine-grained coal from the flowing stream by overlying discard. Axial lifter bars have been employed in some barrels to alleviate this problem. These data were derived for a barrel of 1.8m diameter handling approximately 100t/hour. Barrel diameters up to 2.9m are now available with throughputs approaching 400t/hour.

8.3.4 Fine-grained coal recovery

For the recovery of coal in the size fraction from about 1mm to 0.25mm, it has been common to use Hydrosizers, spiral concentrators and small water-only cyclones.[180] Froth flotation has traditionally been the method employed for separation below 0.25m, but is rarely used because of the high capital and revenue costs, not only of the flotation step but also of

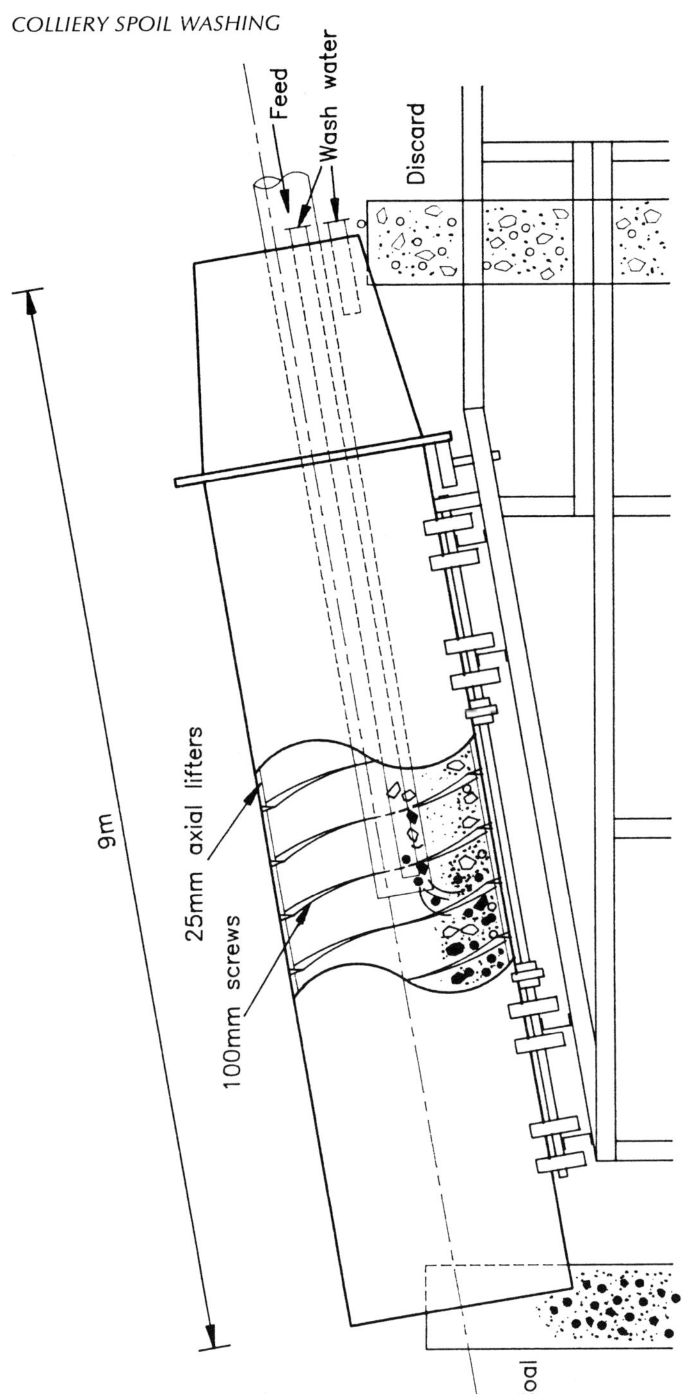

Figure 8.5: Simplified diagram of a barrel washer for coal recovery from colliery spoil

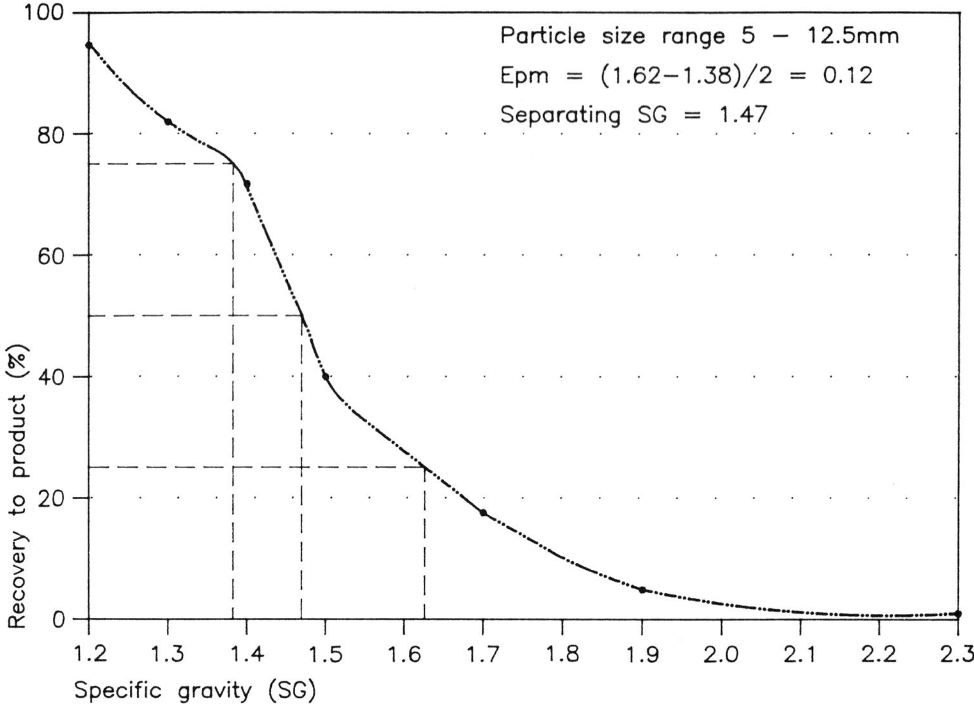

Figure 8.6: Typical partition curve for a barrel washer

Box 8.2: The Epm value

The Ecart Probable Moyen, or Epm value is a measure of the efficiency of a particular piece of equipment used in coal recovery. The Epm value is derived from the partition curve (see Figure 8.6) as follows:

$$\text{Epm} = \frac{\text{SG which will achieve 75\% recovery - SG which will achieve 25\% recovery}}{2}$$

The Epm decreases as the steepness of the partition curve increases, *i.e.* as the separation becomes more efficient.

the associated dewatering facilities. Also, it is often assumed that surface oxidation of the fine-grained particles has occurred to an extent where flotation becomes difficult. However, it is usually possible to achieve good performance with the correct choice of reagents, albeit at slightly higher dosages than those encountered in run-of-mine coal flotation. It may also be necessary to operate at a relatively low slurry concentration, of about 10% w/w solids, to avoid entrainment of slimes.

8.3.5 Solid/liquid separation

Coal product and coarse discard

Since it has been unusual to recover coal of a size <0.25mm, the coal product has been dewatered using a combination of screening and centrifuging. These operations, together with storage on properly drained pads, result in a product which is acceptable in terms of ease of handling. Further moisture reduction by thermal methods may be used if the economics, in terms of extra revenue for increased calorific value, are favourable.

Tailings

In the infancy of colliery spoil washing, use was frequently made of lagoons as a method of storage and dewatering of tailings. Management and engineering constraints on the properties of discard materials often dictate that mechanical means of tailings dewatering are included in modern tip recovery plants. These constraints arise from the high water content and gel-like consistency of lagoon tailings, which limit the after-use of the site due to ground instability, and can produce problems of water seepage and safety risks. Operations such as deep cone thickeners, vacuum filters, filter presses and solid-bowl centrifuges have found application, but are not as common as belt presses (or multi-roll filters) in newer facilities. Newer equipment has the advantages of continuous operation and a greater drop in pressure across the filter *i.e.* they can operate at several atmospheres pressure unlike rotary vacuum separators which can only achieve pressure drops of 1 atmosphere. Flocculant consumption adds a significant operating cost for processes incorporating belt presses, but the tailings moisture content can be reduced to around 30% w/w at rates up to 50t/hour, enabling mixing with the coarse-grained discard prior to formation of the new landforms. If the quantities of tailings are high it may be necessary to consider the addition of cement in order to achieve the required engineering properties.

The use of large open lagoons for the collection and settlement of coal washery tailings is less used in Europe than it was in the past, although where flat land is in abundance it is still an inexpensive option. The need to dewater old lagoon tailings can however prohibit reclamation of these sites due to the high cost of these operations.

8.4 Infrastructure

When assessing the feasibility of incorporating coal recovery into a land reclamation project, the availability of land, transport arrangements for supplies and products, availability of utility services such as water and electricity, and environmental issues, must be considered. The equipment generally requires little space but location is important so as to minimise the nuisance to and disturbance of the environment. Transport of coal must be carefully planned and is often an area of contention when permission is being sought for such recovery operations. Rail transport is preferred, and may be feasible for large volumes of material, even if previously abandoned lines have to be reinstated.

Water requirements are also significant, with quantities of the order of $1m^3$ of make-up water needed for every 10t of feed treated. This quantity tends to be higher for tips with a high content of fine-grained particles.

Power requirements for a large reprocessing centre treating about 500 t/hour are approximately 1000kW. The centres are often operated continuously with a weekly stoppage for maintenance. Labour requirements are in the region of 20 people for the larger centres/works, including managerial, supervisory and laboratory personnel.

Construction of new plant requires a minimum period of six months, and relocation of an existing facility approximately half of this time. Modular construction can reduce these times significantly.

9 STEEL INDUSTRY RAW MATERIALS AND WASTES

Chapter contents

9 STEEL INDUSTRY RAW MATERIALS AND WASTES

9.1 Iron and steel making

9.1.1 Iron and steel

Steel is an alloy of iron and carbon. It may contain up to 1.7% carbon, though generally it contains less than 0.5%. Steel is produced in a variety of forms, for different uses. Alloy steels contain metallic elements other than iron, such as chromium (present at 14% in stainless steel), nickel, vanadium, molybdenum, manganese, cobalt and tungsten. The chemical composition and properties of the main types of iron and steel are shown in Table 9.1.

9.1.2 Iron ore

Iron ores are principally oxides or carbonates of iron. Iron ore has been mined in Europe by both underground and surface mining methods. The main types of iron ore and where they can be found in Europe are shown in Table 9.2. Mining produces waste rock and soil which may be devoid of vegetation because of its poor nutrient status. Spoil from iron ore mining in Lorraine, eastern France contains pyrite and is a source of acid mine drainage (see Section 12.2.3).

Since the 1960s the exploitation of large deposits of high purity iron ore (63-65% Fe content) in western Australia and the Amazon has led to a dramatic reduction in iron ore mining in Europe, with European ore being replaced by imported ore, predominantly from Africa and Brazil. In 1990 the only significant European iron ore producers were France (Lorraine region) and Spain.[124] Production of iron ore is likely to decline further as steel production based on these ores is planned to switch to imported ores or production from scrap.[125]

Some of the contamination of iron and steel sites by non-ferrous metals derives from the metal content of the iron ore. The metals which may be present as impurities in iron ore are shown in Box 9.1.

Table 9.1: Types of iron and steel

Type of iron/steel	Composition	Properties
Pig iron	High in carbon (3-4%) and impurities, including silicon, phosphorus, sulphur and manganese. Much of the carbon is in the form of graphite flakes which are a source of structural weakness.	Brittle, low melting point. Used for manufacture of other types of iron and steel.
Wrought iron	Virtually no carbon. Contains minute, parallel threads of slag.	Ductile, malleable, easily welded, corrosion free. High tensile but low compressive strength.
Cast iron	High in carbon and other elements. Slightly lower in impurities compared to the pig iron from which it is manufactured.	Brittle, cannot be forged or welded, but high compressive strength.
Carbon steels	Similar to wrought iron in that they contain very little carbon: mild steel <0.2%; medium steel 0.3-0.6%; high-carbon steel 0.6-1.7%. No slag threads.	Not as ductile or as easily welded as wrought iron, not corrosion free, but harder wearing and stronger.
Alloy steels	Small amounts of carbon and up to 50% of other metals (e.g. Al, Cr, Co, Mn, Mo, Ni, Ti, W, V).	Wide variety of properties depending on alloy.

9.1.3 Iron smelting

Ancient methods of iron smelting involved heating the iron ore in a charcoal fire within a clay furnace with limited air supply. On completion, the furnace had to be demolished to obtain the 'bloom' of wrought iron. Such furnaces were known as bloomeries. If the supply of air was increased the carbon from the charcoal and silicon from the

Table 9.2: Types of iron ore and principal deposits in Europe

Name	Chemical composition	Physical character	Principal deposits	Purity /impurities
Magnetite	Fe_3O_4	Dense, hard, black masses. Magnetic. Specific gravity 4.9-5.2.	Scandinavia, particularly Sweden.	72.4% Fe when pure, usually deposits contain 55-65% Fe. Purest type of iron ore. Free from sulphur and phosphorus.
Red hematite	Fe_2O_3	Red/black, hard massive lumps to soft powdery deposits. Specific gravity 4.9-5.3.	Sweden, UK (Cumbria), Spain (Bilbao).	70% Fe when pure. Usually deposits contain 60% Fe.
Brown hematite (limonite)	Fe_2O_3 with 8-20% water	Light yellow-brown to red/black. Soft, open, easily, smelted. Specific gravity 4.0-4.7.	France (Lorraine), Belgium, Luxembourg, Germany, Spain.	65-55 % Fe. High phosphorus content.
Carbonate ores (siderite)	$FeCO_3$	Grey/brown.	UK (East Midlands)	48% Fe when pure, deposits average 30%. High phosphorus content, high lime content.

Box 9.1: Non-ferrous metals found in iron ore[22]

Non-ferrous metals which may be present as impurities in iron ore include:

Titanium	Zinc	Nickel	Vanadium
Magnesium	Lead	Arsenic	Chromium
Manganese	Copper	Tin	Boron

iron ore would combine with the iron, lowering its melting point so that the iron became liquid and flowed out of the furnace. This was at first undesirable, as the iron produced, known as pig iron, was brittle and unworkable. However, a method by which this iron could be refined to produce wrought iron was developed in the sixteenth century and production of pig iron from iron ore became the norm, with wrought iron subsequently produced from the pig iron.

The furnace producing pig iron from iron ore, the blast furnace, has changed little since it was first developed, although its size, efficiency and speed of throughput have greatly increased. The largest furnaces in Europe, built since the 1970s, have hearth diameters of around 14m, and working volumes of 4,000-5,000m^3, and can produce over 10,000t of pig iron per day.[125] Figure 9.1 is a diagrammatic section through a blast furnace. Iron ore, coke and limestone are charged into the top of the blast furnace via a bell valve, which prevents loss of gases to the atmosphere during charging. A blast of air, sometimes containing fuel in the form of gas, oil or pulverised coal, heated to 600°C, is introduced near the bottom of the furnace. The coke burns to produce carbon monoxide which reduces the iron ore. Molten iron flows into the bottom of the furnace where it is removed via a tap. The iron may be run into moulds, known as pigs, or may be used directly, in its molten form, for the production of steel or cast iron. Impurities in the iron ore and coke combine with lime from the limestone to produce a molten slag which floats on top of the iron and is tapped off separately.

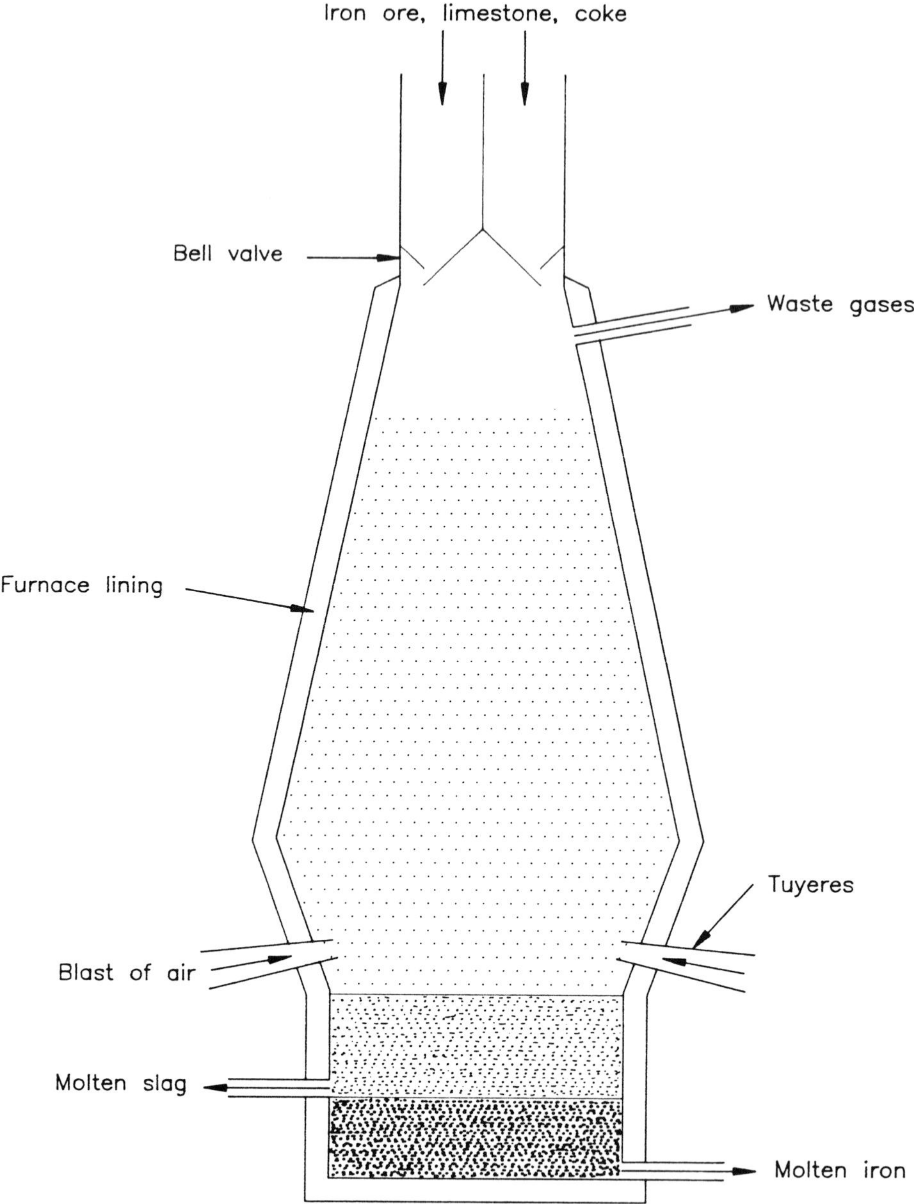

Figure 9.1: The blast furnace

The use of coke, rather than charcoal, for fuel was an important development in iron making, first introduced in the early eighteenth century. Coal could not be used as it softens under heat and would thus choke the furnace. Unlike charcoal, the coke contains sulphur which makes the pig iron produced unsuitable for wrought iron, a disadvantage before steel replaced wrought iron in nearly all its traditional uses.

Blast furnaces are operated continuously, with periodic tapping of slag and iron and charging of raw materials. The furnace is only shut down for renewal of the furnace lining. A modern blast furnace is shown in Photograph 9.1.

Small quantities of iron are also produced by direct reduction processes, which involve the removal of oxygen from iron ore, in the solid state, by gaseous or combustible reductants. Direct reduced iron (DRI) is generally added to electric arc furnaces (see Section 9.1.5) to make steel.

Photograph 9.1: Blast furnaces in Spain (source: Richards, Moorehead and Laing Ltd)

9.1.4 Iron casting

To produce cast iron, pig iron is remelted in a cupola, a furnace similar to the blast furnace, though generally smaller and operated as a batch process. High grade coke, low in ash and sulphur, is used as the fuel and a small quantity of limestone added to form a slag with some of the impurities in the iron. The molten iron is tapped off and run into moulds, formed from sand.

Cast iron has now been largely replaced by steel in most of the many uses it was put to in the nineteenth and early twentieth century. Casting, still commonly carried out with steel, is performed in a foundry.

9.1.5 Steel making

The making of steel is a batch process by which the impurities present in the pig iron are removed. Measured quantities of carbon and other elements are then added to produce steel of the type required.

The oldest steel making process is the Bessemer process, developed in 1850. In this process, molten pig iron was loaded into a vessel, known as a converter (see Figure 9.2(a) and (b)). Air was then blown into the molten iron through a series of tuyeres in the bottom of the converter. The air oxidised the impurities such as carbon, silicon and manganese, and some of the iron. These impurities either left the converter as brown fumes or formed a slag which floated on top of the iron and was removed separately. In the original 'acid' process the converter was lined with silica bricks. Only iron low in sulphur and phosphorus could be used as these elements were not removed by the acid Bessemer process. In the 1880s a 'basic' version of the process was invented by Thomas and Gilchrist. Addition of a lime-rich flux was found to remove the unwanted elements from the steel. The furnace lining had to be made of a basic material (dolomite) to avoid reaction with the slag. A phosphate-rich slag was produced which could be used as a fertiliser. The process allowed steel to be made from the phosphorus-rich ores of France, Belgium, Luxembourg and Germany, where it is known as the Thomas process.

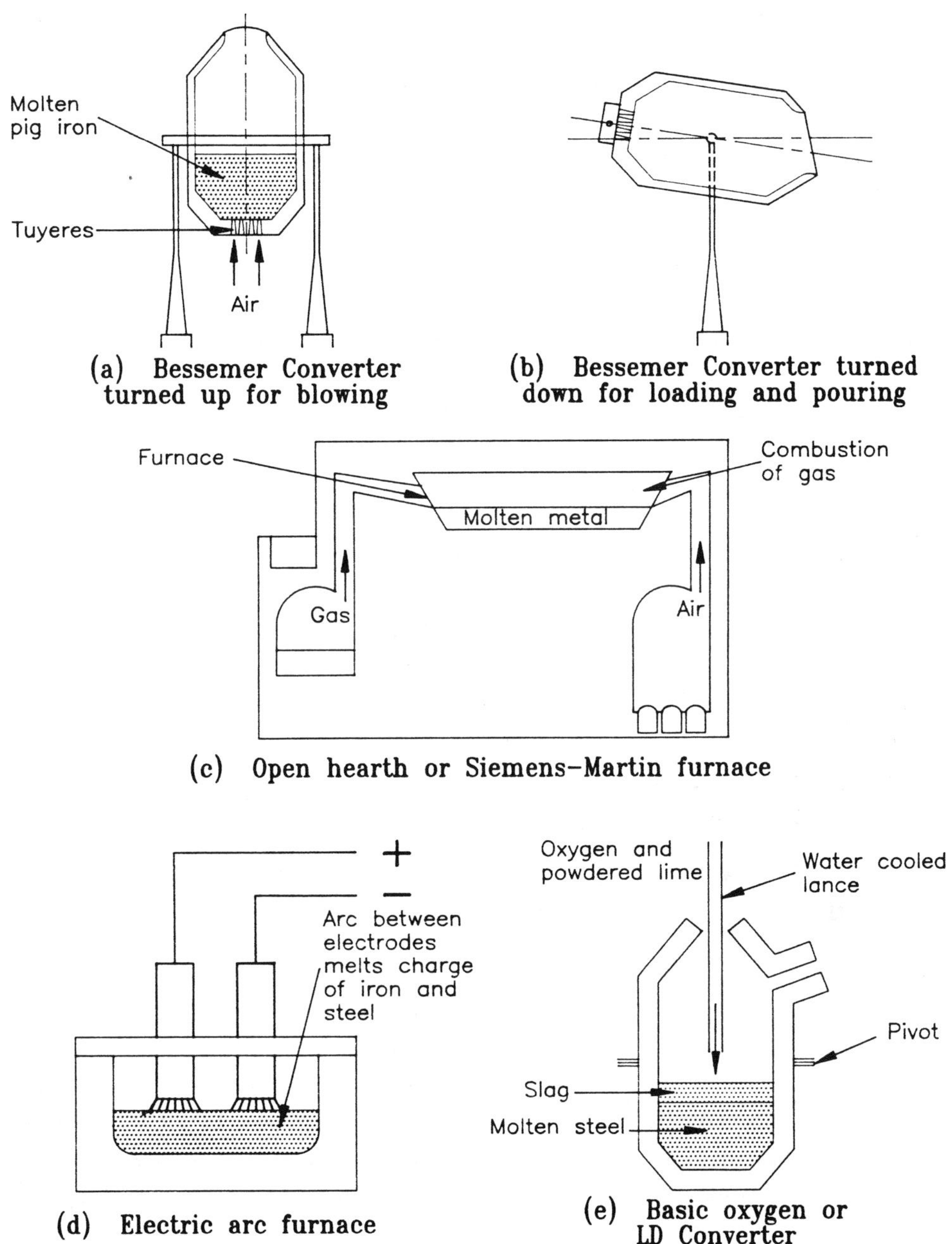

(a) Bessemer Converter turned up for blowing

(b) Bessemer Converter turned down for loading and pouring

(c) Open hearth or Siemens-Martin furnace

(d) Electric arc furnace

(e) Basic oxygen or LD Converter

Figure 9.2: Steel making furnaces

In the first half of the twentieth century the main alternative to the Bessemer processes was open hearth, also known as 'Siemens-Martin', or just 'Martin'. In open hearth steel making the furnace is a shallow, crucible type. Heat to melt the iron is produced by the burning of gaseous fuel within the furnace (see Figure 9.2(c)). Acid and basic processes differ in the nature of the furnace lining and the addition of lime-based flux. The acid process requires low phosphorus pig iron whereas the basic process can use high phosphorus pig iron, but unlike the basic Bessemer process, phosphorus is not essential. Steel scrap as well as pig iron can be used in open hearth steelmaking. In the basic open hearth process fluorspar is added with the limestone as a flux. The fluorspar absorbs sulphur and makes the slag more fluid, but it destroys the value of the slag as a fertiliser as it renders the phosphate in the slag insoluble. Less iron is lost through oxidation with the open hearth than with the Bessemer process. Open hearth steel making is still carried out in Eastern Europe and India.

Pure oxygen became available in large quantities after 1945 and in 1953 a steel making process which used pure oxygen rather than air was developed in Austria. It was called the Linzer-Dusenverfahren (LD) process and is also known as the Basic Oxygen Steel (BOS) process. A converter, similar to that used in Bessemer steel making is used but, rather than air being introduced through the bottom of the vessel, a lance through the top of the vessel blows a blast of pure oxygen with powdered lime on to the charge of molten pig iron (see Figure 9.2(e)). Oxidation of impurities in the pig iron produces heat so no additional heating is required. Up to 30% of the charge can be cold scrap or iron ore. In later modifications of the process oxygen may be blown from the sides or bottom as well as the top of the converter. The process combines the fast production rate of Bessemer with the high quality product of open hearth and it is now the major steel making process. The BOS process is used for making bulk steel at large integrated steel works where molten pig iron from blast furnaces can be fed directly into the converter. In 1991 oxygen steel making accounted for 67.9% of steel production in the European Community.[124]

Electric steel making uses electricity to provide the energy to melt a charge of iron and steel scrap and produce steel of the desired quality. In the electric arc furnace (see Figure 9.2(d)) the metal is placed in the arc between two carbon electrodes. The heat of the arc is sufficient to melt the metal charge and impurities present in the metal form a slag with added lime and fluorspar. In the high frequency induction furnace a high frequency AC current in a copper coil surrounding the furnace induces a similar current in the metal within the furnace. This current is concentrated in the surface of the metal, creating heat which melts the metal. The generation of heat in electric furnaces does not depend on combustion processes so the atmosphere of the furnace can be controlled, enabling conditions to be chemically reducing if necessary. Chemically reducing conditions allow the melting of steels containing easily oxidisable elements such as chromium, manganese and vanadium without loss of these elements, to produce high quality alloy steels. Electric furnaces are generally smaller than oxygen furnaces and can take a cold charge of 100% scrap. Such furnaces can therefore be used at non-integrated steel works where a supply of molten pig iron is not available. In 1991 electric steelmaking accounted for 31.6% of European Community steel production.[124]

9.1.6 Processing of steel

Crude steel from the furnace is first transferred by casting, rolling and/or forging into semi-finished or intermediate products such as ingots, slabs, blooms or billets. These products are known as 'semis' and they may be sold to steel processing companies or may be further processed at the steel-making works. Such further processing may include hot or cold rolling, drawing to form wire, cutting, machining and coating with metallic or plastic materials. The major processes are described in the following sections.[243]

Casting

Liquid steel from the furnace may be cast into ingots by pouring into individual moulds made of pig iron. Once solidified, the ingots of steel are taken from the moulds and placed in soaking pits to be reheated to the temperatures suitable for hot rolling. The soaking pits give an even temperature distribution within the ingots.

Alternatively, continuous casting may be employed. In this method the liquid steel in the ladle is tapped into a vessel known as a tundish. Nozzles in the tundish are then opened to allow the steel to flow into moulds made of copper which are cooled by water. The steel is removed from the mould when it is partly solidified and carried on a series of rollers through sprays of water, to cool it further. The steel is cut into lengths and it is either cooled and inspected, or passed immediately to a reheating furnace prior to hot rolling.

Hot rolling

The ingots, or lengths of steel from continuous casting, are reduced in thickness and sometimes in width, with concurrent elongation, by passage through the rolling mill. Scale, consisting of oxidation products, forms on the surface of the steel during reheating prior to rolling. This scale is broken up by the action of rolling and is washed away by the water used to cool the rolls.

Pickling

After hot rolling the metal cools and a thin scale forms on the metal surface. This scale must be removed by pickling in an acid bath before further processing is carried out. Sulphuric acid or, less commonly, hydrochloric acid is used. Particularly acid-resistant steels, such as stainless steel, require the use of stronger acids, such as mixtures of hydrofluoric and nitric acids. After pickling the steel is rinsed and coated with oil to prevent further scale formation.

Cold rolling

The flat products produced by hot rolling are processed further by cold rolling to produce products such as bright sheet, angles or sections or wire products (by cold drawing or twisting). Oil, or mixtures of oil and water are used to cool the rolls and to provide lubrication. After cold rolling the oil is removed from the steel products by rinsing, electrolytic treatment in an alkaline solution and then washing to remove the alkali.

Coating

Zinc, tin, aluminium, lead, brass, cadmium, paint and various types of plastic may all be applied to steel products as a protective or decorative coating. Coating with zinc, known as galvanising, may be done electrolytically or by dipping the steel in liquid zinc. Coating processes are carried out in a number of stages, several of which involve cleaning the metal surface. These stages include rinsing, alkaline treatment, pickling and oiling.

9.1.7 Integrated steel works

The production of steel is an industry where great increases in efficiency have been achieved through economies of scale, so that increases in steel production since the 1950s have been achieved with larger but fewer steel works. These large works are generally 'integrated'; all the processes involved in steel production, such as smelting of iron ore, burning of limestone to produce lime, coal carbonisation to produce coke, steel making and steel processing, are located on the same site. The products of one process feed directly into the next process, so that, for example, molten pig iron from the blast furnaces feed into the converter to be made into steel (see Section 17.9). In this way heat losses are minimised and efficiency is increased by utilisation of heat or gases generated by one process in other processes. Underground and overground ducts, pipes or conveyor belts transport gases and materials from one part of the site to another.

In many steelmaking towns, such as Seraing near Liege, Belgium, the town and steelworks have grown together. In such places, ducts and conveyors cross residential and commercial areas of the town to connect different sites of the local steel company.

9.2 Wastes from iron and steel making

9.2.1 Introduction

The major processes which take place at iron and steel works, together with their associated raw materials, products and wastes, are shown in Table 9.3.

The principal type of waste from steel making in terms of volume is slag. There are various types of slag, described in Section 9.2.2. Slag has become regarded as a useful by-product rather than a waste material, although this has not always been so, particularly in Britain where large tips of waste slags are found at some former iron and steelworks sites. Flue dusts and refractory linings from furnaces are also significant waste streams, of which deposits may be found at old sites. Coal carbonisation, used to make coke is likely to have been the greatest source of contamination found at steel and iron sites. Coal carbonisation is dealt with in Chapter 10.

9.2.2 The characteristics of slags

Introduction

Slags are synthetic rock-like materials formed by the fusion of the non-iron constituents of iron ore, or the impurities present in the iron, with the limestone flux during steel making. The amount of impurities to be removed is much greater at the iron smelting stage, in the blast furnace, than at the steel making stage, so for a given amount of steel produced from iron ore the volume of blast furnace slag is much greater than the volume of steel slag.

Table 9.3: Processes, raw materials, products and wastes involved in production of steel

Process	Raw materials	Product	Wastes/by-products
Iron ore mining		Iron ore	Waste rock
Iron smelting (in blast furnace)	Iron ore Limestone Coke Air	Pig iron	Blast furnace slag Flue dusts Ash Refractory linings
Steel making	Pig iron and/or steel and iron scrap	Steel	Steel slag Flue dusts Refractory linings
Iron casting	Pig iron Coke Moulding sand	Cast iron	Foundry slag Moulding sand
Processing of steel	Crude steel	Steel products	Spent acids and alkalis, hydroxide sludges, spent plating solutions, mill scale, oils, solvents, paints, non-ferrous metals
Coal carbonisation (see Chapter 10)	Coal	Coke Coal gas	Tars Benzole Naphthalene Ammoniacal liquors Spent oxide Sulphate

Blast furnace slag

Blast furnace slag consists of silicates and alumino-silicates of lime.[146] The major mineral component is melilite. This is a solid solution with the end members gehlenite ($2CaO.Al_2O_3.SiO_2$) and akermonite ($2CaO.MgO.2SiO_2$). Melilite is a stable mineral with good strength properties, responsible for the good engineering properties of blast furnace slag as roadstone, fill material and concrete aggregate.[179, 237] The rate and manner of cooling of molten slag from the blast furnace can be varied to produce three different forms of solid slag, as described in Box 9.2.

The few problems which have arisen through the use of blast furnace slag are associated with the sulphur content of the slag. Blast furnace slag generally contains around 1.5% total sulphur, with over 2% in older slags. The sulphur is derived from the coke used in the furnace. High concentrations of water-soluble sulphate may result in the formation of sulphoaluminates such as ettringite ($3CaO.Al_2O_3.3CaSO_4.30\text{-}32H_2O$).[237] The formation of this secondary mineral is accompanied by an increase in volume of approximately 120%, and may cause ground heave.

High concentrations of sulphate may also result in the chemical attack of concrete (in which sulphoaluminate formation is also involved), particularly if the concrete is placed in contact with the slag in a

Box 9.2: Forms of blast furnace slag[146]

Three different forms of blast furnace slag are produced by different methods of cooling the slag:

Air cooled	Slag which has been allowed to solidify under ambient conditions, producing a crystalline structure similar to that of natural rock. The slag is generally crushed and graded to produce aggregate of the desired size.
Foamed, or expanded	Introduction of water under controlled conditions whilst molten slag is tipped into a container, produces many air pockets within the slag, giving a strong, lightweight aggregate.
Granulated	Slag which has been cooled rapidly with an excess of water to produce glassy sand-like material.

waterlogged situation. The presence of reduced forms of sulphur result in discolouration of the slag and odour problems. A greenish tinge is common and the foul-smelling gas, hydrogen sulphide, may be produced. Leaching of sulphur compounds from blast furnace slags cause the most common water pollution problems associated with the use of slags, although these problems can be largely avoided by not placing the slag in situations where water can drain through them.[83]

Steel slags

Steel slags contain more iron, as free metal and as oxides, than blast furnace slag. This iron gives steel slags a greater density, and the potential to be magnetic. Unlike blast furnace slags steel slags contain very little sulphur. The chemical composition of steel slags, even from the same process, tends to vary more than for blast furnace slags.

Steel slags may derive from acid or basic steel-making processes. Slags from acid processes contain silica with some lime. However, these slags are rarely encountered as basic processes have been responsible for nearly all European steel production in the twentieth century. Basic steel slags are high in lime, and contain calcium silicate, solid solutions of calcium oxide and ferrous oxide, and calcium ferrite.[83, 237] Unlike blast furnace slag some 'free', uncombined lime (CaO) and magnesia (MgO) may also be present, particularly in slags from oxygen steelmaking.[22] These oxides can react with water to form hydroxides, a reaction which is accompanied by a large increase in volume. This reaction can give rise to a greater degree of expansion and is a more common problem associated with the use of slags than the sulphoaluminate expansion mechanism of blast furnace slags (see Box 9.3).

Hydration of calcium oxide has been found to be responsible for the expansion of steel slag over the first year subsequent to production, with hydration of magnesium oxide a slower, longer term reaction, responsible for the continued expansion over a number of years. The heat produced by these hydration reactions may also result in spontaneous heating in slag fill.[236]

Box 9.3: Building damage caused by expansive steel slags

Building damage caused by expansive steel slag was reported in 1969.[64] A single storey building, constructed in 1962, had developed cracks followed by buckling of window frames and interior metal and glass partitions. The damage was found to be caused by the differential movement of the floor slab and interior and exterior columns. The latter rested on spread footings at 1.5m depth whilst the concrete floor slab was cast on compacted fill consisting of slag from a steel mill. Level measurements revealed that the floor slab was rising rather than the footings sinking. The absence of any voids between the floor slab and an underground service tunnel indicated that the movement was caused by swelling of the slag fill. In autoclave tests open hearth slag from the steel mill was found to expand by 10% and it was concluded that this was due to hydration of calcium and magnesium oxides.

Following publication of this paper fourteen other cases of building damage caused by expansive slag were reported in the same journal.[65] The amount of expansion reported was as much as 20%, with rises of 220-370mm beneath floor slabs and 600mm beneath pavement. The causes of the damage were not always recognised, often being wrongly diagnosed as differential settlement or expansive clays.

Slags which are susceptible to expansion can be used in situations where expansion can be tolerated, such as screening bunds, or as aggregate on unsurfaced roads, where steel slag has the advantage of not suffering the discolouration of blast furnace slag. Steel slag has excellent skid resistance properties and is used for roadstone, sealed with an asphalt coating to exclude water.[84, 179] Slag from basic Bessemer (or Thomas) processes, used widely in Germany and France, was rich in phosphate and was valued as a fertiliser. Now that oxygen steelmaking has replaced the Bessemer process, and low phosphate imported ores have replaced the local high phosphate ore, phosphate is added to some steel slag in the Ruhr area so that it can be sold as a fertiliser.[239] Use of steel slag as a fertiliser is described in Box 9.4.

Calcite can be formed if steel slag is placed in waterlogged situations.[83] The highly alkaline pH (10-12) of steel slags may cause surface and groundwater to be alkaline.[225]

9.2.3 Slags as a substrate for plant growth

Because slags have been so widely re-used, large quantities of waste slags have not generally accumulated. The lack of large slag heaps in need of treatment has meant that there has been little study of them as a substrate for plant growth. Where attempts have been made to vegetate slags a principal difficulty has been with those slag heaps which are comprised of large blocks of fused slags. The lack of fine-grained particles has meant that moisture holding capacity is negligible and vegetation has been slow to establish and difficult to maintain. Some very old blast furnace slag heaps have become vegetated naturally (see Photograph 9.2) but where slopes are steep even on old tips vegetation has not established. Old acid slag heaps tend to take on the vegetation characteristics of the surrounding land. Where slag and other wastes from basic steel making

Box 9.4: Basic slag as a fertiliser

Basic slag was used as a fertiliser in many steel making areas. Its principal benefits were:

- alkalinity. Basic slag has a moderate liming effect;
- source of phosphate. The phosphate in slag is only sparingly soluble. Slag was therefore useful as a 'slow release' supply of phosphate, one application often sufficing for 3-4 years on grassland;
- low cost.

These properties are particularly beneficial on soil-forming materials such as colliery spoil or other inert wastes where vegetation establishment is required. Basic slag (often enriched with phosphate) is now imported for fertiliser use into some European countries where it was once a by-product.

Photograph 9.2: A heap of blast furnace slag in South Wales with some natural vegetation (source: Richards, Moorehead and Laing Ltd.)

Photograph 9.3:
White Mullein (*Verbascum lychnitis*) on blast furnace slag (source: Richards, Moorehead and Laing Ltd)

processes has been deposited, naturally established vegetation can be of some ecological value (see Box 9.5).

9.2.4 Assessment of slags

Over the years more and more slag has found a productive use and slags have come to be viewed as a valuable by-product of iron and steel making. Slag is, however, likely to be encountered on derelict iron and steel sites as it will have been used as an aggregate in concrete and as a general fill material. There may also be areas used for disposal of material which was considered to be unsuitable for reuse at the time of its production. Different types of slags and other wastes from the site, such as refractory linings and flue dusts, are likely to be mixed together in such dumps. The assessment of these tips presents a number of problems, especially the difficulty in obtaining representative samples for subsequent analysis. A wide variety of different types of slags of different ages may be present, and cementation of some slags may make it difficult to penetrate with ground investigation equipment. Indeed, in some instances even heavy earth moving equipment may encounter problems in excavating highly cemented slags.

Box 9.5: Wildlife value of steel wastes

The alkaline nature of the slags from steel manufacture can result in colonisation by vegetation of significant ecological value. At one site in North Wales where a steelworks is situated on an estuary, the combination of the calcareous influence of the slag and the marine influence has resulted in the development of an ecologically interesting flora. The plants found include White mullein (*Verbascum lychnitis*) a relatively uncommon plant in the United Kingdom and which is only found in the vicinity of the steelworks on steelworks slag and the waste from the alkali industry (see Photograph 9.3). The wildlife value of the site as a whole is enhanced by the colonisation by flora and fauna of the former steelworks lagoons which have become very rich in bird life and support one of the largest colonies of breeding common terns in the United Kingdom.

Volumetric instability

Although slags present at former steel sites may be over 50 years old this does not mean that reactions causing expansion (described in Section 9.2.2) will no longer occur. This is particularly so if the slag is disturbed, so that large lumps are broken up, exposing unreacted material to water and bringing materials together in new combinations. Old deposits of slags should be disturbed as little as possible, but if they are to be placed in confined situations, such as beneath buildings, an assessment of their volumetric stability should be made.

Assessing the volumetric stability of slags is not a simple matter. The mineralogy of the slags should be determined by techniques such as optical microscopy and X-ray diffraction, and the chemical composition determined by chemical analysis. In particular the amounts of total sulphur, sulphate, free lime and magnesia present should be measured. Specialist techniques, such as thermal analysis, are required to distinguish the oxides, lime (CaO) and magnesia (MgO), from their hydroxides ($Ca(OH)_2$ and $Mg(OH)_2$).[237] Information gained from these investigations can be used by experienced personnel to make an assessment of how likely the slag is to undergo expansive reactions. Accelerated expansion tests should then be carried out on representative samples of the slag material. The simpler test procedures which have been developed for this involve immersion of compacted samples of slag in a water bath which is maintained at an elevated temperature.[82] However, some work suggests that expansion is greater if the slag is in contact with water vapour rather than immersed in a water bath, as in the latter case the hydroxides are formed in solution and tend to fill up the void spaces in the crystal structure, whereas in the former the crystals are pushed apart.[156] Examples of accelerated expansion tests are described in Box 9.6.

If expansion is observed in an accelerated expansion test it indicates that there is a risk that the slag may expand. The precise degree of expansion cannot be predicted, nor can the timescale over which such expansion will occur. During reclamation of a slag bank in Hartlepool, North-East

Box 9.6: Accelerated expansion tests

A test procedure developed in the United Kingdom[238] exposes samples of slag to carefully controlled fluctuating conditions of temperature and humidity within a climatic cabinet. The test is usually carried out over a period of 14 days and, unlike tests where the sample is immersed in water, it detects expansion due to formation of sulphoaluminate type minerals as well as the hydration mechanisms found in steel slag.

An accelerated expansion test which uses steam has also been developed in Germany, where it is used to assess steel slags, from the LD process.[147] The test apparatus has two compartments, the lower of which contains water which is heated to boiling point. The steam generated passes through the upper compartment, which contains a compacted sample of slag. This upper chamber is surrounded by a heated jacket, at 102°C, to prevent condensation of water vapour. A meter measures movements in the upper surface of the slag sample. Expansion has been observed to be hyperbolic, tending towards a maximum value in around 24 hours. Comparison of maximum expansion with free lime content of the slag sample revealed that expansion was minimal, (3% or less) if the free lime content was less than 7% by weight. This work has led to specifications for use of steel slag in road construction in the state of North-Rhine Westphalia.

England, site levels were monitored after earthworks had been carried out to prepare the site for redevelopment. Some of the slag from the bank was known to be expansive as its use in a local road scheme had led to ground heave. The observed rise in levels was generally less than that found in concurrent laboratory real-time expansion tests, but expansion was still continuing after three years.[77, 78] During later development of the site slag was removed over areas where buildings were to be located.[238]

9.2.5 Leaching of metals from slags

Slags contain metals derived from iron ore, coke and limestone. Steel slags may also contain some of the metals used in alloy steels. Metal concentrations of some blast furnace and steel slags are shown in

Table 9.4: Metal content of slags (from Barry, 1985[22])

		Blast furnace slag				Steel slag		
	Sample:	A	B1	B2	C	D	E	F
		TOTAL CONCENTRATIONS (mg/kg)						
Antimony		ND	ND	ND	ND	210	64	220
Arsenic		ND	ND	87	3	ND	ND	ND
Barium		200	100	200	400	700	70	100
Cadmium		10	6	8	9	7	4	9
Chromium		162	65	55	60	420	96	752
Copper		188	90	60	4400	32	17	20
Fluorine		9	1	0.8	0.4	9.5	4.8	0.7
Lead		63	56	48	61	70	38	80
Magnesium		37360	44040	54800	60980	4542	20902	20370
Manganese		18800	2580	2210	3620	44900	12600	43800
Mercury		1.7	ND	ND	ND	ND	ND	ND
Nickel		50	20	30	17	140	50	120
Selenium		24	ND	80	57	ND	ND	16
Thallium		23	95	108	90	53	34	58
Zinc		780	20	25	1160	38	41	54
		EXTRACTABLE CONCENTRATIONS (mg/kg)						
Hot water	Boron	5.5	3.0	3.0	4.0	3.5	3.0	3.5
0.5M acetic acid	Copper	0.3	0.5	0.8	0.6	0.7	-	0.3
	Nickel	0.2	0.4	1.6	1.0	3.0	5.0	0.4
	Zinc	3.6	2.0	3.2	3.4	2.8	8.2	3.2
0.5M ETDA	Copper	0.2	1.3	0.9	1.3	0.1	-	0.7
	Nickel	1.3	0.9	1.4	1.0	1.8	-	2.0
	Zinc	8.9	2.7	4.8	2.9	1.0	-	1.2

Notes: B1 and B2 are 'blind' repeat analyses
ND not detected
\- not determined

Table 9.4. Many of the metals in slags, such as copper, chromium, lead, zinc and nickel, are potentially toxic to people, animals or plants. However, slags have a pH of 10-12 (though this may fall to pH 8 on weathering) and this alkaline pH and the complexing of metals with calcium silicates ensures that these metals are present in insoluble forms. Leaching studies have shown that metals are not readily leached from slags.[83, 127, 141]

9.2.6 Flue dusts

Flue dusts consist of the particulate matter removed from the gases emitted by furnaces of iron and steel production. The coarse-grained material, of a particle size similar to that of sand, trapped by primary dust catchers, is usually returned to the blast furnace after sintering, a process which agglomerates fine-grained particles so they can be used in the furnace without impeding the flow of air. Recycling of flue dusts in this way recovers some of the iron lost in steelmaking and the coke lost in the dust from the blast furnace. However, flue dusts also contain non-ferrous metals so continuous recycling of all dust cannot be practised as it would lead to the accumulation of such metals in the blast furnace.

Fine-grained dusts, collected in bags, scrubbers, precipitators and cyclones may have been disposed of in lagoons, where their physical nature may result in soft ground conditions. Excavation of such fine-grained materials is likely to produce high concentrations of air-borne dust, a health and safety hazard. Chemically, dusts are likely to be enriched in metals such as lead, zinc, nickel, copper, cadmium, chromium, arsenic, alkali metals, and halides.[189, 225]

Flue dusts are often the chief sources of inorganic contamination at iron and steel sites and, unlike slags, their metal contents may be leachable. In a study of the leaching of metals from wastes of steel manufacture,[127] flue dusts from electric steel making were found to be the only waste which gave rise to leachate containing elevated concentrations of metals such as lead, zinc and chromium. The free lime content of the dusts was found to be the major factor determining metal solubility.

Dusts from electric arc furnaces frequently contain up to 20% lead and zinc, originating from the high percentage of scrap metal used in such furnaces. In the USA electric arc furnace dust is classified as a hazardous waste and it must be treated chemically or thermally to remove or stabilise the leachable toxic metals. One such thermal process operating in Europe, the Waelz process, involves heating the dust to high temperatures, with coal and sand. Lead and zinc are vaporised and reclaimed to give a residue containing 60-70% lead and zinc in an oxidised form. Other constituents of the furnace dust form an innocuous slag with the sand.[232] A hydrometallic process involving alkaline leaching, which produces a metallic zinc powder, has also been developed.[256]

9.2.7 Other wastes

The refractory materials used to line furnaces form a significant part of the wastes likely to be found on iron and steel sites. They will almost certainly be contaminated by metals. Refractories from the blast furnace can also contain cyanides.[243]

Sand used for forming moulds for casting of iron and steel, may contain phenolic binders. Large volumes of this sand, used as a fill material has been found at some sites.[129] Foundry and moulding sand may also contain expansive steel slag, giving risks of ground heave.[65]

The processing and treatment of steel results in various types of wastes, for example:[22, 189, 243]

- spent pickling acids (liquors containing, typically, 10% unreacted sulphuric or hydrochloric acid and 5% dissolved iron;[211]
- hydroxide sludges, formed on neutralisation of spent acid solutions, containing iron sulphate or chloride;
- spent plating solutions;
- galvanising scums;

- wastes from metallic coating of steel, which may have used tin, molybdenum, zinc, vanadium, nickel, copper, chromium, cadmium, lead or aluminium;
- emulsified mineral oils from cold rolling of steel;
- waste plasticisers, glues and paints used in coating of steel products.

Substances used in steel production may be present as contaminants at former steel sites. For example, selenium is used as an additive in ferrous metals as it aids casting and improves the machinability of steels. This toxic element, which is chemically similar to sulphur, has been found at elevated concentrations in ground beneath former steel making plant.

Iron and steel sites also contain the wastes common to a wide range of industrial sites. These materials include asbestos, used for lagging of pipes and buildings, ash from coal burning, and waste oils and lubricants.

Areas which were used for storing scrap are likely to be contaminated with materials associated with that scrap, such as paints, oils and non-ferrous metals.

Iron and steel production sites have typically been developed over a long period of time, with several generations of plant following on from each other. In many cases such sites were originally associated with coal or iron ore mines. Shafts and areas of unstable ground may thus be present, sometimes overlain by areas of filled ground several metres deep. Frequently encountered problems, in addition to those caused by the presence of the wastes and raw materials described above, are the presence of massive foundations, underground voids, ducts and pipework. These aspects have been discussed in Chapter 4, under the subjects of demolition and site clearance.

10 COAL CARBONISATION

Chapter contents

10 COAL CARBONISATION

10.1 Introduction

10.1.1 Coke, coal gas and smokeless fuel

The process of coal carbonisation is intimately associated with both the coal and steel industries. Coke, a primary product of the coal carbonisation process, is produced from coal for use in the manufacture of iron and steel, where it provides a fuel and a reductant for the smelting of iron ore. Coke replaced charcoal in the iron-making process in the early eighteenth century (see Section 9.1.3).

In addition to the manufacture of coke, coal carbonisation is used for the production of coal gas ('town gas') and of smokeless fuels for the domestic market. Smokeless solid fuel is produced by low-temperature carbonisation ($<700°C$), whilst coal gasification relies on the high-temperature process ($>900°C$). Coke may be regarded as a by-product of the coal gas industry.

The European coal gas industry has now almost disappeared in favour of natural gas and other, cleaner, alternative sources of energy.

Coal gasification, coke making and smokeless fuel manufacture are all coal carbonisation processes. In these processes coal is heated under oxygen deficient conditions, usually in large batteries of silica-lined ovens, with the subsequent liberation of gaseous combustion products and volatile components, including both aliphatic and aromatic hydrocarbons, oxides of sulphur and nitrogen and other coal impurities.[30] Although coal gas produced from the gasification itself acts as a fuel to fire the ovens, large quantities of gaseous and particulate emissions from the coking and smokeless fuel processes were, in the past, freely liberated into the atmosphere. In many areas this has led to a significant rise in the background concentrations of toxic organic compounds in surface soils.[116, 130]

Coke is thus a purified carbon product. It has a large particle size not dissimilar to that of the coal feedstock, but exists in a highly reduced form, and has robust structural qualities. In the blast furnace coke provides heat, reducing power (in the form of carbon monoxide) and structural support. The long history of the coking and coal carbonisation process has been parallelled by the development of a variety of different designs of carbonisation plant, with significant contributions arising from Germany, the United Kingdom and the Netherlands.[30, 262]

Although the best coking coals are of a low-volatile bituminous nature, other highly volatile and highly sulphurous coals have traditionally been blended with more suitable coal feedstocks for the production of coke. Whilst good coking coals produce significant quantities of by-products, those of a lower quality led to greater emissions to the atmosphere, greater deposition in local areas and larger quantities of by-products and wastes.

10.1.2 The carbonisation process

Coal carbonisation involves the destructive distillation of coal, and during this process as much as 50% of the weight of the coal feedstock is driven off as gaseous and volatile components. These are regarded as by-products of the coking process, and include combustible gases *e.g.* methane and hydrogen, and the lower boiling point components of coal tar *e.g.* benzene and naphthalene.

Older coal carbonisation works burnt coal in horizontal, silica-lined retorts. These were often long pipes with a semi-circular cross section, arranged in two vertical rows of four or more, on either side of a primary combustion chamber where the 'producer gas' was burnt. Gas passes from the retort along a hydraulic main, and here some ammonia, coal tars and phenols are recovered in condensate as by-product liquor.

Larger volumes of tarry condensate are recovered in condenser plant, consisting of vertical water- or air-cooled pipes over which the crude gas is passed. This removes further ammonia and tar in the gas stream.

At town gas works and coking plants gas passes through scrubbing systems, which use dilute acid solutions to remove the remaining ammonia, giving rise to ammoniacal liquor. The liquors associated with gas cleaning are often the source of significant contamination of coal carbonisation sites.

The partially-cleaned gas then passes through purifier beds containing hydrated ferric oxide (or 'bog iron'), where hydrogen sulphide gas was removed.

10.1.3 By-products and environmental considerations

In the past the by-products from coke manufacture often formed the feedstocks for other chemical manufacturing industries, although much of the material was regarded as waste, with little or no intrinsic value. The hazardous nature of these wastes was not fully realised, and consequently the waste disposal practices of the manufacturers were basic, with little or no consideration being given to either the environment or the health and safety of workers at the site. Therefore many sites on which coal processing facilities once stood still contain significant quantities of toxic waste products from their former activities, often indiscriminately deposited in unlined landfills or poorly maintained underground repositories.

The locations and nature of such underground deposits of waste are often identified only after a thorough site investigation. Measures can then be taken to minimise environmental and health risks in the area during the course of a reclamation scheme. Often, however, land reclamation schemes that were undertaken in the past have paid little attention to the adequate removal, containment or treatment of these wastes, and they remain in subterranean deposits beneath newly developed sites. In many

European countries, the presence of an operating gas works in a central location in most towns and cities prior to the use of natural gas, has resulted in a large number of former coal carbonisation sites being redeveloped as part of urban regeneration and development schemes. Although some of these sites remain derelict and/or untreated, there are many that have been reclaimed, and of these a number retain underground deposits of wastes that have been merely covered over in the course of development. The ill-founded philosophy of 'out of sight is out of mind' has been all too commonly adopted in the redevelopment of chemically polluted sites. Coal carbonisation sites are no exception.

Improvements in the 1970s to the design and operation of coal carbonisation facilities have been prompted by the requirement for cleaner, pollution-controlled operations. Previous unwillingness by the operators of coal carbonisation facilities to incorporate new technologies for the purpose of pollution control has been overcome by legislatory pressure to meet increasingly stringent emission standards. What was therefore once viewed as capital expenditure with no foreseeable financial benefits has become a necessary expense to avoid both ongoing pollution and potentially large remedial action costs prior to the sale or redevelopment of redundant sites. Various technologies and equipment are now used in operating coking plants to improve the environmental and output performance of the coking process.[160]

Coking and coal gas production are industries which originated in the nineteenth century, and, like coal and steel production themselves, they have left behind a legacy of industrial dereliction and environmental damage. Derelict land from the closure of coal carbonisation facilities continues to be created throughout Europe, as the requirements for coal products and by-products decline in the face of new technologies and other changes to industry.

The fundamental environmental consequence of the coal carbonisation process is that it involves the separation and concentration of various chemical fractions, from their relatively low concentration within the coal-carbon matrix to several by-product chemical streams of highly elevated concentrations. In addition, some of these by-product chemicals are of relatively high mammalian toxicity, and therefore pose a risk to human and animal health, and to the environment in general.

10.2 Nature of by-products and wastes

10.2.1 Introduction

There are a number of categories of by-products that arise from the various coal carbonisation processes.

The major by-product streams from carbonisation can be placed into five categories:

- liquid effluents;
- high boiling point tars;
- spent oxide;
- acidic sludges;
- benzole still bottoms.

The chemical nature of all of these waste streams is dependent on the nature of the coal used to feed the plant and on the type of process employed. The following gives a broad account of the composition of these by-product streams.

10.2.2 Liquid effluent

Ammonia, a simple though potentially toxic chemical combination of nitrogen and hydrogen, is a major constituent of the reducing process of coal carbonisation. Ammonia was removed principally as a waste product by condensation and scrubbing liquors, which were often produced in large volumes during coking and gasification. Such liquors often contained in the region of 1 to 2% ammonia.

Box 10.1 lists the compounds found in a typical ammoniacal liquor. Ammoniacal liquors are usually alkaline, and in addition to the substances shown in Box 10.1, they may contain significant quantities of other potentially toxic chemical species, including pyridine and other heterocyclic nitrogen compounds. Generally they contain high concentrations of inorganic and organic compounds, and therefore have high biological and chemical oxygen demands (see Box 10.2). Ammoniacal liquors are therefore of great significance as potential polluters.

In addition to the scrubbing and condensation processes described in Section 10.1.1, ammoniacal liquors amended with ammonium sulphate and sulphur, or alternatively, alkaline ferrous sulphate liquors, have been used to remove further cyanide from the coal gas. Significant quantities of ferrocyanide and ammonium thiocyanate are produced at this stage.

The waste disposal practices of former carbonisation plants are likely to have included indiscriminate discharge into local watercourses, and even to groundwater. Later effluent management practices may have involved the storage of liquors in holding tanks prior to collection and disposal off site. Where a coal processing plant was closed, undisposed liquors may

Box 10.1: Chemical components of a typical ammoniacal liquor from coal carbonisation

Phenols	e.g. phenol, cresol, dimethylphenol
Cyanide	as HCN
Thiocyanate	as SCN^-
Sulphide	e.g. H_2S in solution
Thiosulphate	as $H_2S_2O_3$
Ammonia	e.g. NH_4^+ in solution or fixed ammonia
Carbon dioxide	as CO_2
Light aromatics	e.g. Naphthalene, methylnaphthalene, benzene
Chloride	as HCl

Box 10.2: Definition of biological oxygen demand and chemical oxygen demand

Organic molecules can be oxidised, *i.e.* combine with oxygen and donate electrons, either through biological processes (mediated by enzymes) or by chemical oxidation (facilitated by oxidising agents). Oxidation can give rise to the mineralisation of an organic molecule, producing carbon dioxide and water.

Biological Oxygen Demand (BOD) provides an estimate of the amenability of organic materials in solution to biological oxidation. Thus, the BOD test involves the dilution of a liquid sample with aerated, distilled water, and the measurement of dissolved oxygen in each of two identical sub samples; one immediately after dilution and the other after seeding with aerobic bacteria followed by a period of incubation at 20°C. This incubation is typically for five days, in which case the notation BOD_5 is used. The BOD test, by quantifying oxygen consumption, measures oxidisable organic matter, oxidisable nitrogen compounds and some reducing compounds *e.g.* sulphides. The latter is due to chemical rather than biological reactions.

Chemical Oxygen Demand (COD) measures the chemical oxidation of organic material in a solution. The method uses boiling acidified potassium dichromate to oxidise the material, and the excess dichromate remaining after the reaction is measured by titration. Some chemical species, notably chloride and ammonia, can give anomalously high COD values, however, and in addition some organic molecules *e.g.* benzene and toluene, are incompletely oxidised.

Results for BOD and COD are expressed in mg/litre.

Although both tests have limitations, as outlined, they provide useful indications of the general nature of organic contamination in a liquid sample. The testing of BOD and COD on the same sample can, in addition, provide a useful estimation of the proportion of organic contamination that is amenable to biodegradation. This can be expressed as the COD:BOD ratio.

have been left on site, and these may remain in underground tanks and pipes for many years. Leakage of liquor from both above- and below-ground structures would have invariably led to local contamination and possibly the accumulation of residual chemicals in spatially distinct soil and groundwater. Changes in the local soil and groundwater conditions, during, for instance, a reclamation operation, structural development or site investigation, may result in the spread or redistribution of contamination (see Section 4.5).

10.2.3 Tars

Tar was often a valuable by-product at coking works and associated facilities. Tar was removed by either condensation or gravity separation from the liquors or by tar extractors used for purifying the gas stream.

In all but some of the more recent coal carbonisation plants, tars were generally stored in structures below ground, and were often transported from one part of a site to another via large underground pipes. Tar pipes, wells, pits and slumps are common features of former coal processing facilities, and are often a cause for concern during reclamation works due to the problems associated with the handling and disposal of the residues contained within them.

More recent coal carbonisation facilities were designed with above-ground tar storage tanks, from where crude tar could be collected and transported off site for use elsewhere.

Tars are a complex and diverse mixture of organic compounds. Amongst the more toxic and harmful to the environment are the PAHs (polyaromatic hydrocarbons). A great number of this diverse group of hydrocarbons are present at high concentrations in coal tars. Box 10.3 gives details of some of these compounds.

Coal tars are an important product of coal carbonisation. Tars are distilled and refined to produce a variety of products, such as

Box 10.3: Major chemical components of coal tar

Polyaromatic hydrocarbons (PAHs) are large organic compounds, usually of low aqueous solubility. Many are known to be carcinogens. PAHs have a basic structure of two or more benzene rings. Analysis of coal tar for PAHs is likely to focus on the following members of this group:

Compound	Number of rings in structure	Typical composition (% of coal tar)
Naphthalene	2	5 - 15
Acenaphthene	2	
Fluorene	3	5
Anthracene	3	10 - 20
Phenanthrene	3	
Fluoranthene	4	
Pyrene	4	
Chrysene	4	
Benzofluoranthene	5	25 - 40
Benzopyrene	5	
Perylene	5	
Benzoperylene	6	
Coronene	7	

Other components of coal tar, besides PAHs, include phenols (monohydric and polyhydric), benzene, toluene, xylene, tar acids, tar bases and a variety of compounds that comprise pitch. Pitch is the high boiling point fraction of coal tar.

weatherproofing and road building materials. The sites of these tar distilleries are also important sources of wastes and potential contamination of ground. Similar precautions to those at coal carbonisation facilities should be adopted at tar distilleries.

Low quality tars and tar residues were traditionally tipped in discrete areas of a given coal products site, often with little regard to the future use of these areas. In many instances these tar dumps or low-technology landfills now present significant constraints on future development due to their large size, high toxicity and low potential for on-site decontamination of the waste material.

10.2.4 Spent oxide

During the process of town gas production and coke making the gas produced was often fed over a series of purifier beds in order to remove toxic and foul smelling hydrogen sulphide prior to sale and use as a fuel. Hydrogen sulphide was often present in very high concentrations, and had to be reduced to a concentration below 1mg/l in the gas distributed to domestic users. These purifiers contained moist iron oxide which reacted with the hydrogen sulphide to form iron sulphide.

Although the resulting iron sulphide material could be reoxygenated by exposure to air, eventually the material was discarded as spent oxide. In addition to the iron sulphide the oxide also accumulated contaminants such as cyanides and elemental sulphur.

Spent oxide produces the distinct 'gassy' odour typical of coal carbonisation sites, and was often tipped on site with little regard to safe disposal practices. Its excavation and disposal during the course of reclamation of a site can be therefore be problematic, not only because of its chemical nature, but also due to its odorous properties.

Spent oxide, present to a greater or lesser extent at old carbonisation sites, is often mixed in with other wastes and fill materials. The appearance of spent oxide varies, but often resembles clinker and slag material, though it has a characteristic blue-grey colour, due to the presence of complex cyanides. Box 10.4 gives the chemical composition of a typical spent oxide.

10.2.5 Acidic sludges

Acid tars and sludges are derived from the acid washing of crude tar oils resulting from the coking process. Acid washing results in the extraction of a variety of impurities including aromatics, hydrocarbons, and sulphur and nitrogenous compounds.

Mineral acids, particularly sulphuric acid, were widely used as washing and scrubbing liquors at coal carbonisation facilities. Sulphuric acid was particularly popular for the chemical washing of benzole, and for the conversion of ammonia in by-product liquors to ammonium sulphate.

Acid sludges are often of very viscous, which causes handling difficulties. This viscosity, combined with their complex and toxic chemical nature, and their low pH, makes acid sludges a waste product of considerable potential environmental impact, requiring special treatment or disposal to reduce the risk of contamination of surrounding ground.

Box 10.4: Typical chemical composition of spent oxide (adapted from Environmental Resources Limited, 1987[86])

Compound/Chemical	Concentration range
Free sulphur	35-60 %
Sulphate (SO_4^{2-})	1-3 %
Cyanide (total)	2-5 %
Thiocyanate	0.1-0.5 %
Iron	2-10 %
pH	2-6
Metals e.g. Cd, Pb, Cr	2-100 mg/kg

Treatment of these acidic residues during operation of a carbonisation plant was not often practised, and they were simply tipped into pits on or near the plant. Any treatment of the wastes was invariably limited to either the burning of the residues in the ovens on site, or hydrolysing them with hot water. The latter included neutralising the acidic water phase with lime. This process resulted in two waste streams (neutral effluent and dilute tar), which remained a problem for disposal. The high cost of hot water hydrolysis and neutralisation tended to dissuade operators from practising this method. Acid tar lagoons are thus still a feature of some industrial sites, and treatment remains as inefficient and costly as ever. Such treatment methods, which may include excavation and incineration, are discussed in detail in Chapter 11.

10.2.6 Benzole

Benzole was a valuable crude product of coke manufacture, produced in countercurrent scrubber towers, where a mineral wash oil was used to absorb the benzole hydrocarbons. The benzole hydrocarbons were then recovered in a benzole distillation plant.

As with other by-products of the coal carbonisation process, the chemical composition of benzole differs with coal feedstock and separation and distillation processes, but essentially the main components are benzene, toluene, xylene and related compounds (often referred to as BTEX). Impurities include phenols, carbon disulphide, naphthalene, oils and paraffins.

The liquid and readily water-miscible nature of these compounds, combined with their toxicity, makes them significant contaminants of former coal carbonisation sites.

10.2.7 Coal and coke

Coal and coke are also found in the vicinity of coal carbonisation plants. Whilst these materials are not of great significance in terms of their toxicity, they do constitute a combustion risk at such sites, particularly in areas of former coal and coke stocking. Site investigations into the nature of ground conditions must therefore take into account the calorific value of soil containing quantities of potentially combustible material. However, if the quantities of coal and coke materials at a given site are not sufficient to warrant economically viable recovery (see Section 8.2), there may still be a strong case for removal prior to site development if by doing so the area will be brought into a state of low combustibility.

10.3 Background levels of contamination

Coal carbonisation facilities are usually situated in areas of either high urban density or dense industrialisation, where soil concentrations of coal-related contaminants are already elevated. Coal carbonisation will further increase the concentrations of these soil contaminants over an area far wider than that of the former carbonisation facility alone because of atmospheric discharges and deposition.

The elevated soil concentrations of coal-related contaminants common in urban areas can result in misinterpretation of data from a site investigation if concentrations are being compared with those found in rural locations (see Section 2.6.5). Background concentrations of coal-derived contamination in the locality of a carbonisation facility should therefore be taken into account when setting reclamation standards (see Section 11.9). Elevated concentrations of PAHs in urban areas may derive from the burning of coal and other fossil fuels for domestic and transport purposes, as well as industrial uses.

During the active life of a coal carbonisation plant, and especially those that operated with little or no regard for the release of emissions to air, potentially toxic compounds including sulphur dioxide and PAHs were almost continuously pumped into the atmosphere. The soluble and particulate nature of such contaminants meant that they were destined to become deposited on the land in due course.[130]

Such deposition of airborne pollution is prevalent in areas where highly sulphurous or bituminous coal was utilised, and where the situation of the site is such that high annual rainfall and/or low levels of prevailing winds have optimised the potential for fallout in the vicinity. The soils around the brown coalfields and associated coal processing areas of eastern Germany, such as those in the Leipzig-Halle region, are, for instance, highly acidic because of the deposition of sulphuric acid from the airborne sulphur dioxide of coal processing origin (see Section 17.7). The effects of the resulting acidity on the mobilisation of toxic metals *e.g.* aluminium from soils, is well documented,[81] and has resulted in widespread deterioration in soil, surface and groundwater quality, and in the corresponding depletion of vegetation and wildlife.

Metals too were deposited from the atmosphere in the form of fly ash. Some of these metals are toxic, and occur as both background contaminants, as well as components of bulk wastes, such as spent oxide. Such metalliferous contaminants include chromium, lead, mercury, cadmium, copper and arsenic.

10.4 Structures and materials on site

Coke works and associated industrial coal processing facilities were often composed of large structures in which the various processes were carried out. Coke ovens, distillation towers, gas purifier beds and venting chimneys are structures which are all closely associated with the generation of contaminated materials, and as such require special regard during the decommissioning and demolition of these facilities. These matters are discussed in Chapter 4 in more detail. The handling of contaminated building materials is as important an issue as the handling of toxic wastes, and care must be exercised in the disposal, retention or reuse of such materials during redevelopment work. Analysis of brick from the chimneys which once vented emissions from the Siemens-Martin steel furnaces of the Riesa steelworks in Germany, showed high concentrations of some potentially toxic metals up to 60mm into the brickwork lining the chimneys (see Section 17.7).

Therefore, demolition of the structures present at a cokery is not limited to problems of scale and the resilience of building materials. Preliminary assessments of contaminated structures should be made prior to demolition (see Box 4.5). This can be done by a visual assessment, or preferably, in addition, by the collection of samples for appropriate chemical analysis. Visible contamination may include the blue staining of materials by ferrocyanides (sometimes called 'Berliner Blue' or 'Blue Billy'), the retention of spent oxide in old purifier boxes, or bright yellow deposits of elemental sulphur. These materials need to be accounted for in terms of both risk and cost in the overall plans for remediation of a given site.

10.5 Environmental contamination

10.5.1 Introduction

The complex chemical nature of coal gives rise to the array of compounds produced during the carbonisation process. It is this mixture of chemical elements and compounds that produces the unique circumstances to be found at coal carbonisation sites, and gives cause for the special measures that are required in order to restore the site to a safe and appropriate after-use.

Most nationally adopted standards and guidelines for contaminated land in Europe take into account the contaminants typical of coking, gas and smokeless fuel works. In the United Kingdom, for example, the frequently used ICRCL (Interdepartmental Committee for the Reclamation of Contaminated Land) guidelines for contaminated soils, are based on common coal carbonisation contaminants. Such guidelines, whilst forming a nationally accepted standard for the reclamation of contaminated soils of all types, may not cover all the contaminants present at former coal carbonisation sites. Chemical guidelines for the decontamination of polluted soils are published in many European countries, although nationally adopted guidelines are less common than regional standards. In many cases soil chemical standards are still set on a case by case basis, even where local guidelines exist.

The wastes generated and frequently present at these sites may have the following effects:

- lowering of groundwater quality;
- direct toxicity;
- degradation of building materials;
- difficult ground conditions.

At some coking works large volumes of effluents were pumped to settling tanks or lagoons, where particulate and heavy materials were allowed to settle out under gravity, and the clarified effluent released either directly to the sewers or further treated prior to disposal. At coke works in some mining regions this practice has led to the leakage of effluent waters into underground mine workings.

Settled materials in primary treatment lagoons build up during the course of operations, and the residues are likely to be of a contaminated nature, requiring treatment or disposal at a later date. Such residues must be carefully considered as potentially hazardous material during a reclamation scheme, and should be treated accordingly *i.e.* by sampling, analysis and, where necessary, remedial treatment.

10.5.2 Groundwater quality

Introduction

In general those chemicals of high to medium aqueous solubility such as soluble components of former liquid effluents, including ammoniacal liquors, are of greatest concern to ground and surface water quality at coal carbonisation sites, particularly where water is extracted to supply potable water or water for other industries.

The solubility and mobility of contaminants in bodies of water is, however, only one part of a much more complicated equation, as aqueous solubility is often strongly related to other factors such as pH, microbial action and soil disturbance.[246]

Inorganic contaminants

Coal, the parent material of the coke making industry, contains a variety of chemical elements (Table 5.4). Amongst these elements are many toxic heavy metals, and other hazardous components such as arsenic, which are associated with the pyritic materials present in coals.

Heavy metals such as lead and mercury tend to be far more soluble under acidic (low pH) than alkaline conditions. At coal carbonisation sites, where acids were regularly used and disposed of, soil and groundwater pHs are often low. In such conditions metal solubilities are high, giving rise to a high potential for metal leaching and dispersion over a wide area. Where the concentrations of soluble metals are high, little or no vegetation may exist. In this way sparse or limited vegetation in an area can indicate the possible presence of phytotoxic (plant toxic) metals or other chemicals in soil water. Plant communities tolerant of metals may also develop. An experienced botanist will be able to identify such metal tolerant plant communities, and therefore highlight areas of potential contamination.

Inorganic contaminants of ground and surface waters in the region of coking works and associated facilities are not limited to metals. Sulphates and other anions, including chloride associated with coal ash residues, are often present, derived from the metal salts in waste materials.

Chemical attack of building materials

Sulphate solubility is enhanced by low pH, and increasing concentrations tend to have correspondingly deleterious effects on concrete building materials. Sulphate reacts with cement-based materials causing structural decline and associated risk, and as such can be a significant contaminant at coal carbonisation sites[177] (see Section 2.6.5). Similarly, phenols, as significant organic contaminants of soil and groundwater at many coal carbonisation sites, can produce adverse effects upon both concrete and plastic building materials in contact with contaminated materials.[224] Excessive acidity in contaminated ground may also cause structural problems if the ground is not decontaminated or target building materials protected.[112]

Organic contaminants

Organic contaminants in water present a different type of problem from those posed by metals and other inorganic substances. Many organic wastes from coal processing are only of limited solubility in water. The constituents of coal tars are, for instance, of varying aqueous solubilities, from the simple aromatic compounds of phenols, benzene and xylene, which are all relatively soluble, to the polyaromatics (PAHs), many of which are practically insoluble. Insolubility does not however exclude them from causing contamination, and organic liquids which are less dense than water can form immiscible surface layers on groundwater. The organic liquid will then flow under its own hydrostatic head in the same direction as the groundwater. Tar oils and other mineral oils can cause such phenomena to occur, and lead to severe contamination of subsurface water.

The differing aqueous solubilities of the organic compounds of coal carbonisation wastes, is perhaps one of the most significant barriers to their successful remedial treatment, and the reason why such materials are often dealt with by containment. Cyanide, for example, a common contaminant at former coal carbonisation sites, exists in many different chemical forms, from the extremely toxic hydrogen cyanide gas, through free cyanide in solution, to the commonplace metal-cyanide complexes which are relatively insoluble and of lower toxicity. The latter forms of cyanide are therefore resistant to leaching, which gives them a lower potential to cause more widespread pollution, but also makes their removal from the soil matrix difficult.[97]

In addition to the chemistry of the soil and groundwater, the underlying hydrogeology has important consequences for the spread of a groundwater contamination plume. The containment and treatment of such groundwater contamination arising from coal carbonisation sites is discussed in Chapters 11 and 12.

10.5.3 Direct toxicity

There are many chemicals that produce toxic effects in animals and plants when they are presented directly to the target organism in unnaturally high concentrations. At sites of current or former coal carbonisation activity, high concentrations of the most ubiquitous of chemical species, such as sulphates, ammonium and sulphur, can produce toxic effects. Table 10.1 summarises general information on the toxic effects of some coal carbonisation by-products and includes the nature of the chemical, its likely form at coal carbonisation sites, and its mammalian toxicity.

The assessment of the risk of toxic substances causing harm is not a simple matter; not least because harmful effects may often only be seen many years after exposure, following a period or periods of continuous or intermittent exposure. The exposure of an organism to carcinogenic compounds, such as coal tar PAHs, may consequently not reveal the symptoms of toxicity *i.e.* tumour growth, until some years after a single period of exposure. A similar scenario exists for teratogenic and mutagenic effects.

Table 10.1: Summary of the toxicity of major coal carbonisation by-products

By-product and potential contaminant	Appearance at former coal carbonisation sites	Toxicity and hazards
Coal tars	Black, viscous with typical tarry, phenolic odour. Often in buried deposits or tanks.	Lighter fractions may be absorbed through skin. Immediate toxic effects unlikely, but many components are carcinogenic. Tar has a high calorific value and may combust.
Spent oxide	Often mixed with other wastes and soils. Typical 'gassy' odour. Some may have a clinker-like appearance, with brown, yellow and blue coloration.	Direct skin contact may cause irritation and burning. Dusts may result in inhalation and eye contact, which may cause irritation and/or allergic reaction. Toxic components include sulphur compounds and cyanide.
Light tar oils and phenols	Typically present as liquors with an oily sheen and oily or phenolic odour e.g. in groundwater. More concentrated liquids are viscous and tend to coat soil and stone particles.	Skin contact can cause irritation, and vapour inhalation may produce headaches and drowsiness. Liquids can be corrosive to plastics and concrete. Long-term exposure can cause damage to central nervous system.
Ammoniacal liquors	May be present in underground tanks and voids, or mixed with soils and groundwaters. Phenolic and ammoniacal odours.	Irritation and burning following skin and eye contact. Vapours can produce nausea and asphyxiation in confined spaces. Liquors may attack concrete and plastics underground.
Cyanide residues	Cyanide exists in various forms at coal carbonisation sites e.g. as the residues on stonework and in spent oxide, or as hydrogen cyanide gas, or in liquids.	Free cyanide causes respiratory failure. However, hydrogen cyanide gas, the most toxic form, is only likely to occur in confined spaces. Complex cyanides are less toxic but can cause skin irritation.
Sulphide residues	Associated with various coal carbonisation wastes as solid metal sulphides (principally brown iron sulphide) and gaseous hydrogen sulphide, typified by an aroma like rotten eggs.	Solid sulphide residues may be corrosive and cause skin irritation. Hydrogen sulphide gas is highly toxic at low concentrations.

For those involved in the remediation of contaminated sites therefore, there are two major considerations concerning direct toxicity:

- human contact with toxic materials;
- plant and animal contact with toxic materials following reclamation.

The first of these considerations is critical before, during and after reclamation. Contact may include the inhalation of contaminated dusts, skin contact and direct ingestion. The second must take into account the continued presence of contaminated materials on site after reclamation, for example contaminants within a containment cell, or as residual concentrations of contaminants remaining in the soil after decontamination treatment.

10.6 Reclamation of coal carbonisation facilities

The information provided in this chapter highlights the complexity of factors affecting the reclamation of coal carbonisation sites. In many ways they are amongst the most challenging sites to the developer because of:

- the large structures present at the sites;
- the indiscriminate handling and disposal of toxic wastes at older sites, and the complex mixtures of contaminants that are likely to occur in soil and groundwater;
- the location of many gas works sites in areas of high population density.

The last point here draws attention to the often valuable development potential of these sites, despite their constraints. Thus, although the costs of redevelopment of coal carbonisation sites will often be increased due to the assessment and treatment of contamination in addition to the demolition, site preparation and construction costs, these extra costs may be offset by the value of the land.

The options for clean-up of contaminated coal products sites are limited by:

- nature of the contamination;
- characteristics of the geology and hydrogeology of the site;
- topography of and access to the site;
- proximity of treatment or disposal facilities for hazardous wastes;
- cost.

These factors, and the methods suitable for the treatment of contaminated materials, are fully discussed in Chapter 11. An appreciation of the risk of contamination should however be gained at the outset of site works. This will include the potential for encountering contaminated building materials, as well as contaminated ground and groundwater. These possibilities are best assessed at an early stage in the reclamation and redevelopment works, by the undertaking of carefully planned and executed investigations (see Chapter 2).

In many cases the environmental liability attached to coal carbonisation sites is significant, and national and regional legislation may be unclear about who is responsible for any pollution arising from a site. In this respect, however, the EC directives on groundwater quality and protection (see Section 12.6.2 and Boxes 12.4 and 12.5) are applicable throughout the Community, and it is the duty of a developer to ensure that the quality of groundwater is not adversely affected by:

- the undisturbed site;
- the site during ground engineering or remedial works.

The varied nature of toxic contaminants at coal carbonisation sites makes them a challenge to land reclaimers. In this way these sites require special regard to the reduction of on-site health and safety risks to ensure the future well-being of those who use them, and the protection of the quality of local ground and surface waters.

11 THE TREATMENT OF CONTAMINATED SOILS

Chapter contents

continued...

11 THE TREATMENT OF CONTAMINATED SOILS

11.1 Introduction

11.1.1 Contaminated soil in Europe

In the latter quarter of the twentieth century there have been rapid advances in the development of techniques for the decontamination of chemically polluted soils and associated groundwaters. Environmental disasters, pressure group activity, and national governmental and European policies and legislation, have focused attention on the need for safe options for dealing with potentially toxic wastes and contaminated materials. Both soil and water have been identified as invaluable natural resources, which should be protected.

Contaminated soil presents a challenge to land reclaimers and developers, and the risks associated with the presence of potentially toxic chemicals in soil and water need to be removed in a manner that retains as much of the indigenous materials as possible. The need to retain uncontaminated material on site has led to an increase in the use of methods to decontaminate soils and groundwaters instead of using engineered systems of removal, burial or containment.

In the European Community there are many thousands of sites which are known to have been contaminated by industrial activities. In Germany alone, for example, there are thought to be between 70,000 and 90,000 contaminated sites,[35] a significant proportion of which are associated with coal and steel production and processing.

Each country has its own approach to the classification and treatment of contaminated sites, and to waste management and disposal practices,[102] although current trends are towards a common approach, where waste producers, site owners and central and regional governments have increasing commitments towards the avoidance and clean-up of contaminated land.

11.1.2 The options available

The activities of the coal and steel industries have produced a diversity of potential chemical soil contaminants, as has been established in earlier chapters. Soils are themselves highly complex media, often displaying great variation in chemical, physical and biological characteristics. No one technique for the remediation of contaminated soil is therefore ever likely to be applicable to more than a relatively small proportion of all contaminated land situations. At many contaminated sites a combination of remedial techniques is often the most appropriate way to deal with contaminated material.

It is a primary objective of those responsible for the reclamation of sites containing toxic materials to ensure not only that these materials are dealt with in such a way as to reduce risk, but also to implement schemes that aim to conserve uncontaminated natural materials as far as possible. The treatment of contaminated land has thus undergone a change in emphasis in order to:

- reduce the amount of toxic materials being dealt with by confinement in terrestrial repositories;
- increase the separation of contaminants from the matrix which they are polluting, so reducing the volumes of waste generated, and conserving soil and water.

In any reclamation operation there are site specific constraints such as the availability of materials, the location, area and topography of the site and, of course, cost. Such factors as these define, at an early stage in the design, the methods that may be applicable to the site.

The methods available divide broadly into:

- traditional methods of isolation or removal of contaminated materials;
- decontamination techniques which remove contaminants from soil and water.

The types of isolation and removal techniques available are outlined in Table 11.1. Figure 11.1 shows the range of treatments for contamination that are available, and their applicability to various target contaminants. Decontamination treatments can be broadly divided into three categories for soil and two for water, as illustrated in Figure 11.2.

Table 11.1: Typical options for the isolation and/or removal of contaminated ground

Method	Characteristics
Isolation	
Capping	Impermeable layer covering the contaminated ground to reduce rainfall infiltration.
Vertical barriers	Impermeable, vertical, subterranean barriers to minimise migration of contamination and pollution movement in groundwater.
Diversion trenches	Drainage systems to intercept water and/or pollution from contaminated ground.
Break layers	Layers of single-sized stone above saturated zone of contamination to prevent upward movement of contaminants by capillary action.
Horizontal barriers	Impermeable barriers beneath waste or contaminated ground to reduce downward movement of contaminants.
Removal	
Excavation	Physical excavation of contaminated ground and removal to a suitable landfill or treatment plant.
Total containment	Excavation of material and replacement in a purpose built impervious cell.

Target compounds / Treatment technology	Vapour extraction	*In situ* bioremediation	*Ex situ* bioremediation	Soil washing	Containment	Stabilisation/ solidification	Thermal treatment	Vitrification	Solvent extraction	Pump and treat	Leaching
Metals (fines and soluble)	✗	✗	✗	✓	✓	✓	○	✓	✗	○	✓
Metals (larger particulate)	✗	✗	✗	✓	✓	✓	✗	✓	✗	✗	✗
Volatile organic compounds	✓	✓	○	✗	✗	✗	○	○	✓	✓	✗
Semi volatile organic compounds	○	✓	✓	✓	✗	✗	✓	○	✓	✓	✗
Halogenated organics	✓	○	✓	✓	✗	✗	✓	○	✓	✓	✗
Oil hydrocarbons	○	✓	✓	✓	✗	✗	✓	✗	✓	✓	✗
Coal tars	✗	○	○	✓	✓	✓	✓	✗	✓	○	✗
Asbestos	✗	✗	✗	✗	✓	✓	✗	✓	✗	✗	✗
Coal	✗	✗	✗	✓	✓	✗	✗	✗	✗	✗	✗
Dioxins	✗	✗	✗	✓	✓	✓	✓	✓	✓	✗	✗

✗ : Inappropriate in most cases

✓ : Appropriate in many cases

○ : Of some potential under certain circumstances

Figure 11.1: Target contaminants of soil and groundwater, and the treatment technologies appropriate for these contaminants

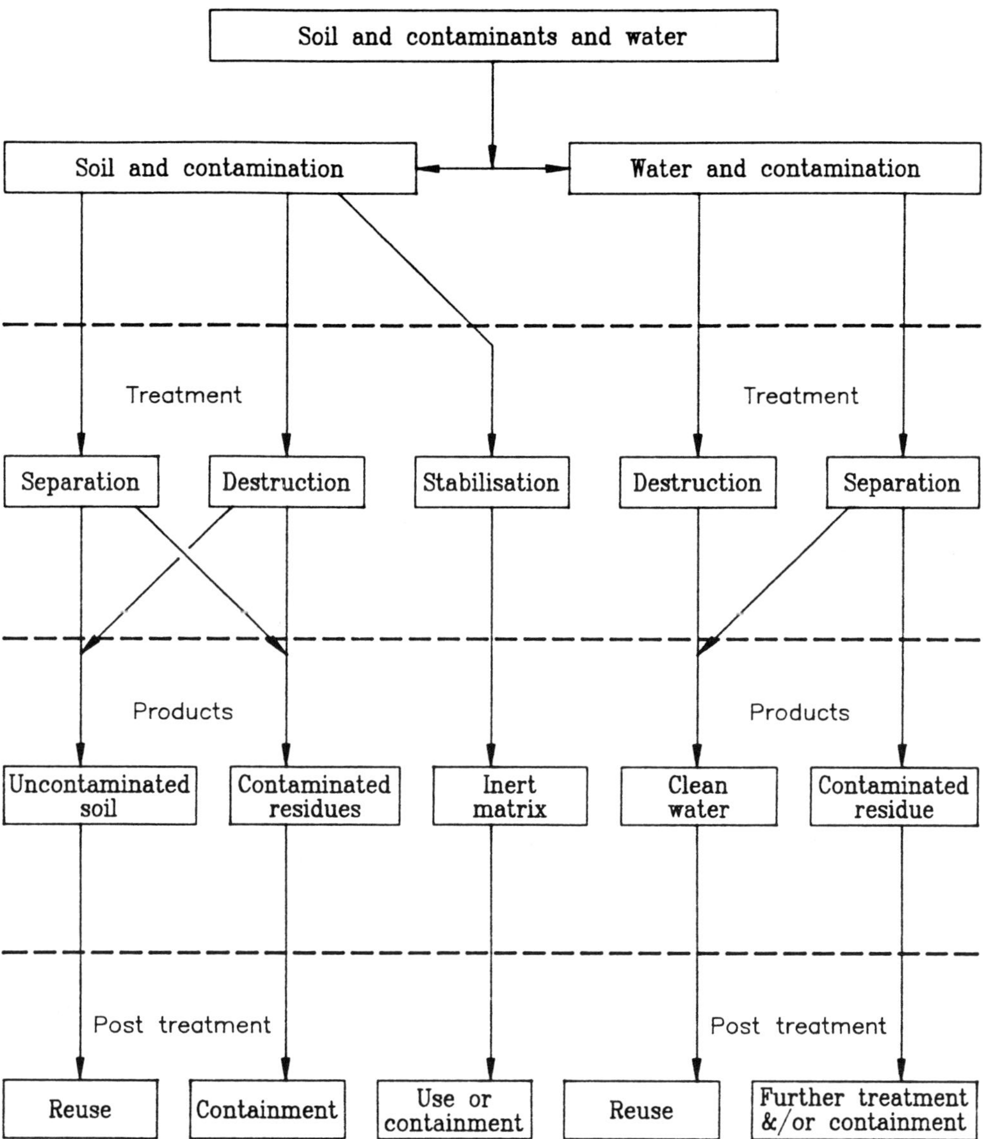

Figure 11.2: Application of treatment technologies to contaminated materials and the implications for product formation and further treatment

Whether a decontamination treatment rather than more traditional techniques is applicable is affected by:

- the nature of the contamination present;
- the nature of the soil materials present;
- target concentrations required after decontamination;
- the volume of materials to be treated;
- site size and access;
- availability of technology locally;
- cost effectiveness of decontamination treatment compared with excavation and containment techniques;
- willingness of the reclamation team to use new and emerging technologies.

The various techniques available for treating contaminated soils are described in subsequent sections as follows:

- removal and disposal of contaminated materials;
- *in-situ* isolation;
- cover systems;
- grouted horizontal barriers;
- thermal techniques of decontamination.

11.1.3 European Community Directives on waste

The creation and treatment of contaminated materials in the European Community is addressed within the Council Directive on waste (75/442/EEC), originally implemented in 1975, and subsequently amended in 1991 (91/156/EEC). This Directive recognises the following important points:

- waste disposal must aim to protect human health and the environment from damage during the collection, transport, treatment, tipping and storage of wastes;

- recovery and recycling of materials is encouraged, including composting and biological treatments;
- trade and competition in waste handling operations should not be affected where disposal is necessary;
- systems involving the use of permits should be implemented to control the storage, tipping and treatment of wastes on behalf of third parties;
- the principle of the 'polluter pays' *i.e.* polluters will pay fines and/or clean-up costs, is established.

Within the Waste Framework Directive the following Articles are of particular relevance to contaminated soils and residues encountered at steelworks, colliery and coal carbonisation sites:

- Article 3 promotes the use of waste as a source of energy, and the recovery of materials from wastes during reclamation;
- Article 4 promotes the use of safe methods of disposal or recovery in order to protect soil, water, air, plants and animals;
- Article 5 requires the setting up of a network of disposal installations which takes account of the 'best available technology not involving excessive costs'.

The following sections within this chapter outline some of the technologies indicated in Article 5 of the Waste Framework Directive, and the ways in which these can be used to overcome the risks associated with contaminated materials, during the reclamation of coal processing and steelworks sites.

11.2 Removal and disposal of contaminated materials

11.2.1 Landfill - past and present

The removal and disposal of materials to landfill should be considered in reclamation design for two reasons:

- past landfill practices at industrial sites often entailed burial of wastes within or near to the site. In this case landfills can be encountered as obstacles to reclamation at a given site, notably in terms of ground stability and potential contamination;
- landfill is an option for the disposal of contaminated materials arising during the course of a reclamation scheme.

A reclamation scheme may involve the design and construction of a new landfill at the site itself, in order to dispose of unwanted or difficult materials, or such materials may be transported off-site to existing landfill facilities.

Whilst this chapter deals with the techniques of ground treatment, and thus the use of landfill as a means of disposal rather than an obstacle to reclamation, the following review of landfill technology will serve to illustrate factors that will influence the impact of old landfill sites on a reclamation operation.

11.2.2 Hazardous waste landfilling in Europe

The only universal option for eliminating risk from the presence of contaminated materials at a given site is by physical removal of the materials by excavation. This has been widely practised throughout Europe in the past, and will continue to be used. Nevertheless, many countries which are densely populated *e.g.* the Netherlands, and/or lack deep soils and natural depressions above bed rock *e.g.* Finland, have restricted space suitable for landfill. These factors, together with the move towards decontamination technologies, have led to a decline in the availability of landfill space. In the Netherlands, for example, where a rigorous policy of soil protection and decontamination has been implemented, only one landfill remained for the disposal of toxic materials in the early 1990s. The price of such disposal is high.

The removal from a site of the toxic material posing risks to the environment, human health or other targets may therefore remove the potential hazards but the transport of such materials simply transfers an equal or greater risk to another site if care is not taken to properly confine or treat the material. In many countries, hazardous wastes such as contaminated soils arising from former coking works sites, are removed to special landfill sites, where provision has been made to control the movement of toxic chemicals outside a defined area. Such controlled landfills are however relatively new facilities, and many countries either have no such landfills available, and/or possess a legacy of hazardous waste landfills where the precautions taken to avoid contamination of the surrounding ground were inadequate. In Italy, for example, a number of illegal hazardous waste landfills have been discovered in the Liguria region, where there is a shortage of landfill sites licensed to accept controlled wastes.[10]

The unspoken tenet of controlled hazardous waste landfilling is that the materials so deposited will remain in a confined volume of ground for ever. No guarantee can be given with any degree of confidence that this will be so, because landfills and underground repositories are generally not failsafe (see Box 11.1). Many landfills only have design lives for sub-surface containment in the order of 20-30 years. Thus, with existing technologies for landfill design, and the often incomplete chemical screening of materials prior to disposal, no landfill operations are risk-free.

The precautions taken to control the movement of hazardous substances, and thus reduce the risks associated with the new placement of materials in landfills, are reflected in the relatively high price paid for the tipping of material. The lack of alternatives have maintained landfilling in the

Box 11.1: The concept of the terms 'design life' and 'failsafe'

The 'design life' of a structure refers to the time within which it will satisfy the criteria with which is was designed to cope. In the case of an engineered system of containment for contaminated materials, this means the time within which the system will, with a high degree of statistical probability, maintain its integrity. A 'failsafe system' incorporates means of detecting failures of a containment at an early stage so that corrective action can be taken prior to a pollution incident.

waste management market place, although this situation is changing in the face of new technologies. It is because of price advantage and the large volumes of waste being produced that disposal to landfill is still, quantitatively, the most important form of toxic waste management in the world.

11.2.3 Containment practices

Landfills are often created in areas of natural or Man-made ground depressions. In the past little attention was paid to the nature of the natural geology of these areas, and thus landfills are found in geological materials that are totally unsuitable for the retention of liquid or gaseous materials within the body of the fill. Pits where sand and gravel were once extracted, for example, have become sites for the tipping of toxic wastes, with the consequence that leachate and gas are able to migrate through the permeable strata, sometimes over considerable distances, leading to the pollution of previously uncontaminated land or bodies of water.

Clearly, a waste landfill, particularly that for toxic waste, must be located so that migration and pollution are minimised. Therefore, these tips should be sited in relatively impermeable ground, such as in stiff clay and, if newly constructed, should include the following design features:

- impermeable lining materials;
- peripheral monitoring wells for groundwater and gas sampling and analysis;
- provision of leachate collection facilities within the body of the landfill;
- gas venting or collection measures;
- adequate surface drainage features.

In addition to these design features there should be tight control exercised over the wastes that are deposited. Thus, sites should be licensed to receive only certain types of compatible wastes, and should be operated by the owner in such a way as to ensure that this is adhered to. In the past, in countries operating site licensing schemes for hazardous waste disposal, such as the United Kingdom and Germany, there have been serious breaches of landfill licence conditions, leading to later pollution and site restoration problems. A landfill operator may be prosecuted for the breach of such conditions.

Landfills are chemically and biologically active structures during the course of their operation, and usually for a period of many years after the tipping of wastes has ceased.

11.2.4 Landfill gas and leachate

Whether a landfill operates under strict control or not it is likely to contain a variety of materials derived from different sources. These materials, depending upon their physical and chemical components, will interact both with each other and with a variety of associated microorganisms. These interactions make the accurate prediction of the life cycle and environmental impact of a landfill difficult to achieve.

In general, two products of the microbial decomposition process are of concern at landfill sites:

- landfill gas, likely to contain carbon dioxide, methane and possibly other toxic, asphyxiant and inflammable gases and vapours;
- leachate, deriving from water entering the landfill and other liquid, soluble and suspended wastes (and degradation products) of varying toxicity and corrosiveness.

It is these two products of the landfill process that make the provision of adequate containment and control measures such a critical issue at actively gassing landfill sites. Such measures may include, in addition to containment systems, gas venting, gas pumping and combustion, and leachate pumping for recirculation, disposal or treatment.

11.2.5 Reclamation and disposal

In the case of disposal to landfill as part of a reclamation scheme, it is often left to a works contractor and landfill operator to ensure the safe disposal of contaminated wastes. It is however the responsibility of those in direct control of the reclamation scheme to ensure that the disposal of the wastes is in the hands of organisations that are applying the following criteria:

- adequate chemical characterisation of the waste prior to landfilling;
- strict health and safety precautions for those handling the materials between excavation and disposal;
- disposal of the wastes to a landfill that is of suitable design quality, to ensure proper containment of the materials, with the necessary waste disposal licensing.

The design and operation of an on-site landfill during the course of a land reclamation scheme should conform to similar criteria. The choice of landfill design will be dependent on the nature of the site, the development plans and the type of materials to be contained. The latter will particularly influence the choice of Man-made polymers that may be used to line the landfill, which could be adversely affected by organic constituents of toxic waste from, for example, coal carbonisation by-product residues. The manufacturers of such geosynthetic polymers should provide information and data on the use of these materials in chemically disturbed conditions, and materials without the correct specification should not be used.

11.3 *In situ* isolation techniques

11.3.1 Barriers to pollution

Where contaminated materials are present at a site, there may be the potential for containment of the contaminants without recourse to excavation.

A variety of techniques are available to isolate adjacent ground from contaminated materials. These techniques usually take the form of a vertical barrier, incorporated after the excavation of a deep trench.

Vertical barriers have limited applicability because they can only prevent lateral movement of contaminants. Such barriers are suitable for use where an impermeable material *e.g.* rock or clay, exists immediately below the waste, and there is the potential for placing vertical barriers around the waste to the depth of the impermeable material. However, in such situations a perched water table may be created within the contaminated ground, resulting in contaminated water at or near the surface, and waterlogged and unstable ground.[13] Such situations can be treated with biological soil and groundwater decontamination techniques (discussed further in Section 11.8).

Materials and techniques used in the construction of vertical cut-off barriers include:

- geosynthetics and associated polymer sheeting;
- sheet piles, and associated displacement barriers;
- slurry and grouted walls;
- open or gravel filled trenches with incorporated drainage and leachate collection facilities.

Figure 11.3 shows a vertical barrier system which uses a geosynthetic membrane.

11.3.2 Geosynthetic membranes

Proprietary synthetic membranes, with specifications geared towards their use in contaminated materials, are available from a variety of sources. They are often available in rolls of sheeting, with appropriate bonding materials to seal joints together. Many of these materials are plastic polymers, which, although durable, are prone to tearing, puncturing and abrasion. Such membranes should therefore be installed so that they are in contact with materials which will not puncture them and which will protect them against damage. These materials include gravel, sand or bentonite clay.

Detailed design of vertical barriers must take into account the practicality of installing geosynthetic membranes. A system which is easy to install will minimise the potential for damage to the isolating barrier material.

The careful supervision of installation of geosynthetics is essential in order to ensure the integrity of the finished vertical barrier. A badly installed confinement system may produce no better control of pollution migration than existed before installation, and may even lead to worse problems occurring locally.

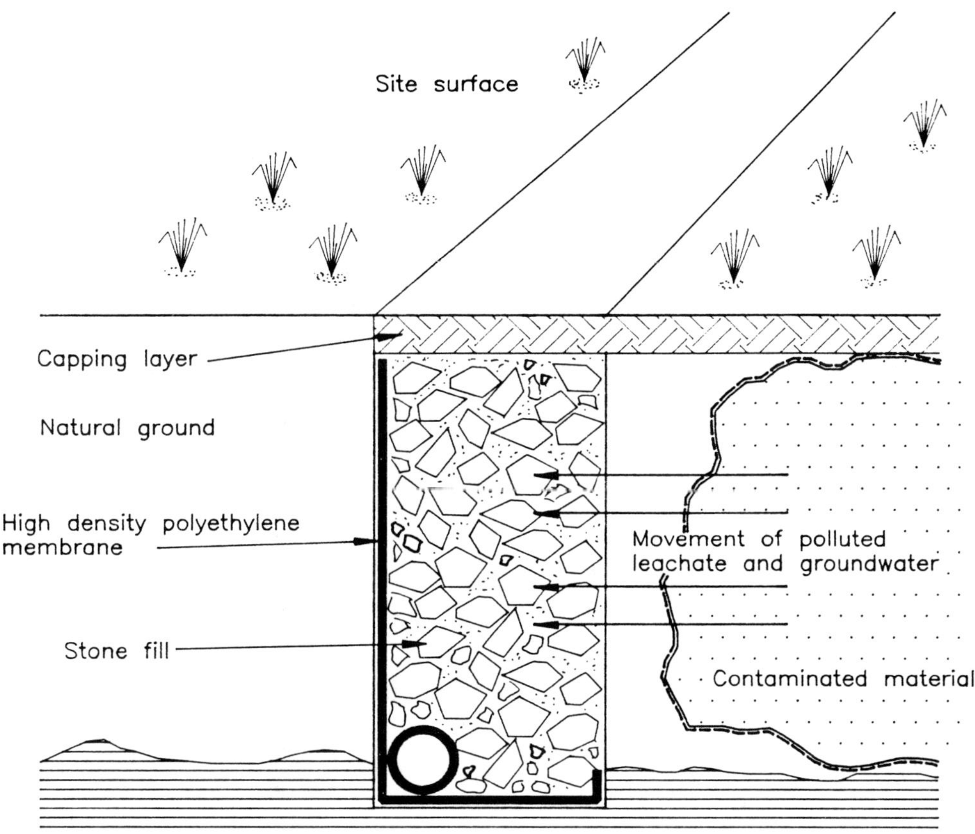

Figure 11.3: Cross section through a typical deep vertical barrier system, using high density polyethylene sheets

To ensure that geosynthetic lining material are impervious when laid damaged sheets must be discarded and seams checked for leaks. Various methods to test the latter are available *e.g.* ultrasonic and conductivity meters.

11.3.3 Sheet piles

Sheet piles, whilst being more robust than geosynthetics, are considerably more expensive. Piles can generally only be pushed into soft ground if there are no obstructions. They are therefore usually vibrated into the ground using special equipment.

Sheet piles fit together by their shaped, interlinking edges, and are not normally completely water tight, unless used in combination with other materials. Whereas sheet piling has been used for the confinement of contaminated materials, it is not generally the most suitable method, because of the permeability of the piles to liquids and their high material and installation costs.

Piling may be used to facilitate the installation of other barriers. For example, steel piles may be inserted and withdrawn, whilst, at the same time a cementitious grout is injected into the piled void.

11.3.4 Slurry walls and grouted barriers

Slurry walls can be used as barriers in situations where gas and leachate migration from landfill requires control.

The installation of slurry walls involves the excavation of a trench, which is accompanied by simultaneous filling of the trench space with a slurry (see Figure 11.4). The slurry prevents the collapse of the sides of a deep trench. The trench may be filled either with the final slurry compound or with a temporary support slurry which is then replaced by displacement with an alternative slurry which sets to form the barrier. During the placement of slurry materials, air spaces must not form if the final structure is to adequately control contaminant migration.

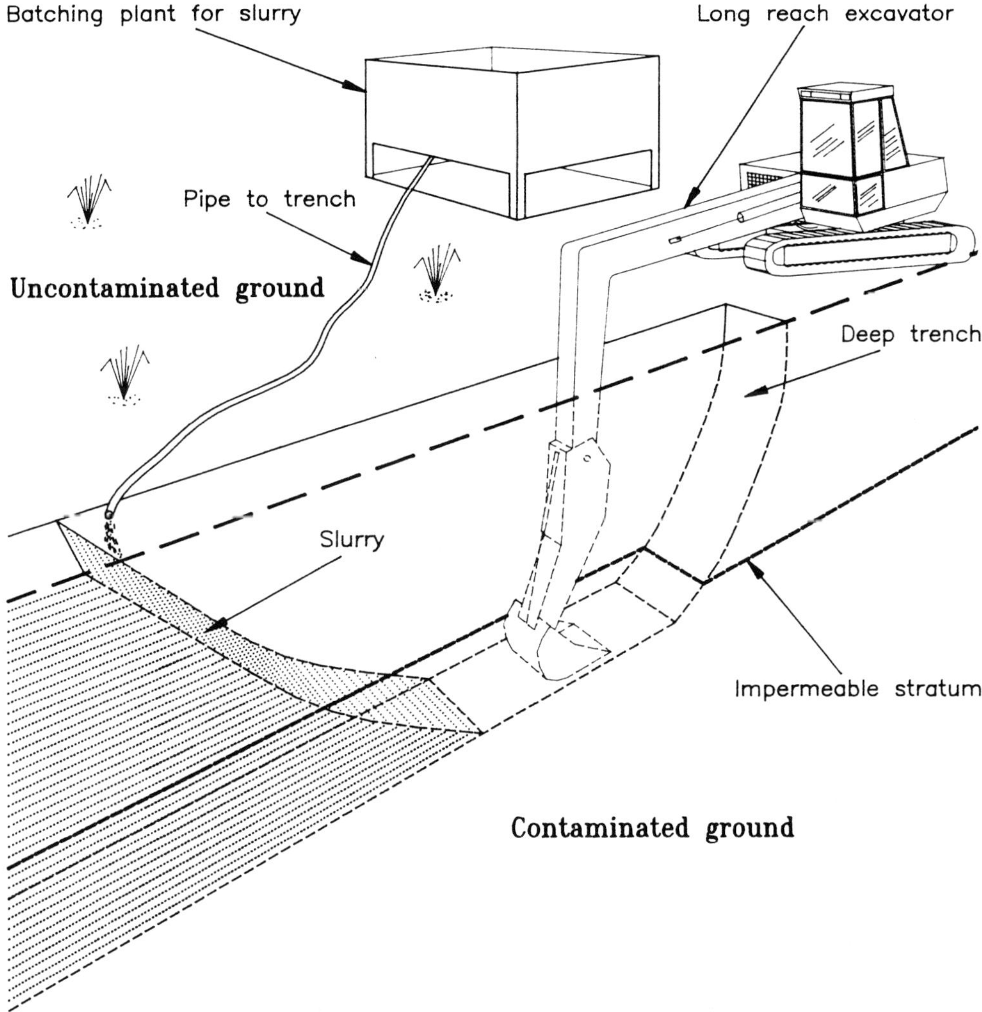

Figure 11.4: Construction of a slurry wall for vertical containment of polluting materials

Slurry materials may include clays (such as bentonite), clay-soil mixtures and clay-cement mixtures. New slurry materials and high density, hydrophobic compounds increase the number of options to be considered. The success of a barrier is determined by its final permeability and integrity. Permeabilities may be expected to be in the region of those for stiff clays ($<10^{-9}$m/s).

Grouted barriers can be introduced behind a high pressure air or water jet or, alternatively, directly into the ground. The grout saturates voids and fissures to reduce lateral permeability.

Future developments in this field are likely to include methods to confine contaminated materials within the ground by the use of *in situ* vertical and horizontal cut-off materials. Such techniques may include the *in situ* formation of impermeable polymeric compounds by the injection of polysaccharides mixtures. Similarly, the stimulation of the growth of soil microorganisms, and the associated formation of cellular material and polysaccharides within soil pore spaces, could provide a means by which the permeability of soil materials may be reduced.

11.3.5 Trench systems

In some instances the construction of relatively simple trench systems for intercepting pollution as liquid leachate or gaseous emissions, can be efficient and cost-effective solutions to pollution problems.

Trenches have often been used around the perimeter of actively gassing municipal landfill sites. These trenches intercept gas migrating towards adjacent ground and vent it to atmosphere in a controlled manner, via the path of least resistance. Such passive venting trenches are usually gravel filled, and incorporate a drain at the base to divert infiltrating rainfall and run-off water towards a sump, thus avoiding the accumulation of water within the trench. Gravel-filled trenches may be covered to prevent the infiltration of fine-grained particulate matter into the trench, and venting pipes installed through the cover material to allow the release of gases to the atmosphere.

In a similar fashion, gravel-filled or open trenches containing a drainage system and sump can serve to control the migration of leachate from areas of contaminated wastes (see Figure 11.5). Trenches such as these have been used to avoid pollution breaching site boundaries in shallow ground. Leachate passing into a drainage collection system may be removed by the incorporation of a simple collection sump from where contaminated fluids can be pumped into containers for disposal or treatment off-site. Alternatively, leachate can be diverted to an on-site treatment plant; for example a biological reactor.

11.3.6 Costs versus effectiveness

The costs of piled, grouted and slurry barriers can be high, especially in ground requiring the construction of deep barriers. In addition these methods may not always be completely reliable, and can have design lives that are limited by the corrosive effects of the substances that they are designed to contain.

The most effective designs are often those which integrate two or more systems. An example of this would be the incorporation of a polymer sheet membrane within a slurry wall.

The high capital cost of integrated systems may be offset by their longevity compared to single systems.

The choice and correct installation of the most suitable materials is crucial to the successful implementation of a barrier system to control pollution. The costing of a barrier scheme should always allow for long-term monitoring in order to detect any failures of the system (see Box 11.2).

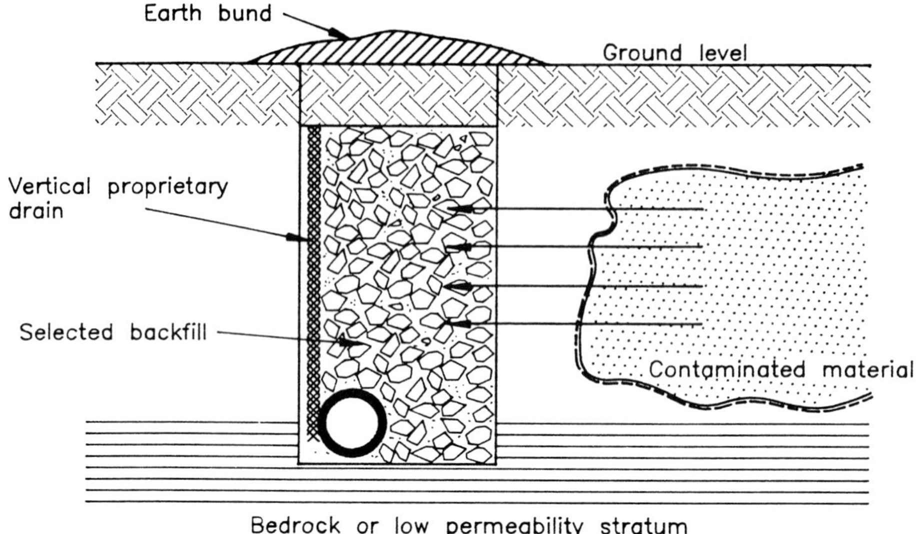

(a) Filled trench

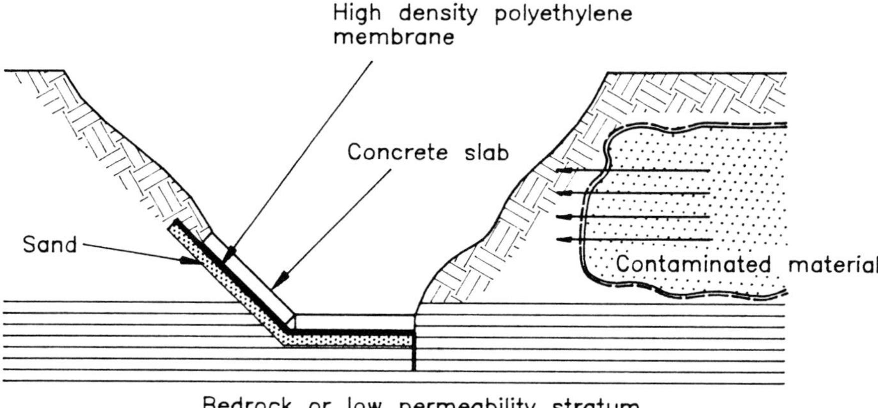

(b) Open trench

Figure 11.5: Examples of cross sections through shallow filled and open trench systems designed to control the migration of pollution

Box 11.2: Cost considerations in the design of containment systems for pollution control

The costs of implementation of physical containment systems, including vertical barriers and cover systems, include many features other than those relating to civil engineering construction and materials. Management costs may include:

- weed control in trenches;
- vegetation management above a containment to avoid deep-rooted species;
- drainage pipe and channel clearance;
- emptying of drainage collection sumps.

Monitoring costs may include:

- installation of borehole stand pipes;
- periodic sampling of stand pipes for gas, groundwater or leachate;
- chemical analysis of samples;
- documentation of monitoring results.

11.4 Site cover systems

11.4.1 Introduction

One of the most frequently used methods of dealing with contaminated land has been by the covering of materials with a low permeability or impervious capping layer. These methods must however fully consider the containment and/or interception of contaminated materials below ground level, as cover systems only restrict movement of materials through the ground surface, and not laterally or below the contaminated ground. The advantages of capping a site are that:

- infiltration of rainwater and/or other liquid inputs *e.g.* flood waters, through the materials is reduced;

- the capping layer forms a barrier between users of the site and the contamination beneath;
- the need to remove contaminated materials from site or to treat them is deferred.

The use of high permeability break layers can assist in the prevention of upward capillary movement of buried contamination towards the surface of the ground.

Cover systems are used for dealing with contaminated land in housing areas in the Netherlands.[155] The need to reduce excavation and removal of contaminated ground for environmental and cost reasons is often a stimulus for the use of cover systems.

Any covering system is likely to include the following elements:

- sealing system *e.g.* clay;
- surface substrate *e.g.* top soil;
- gas control and drainage layer;
- vegetation cover *e.g.* grass, (where applicable);
- long-term management.

In the case of horizontal systems for the control of vertical movements of liquids, attention must always be paid first to the characteristics of the site in relation to groundwater movements and, particularly, changes in levels. In addition, following the implementation of a system, a strategy for monitoring the site is required in order to assess the integrity of the cover, so that any breaches can be detected and repaired quickly.

11.4.2 Capping layer

Capping materials

One of the most critical issues in constructing capping systems is that of the permeability of the materials used. Clay has often been used as a cap material due to its low permeability, high plasticity and moisture retaining properties. Clays vary greatly in structure and properties however, and care should be taken in the construction of new clay caps so that only clays which have low and homogeneous permeabilities are used. The use of such clays will ensure adequate containment of the materials beneath, whilst substantially reducing the infiltration of water from the surface.

Similarly, the installation of a clay cap must be undertaken with adequate compaction, in order to avoid the incorporation of air spaces and breaches of the continuity of the material. A good quality, low permeability clay, will not be effective if the material is incorrectly or too thinly laid. Clay capping layers are usually required to have a permeability of 10^{-9}m/s or less.

If site conditions permit and materials are readily available, covering systems should incorporate a levelling layer of placed material, so avoiding major disturbance of underlying contaminated ground. The use of a levelling layer can reduce irregularities in site topography facilitating the placement of further layers of the capping system. Figure 11.6 shows an example of a cover system incorporating a clay cap.

Clay layers should be laid to agreed tolerances, so that a minimum depth of clay can be guaranteed. This is achieved in practice by using levelling markers at regular intervals across the site, and using these to guide the placement of clay. The final levels on the clay surface, and the falls that are allowed, should aim to avoid ponding at the surface during periods of rainfall.

Within levelling layers gas and drainage control measures can be incorporated.[107] These can divert gas and leachate away from the contaminated ground beneath the cover, and from the site surface, to areas where they can either be released or treated.

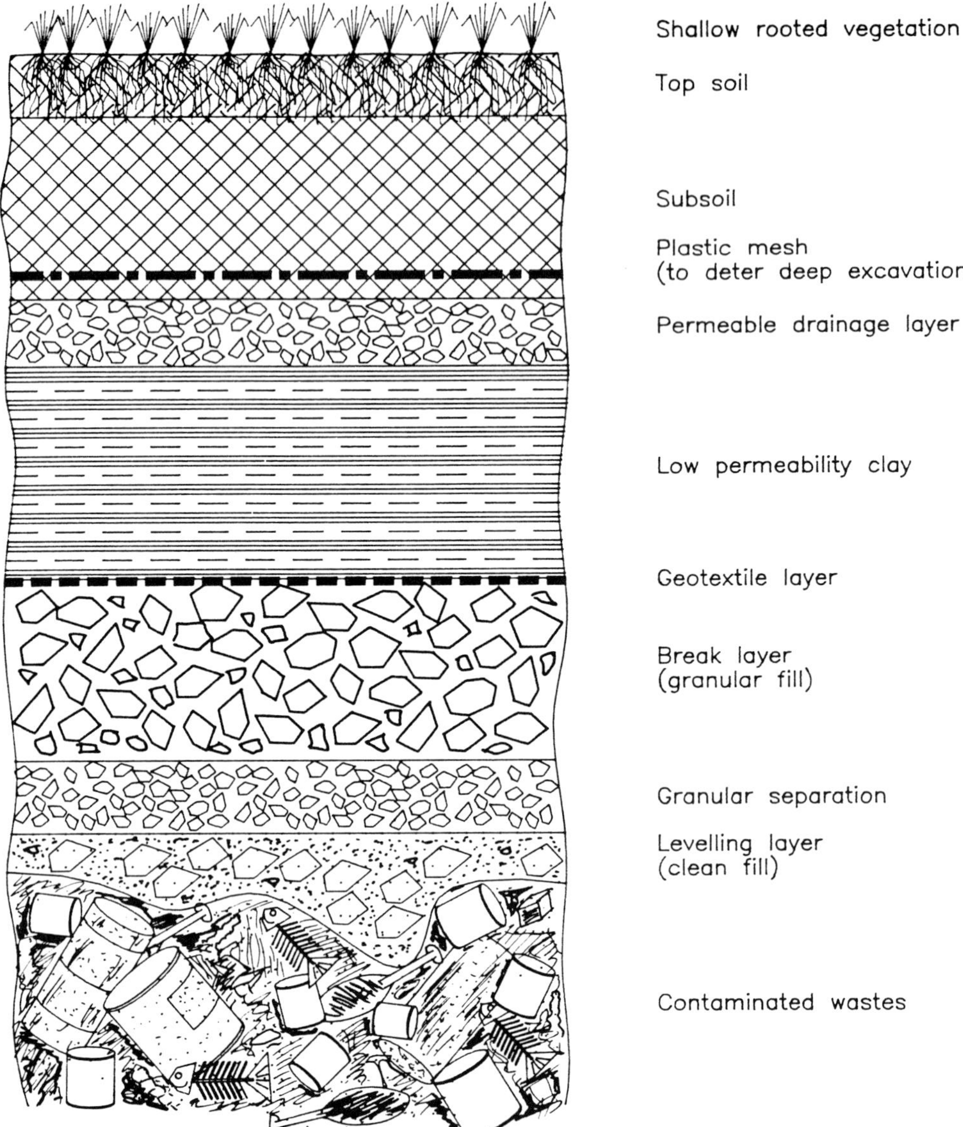

Figure 11.6: Example of a cover system for contaminated ground incorporating a clay cap and break layer (not to scale)

Breaches of capping structures

Clay should always be kept damp during placement, and should remain damp during the design life of a cap, to avoid desiccation and cracking. The latter is usually achieved by placing an adequate layer of clean soil above the cap. In addition any capping material that has the potential to fracture must be used with due attention to the possibilities of ground movement. In areas where differential settlement *e.g.* newly completed landfills, or intense geological movement *e.g.* earthquake activity, is likely, capping may not achieve the required degree of robustness. In such cases, fractures can lead to seepages of contaminated liquids and gases, and the need for remedial measures.

Similarly, breaches of caps can occur through the action of plant roots or burrowing animals. Accidental excavation of capping strata can be avoided by the use of warning grids buried above the cap (see Figure 11.6).

The use of artificial lining materials such as high density polyethylene, can improve the durability of a sealing system and may be used with or as an alternative to clay liners. The comments in Section 11.3.2 regarding the choice of geosynthetic membranes also apply here.

11.4.3 Break layers

The incorporation of stone, gravel or rubble into a cover system can improve performance by forming a stratum through which capillary action cannot occur (see Figure 11.6). Even in the best laid clay-sealed covers upward movement of soluble or liquid contaminants can occur through micropores in the capping layer, and thence through the cover soil, by capillary action. By maximising the spaces between the particles in an artificially created, competent ground layer, capillary movement can be minimised, so reducing the risk of upward contaminant migration.

Such break layers should lie above the highest ground or perched water level, and consist of material that is not prone to erosion and fracturing. Good quality materials, such as hard rock fragments, rather than mixed building rubble of non-uniform or uncertain composition should be used to ensure that degradation of the break layer does not occur. Material of a single particle size will give most void spaces. In addition, geotextile fabrics can be used to avoid clogging of break layers by the infiltration of fine-grained particles from the upper layers of the cover system.

11.5 Grouted horizontal barriers

Vertical barrier systems of containment have been discussed in Section 11.4. If there are no naturally occurring horizontal barriers beneath a contaminated site there are options for creating such barriers at depth using grouting techniques.

These methods are generally both difficult and costly to apply. The techniques fall into three categories:

- jet grouting;
- chemical grouting;
- claquage grouting.

These methods are summarised in Box 11.3.

11.6 Total containment

In circumstances involving small volumes of contaminated ground, it may be possible to excavate material and replace it within a purpose-built containment cell. Such a cell may be constructed of clay or bonded geosynthetic sheeting with adequate provision for drainage around the contained volume. This option avoids transport of contaminated materials off site, but requires a programme of long-term monitoring, due to the

***Box 11.3*:** Horizontal grouting techniques

Jet grouting uses high pressure, water jets to cut voids in ground, which cement and bentonite slurry is then used to infill, in order to create a barrier. The technique relies on the stability of ground above the created voids and the nature of the ground removed under pressure.

Chemical grouting is designed to fill void spaces in existing ground. Thus, slurry is pumped into the ground in the region of treatment *e.g.* adjacent to or within contaminated soil or buried materials.

Claquage grouting uses high pressure grout injection to cause fracturing of the ground prior to grout infiltration.

possibility of containment damage and the limited design life of the system (see Box 11.1).

The option for total containment will be limited by the qualities and types of contaminated materials involved, and by local soil and groundwater conditions. In addition, the nature and location of sub-surface containment cells must be well documented and protected in order to avoid accidental disruption at a later date *e.g.* during subsequent site redevelopment.

11.7 Thermal techniques of decontamination

11.7.1 Types of thermal treatment

Contamination by most organic and some inorganic compounds within soils and slurries can be largely destroyed by thermal treatment. Thermal treatment techniques fall into three broad categories:

- incineration;
- two-stage thermal treatment;
- vitrification.

Thermal treatments in general are useful disposal techniques for materials that are contaminated with highly toxic chemicals which are not separable, recoverable or treatable by other means, and which present too great a risk to be contained within a landfill or other containment system. Such compounds may include highly chlorinated hydrocarbons, such as PCBs.

Solid residues from thermal treatments, including bottom ash and fly ash, cannot be treated as inert waste until fully chemically characterised, as they may contain high concentrations of heavy metals and toxic combustion products. Some metals and other inorganic substances, such as mercury, arsenic, lead and cyanide may, however, volatilise during incineration, and therefore not remain at high concentrations in the solid residues. However, if emission control measures are not taken such volatilisation transfers pollution to the medium of air.

11.7.2 Incineration

Total incineration destroys not only the contaminant but also the structure of the contaminated matrix. In the case of soils particularly, incineration therefore constitutes a destructive technique for decontamination which takes no account of the retention of soil structure and function.[35]

Most European countries operate large incineration facilities to handle special wastes, including contaminated soil. A large incinerator can accept in the region of 100,000t/year.

Although incineration is a destructive technology, it does produce residual wastes in the form of ash and gaseous emissions. The latter have caused concern in the proximity of incineration plants handling toxic wastes, especially regarding the deposition of dioxins. The tightening of environmental legislation in Europe, and the improvement of pollution control technology in the industry, is leading to improvements in materials handling, de-ashing and the control of emissions to air, in both new and existing plant.

Various forms of incinerator plant exist, and these include:

- rotary kilns;
- plasma arc furnaces;
- fluidised bed incinerators;
- cement kilns;
- multiple hearth incinerators;
- circulating bed combustors.

The suitability of a contaminated waste to a particular type of incinerator should be approved by the operator. A variety of wastes from coal and steel facilities may be treated by thermal methods. Fluidised bed incineration for example, has been used to treat waste tar from coking plant operations, at a throughput of approximately 450 t/h.[36]

Whilst incineration is often carried out at permanent facilities, mobile units are also available. Although these have the advantage of not requiring the transport of the wastes over large distances, in practice it may be difficult to obtain permission from a local authority and/or environmental regulation authority to operate such plant in some areas. Infrared and electric units are available as mobile incinerators.

11.7.3 Two-stage thermal treatment

Introduction

The two-stage thermal treatment of volatile contaminants can be achieved by systems that release the contaminants under heating and then thermally destroy them in the gaseous phase.

Rotary kiln incineration

Rotary kiln incineration is a commonly used thermal treatment system, which can also be adapted for two-stage treatment of contaminated materials at lower temperatures than other methods. The rotary kiln consists of a long inclined tube, lined with refractory material, which is slowly rotated whilst the material within is heated, either through the shell of the vessel or via pipes within the kiln. The exhaust gases, containing volatile fractions of the material undergoing treatment (including water and contaminants), are collected and passed into a secondary chamber for thermal treatment of the gases and fine-grained particulate material.[210]

Rotary kilns, which generally have a high throughput, are able to accept waste materials of varying consistency, including building wastes, although clays can form clods during treatment.[12]

Wet oxidation

Wet air oxidation is a decontamination treatment involving the exposure of oxidisable contaminants to gaseous oxygen, in the presence of water, at elevated temperatures. It can be use to treat soils and effluents contaminated with a variety of organic compounds, including coal tars.[228] These contaminants are largely converted to soluble, non-toxic, biodegradable residues in the resulting condensate from the reactor vessel, and can subsequently be treated or disposed of.

Evaporation and condensation

Other two-stage thermal treatment systems also exist. One such system, developed in the USA, vaporises volatile contaminants into a gas stream, which is then condensed to produce a liquid for second stage treatment. This has been developed as a transportable plant.[233]

11.7.4 Vitrification

The vitrification process applies intense heat to contaminated solid wastes, which are pre-mixed with silica particles or other suitable materials *e.g.* sand, limestone, alumina, fly ash. The heat not only drives off volatile and combustible components but in addition causes compounds with extremely high boiling points (including the silica) to melt and fuse on cooling, so forming a solid, glass-like end product as the residue from the process. The glass product can be produced as small particles (frit) or as cast blocks. This vitrified product is chemically inert, and of potential use as a construction material *e.g.* in coastal defences. Vitrification may therefore be seen as a waste recycling activity.

The option to vitrify waste is limited, as with most techniques, by the chemistry of the materials. Some wastes *e.g.* those that are highly alkaline, may not be suitable for processing.

Vitrification for the treatment of contaminated soils is a relatively infrequently used process, and the availability of equipment in Europe is somewhat limited. It has the potential to become more widely used, however, and may be of particular use for smaller volumes of highly toxic materials. Typically a vitrification plant may have the capacity to treat 300-500 tonnes of contaminated material per day.[159]

11.7.5 Application and cost

Thermal treatment is a relatively expensive technique in relation to other waste disposal and treatment options. This is principally due to the high capital cost of the treatment plant and the large energy inputs needed to maintain the high temperatures required (in the range 800-2500°C). Thermal treatment of contaminated solid materials, such as soils containing a high proportion of inert inorganic matter, tends to be a relatively inefficient processes because heat must be transferred throughout the mass of the waste. Organic soils can be less expensive to incinerate due to the higher proportion of combustible material present, so that some of the heat required is produced by the waste material rather than the fuel.

The cost of incineration depends to a great extent upon the type of waste, as precautions during the processing and the subsequent treatment and disposal of residues will differ according to the original nature of the contaminated materials. This means that the operator of the incinerator will require a detailed description of both the physical and chemical nature of the wastes prior to the quotation of prices and acceptance of the material.

As the cost of thermal treatment is high, and the process is largely destructive, such treatment is best limited to small volumes of highly contaminated materials. The advantage of such a disposal route is that once agreement has been reached with the operator all that is required is the safe removal and transport of the contaminated materials to the thermal treatment plant. This can be undertaken relatively quickly, thus removing the source of risk from the site during the reclamation. However, the location of the thermal treatment plant in relation to a given contaminated site may limit the sites to which this treatment technology can be applied.

Box 11.4 lists some further types of thermal treatment, which are currently receiving attention at field scale.

11.8 Non-destructive decontamination techniques

11.8.1 Introduction

There are a number of technologies for the treatment of contaminated materials which fall under the general classification of non-destructive decontamination techniques, as they do not completely destroy the soil. These can be subdivided into the following categories:

- solidification and stabilisation;
- biological;
- separation.

The principle of these techniques is to immobilise, degrade or separate the contaminant in the soil or water which they pollute. The nature of the

Box 11.4: Other thermal treatment systems

Amongst recently developed thermal treatment methods are:

- Thermal stripping of volatile contaminants;
- Radio frequency heating of soils to volatilise contaminants at temperatures of up to 600°C;
- Electric heating of soil *in situ*, using buried electrodes to volatilise contaminants;
- *In situ* vitrification, using graphite electrodes and massive electrical current;[93, 94]
- Infrared incineration;
- Advanced electric reactor.

These technologies are either of limited availability or only on trial at pilot scale. They are available, as with most thermal treatment technology, through specialised contractors who should be consulted fully to ensure the suitability of the techniques to the waste in question. The technical details of some of these techniques have been summarised.[12]

contaminant(s) and the soil, and their interaction, are critical factors in determining the applicability of the technology. However, once the suitability of these techniques has been established for a particular site, they offer powerful tools for the reclamation process.

Decontamination processes may be applied either to excavated or extracted materials (*ex situ*), or within the relatively undisturbed soil-groundwater matrix (*in situ*). There are a number of advantages to the treatment of contaminated soil and groundwater without recourse to excavation of the materials:

- less chance of disturbance of the ground exacerbating pollution outside the contaminated area;
- the potential for undertaking treatment whilst a development proceeds at the site surface;
- cost and time savings by avoiding the excavation of materials;
- simultaneous control of pollution plumes during treatment.

Whether a given decontamination treatment is undertaken *in situ* or *ex situ* depends on:

- the limitations of the method;
- the ground conditions at the site where treatment is required;
- the type of contamination;
- relative costs of the available options.

11.8.2 Solidification and stabilisation

Introduction

In some environments chemical contaminants become naturally bound to indigenous soil particles in such a way as to render them immobile, and therefore less polluting. Clays, for instance, are important constituents of soil in this way, and without them many soil chemicals would be washed out of the soil by leaching.[102, 229] This property of binding has been utilised to good effect in artificially created systems in order to form complexes between contaminant chemicals and a matrix of soil and other materials. The solidification and stabilisation of wastes have thus become important options for the treatment of toxic materials, limiting the solubility or mobility of the contaminants concerned.

In practice the process of solidification for hazardous wastes also involves stabilisation processes. Here, the two processes have been separated for the purposes of illustration.

Solidification

The fixing of contaminants within a resistant solid matrix is termed solidification. The end product, often termed a monolith, has a high degree of structural integrity. In cases where solidification occurs by the physical trapping of contaminants within the pore spaces of a solid, rather than by chemical fixation reactions, the process is known as microencapsulation.[188] These processes may use concrete-based matrices or thermoplastics, such as asphaltic or polymeric compounds.

These methods are likely to include two principal stages:

- minimisation of contaminant solubility;
- cementation of the contaminant(s) by the use of a binder.

Unless contamination is already in a solid form *e.g.* dusts or metal particles, the initial stage of a solidification treatment may include the separation of the contaminant into a solid phase. In the case of organic contaminants this may include the use of clay minerals to form bonds with the contaminant molecules.

The binder materials are usually cementitious, and can include Portland cement, powdered slags, power station fly ash and kiln dust, in varying proportions. The contaminated materials are mixed with these dry binders, and are then hydrated to form of a slurry or paste. This mixture is then cured in appropriate moulds to form the solidified matrix.

The choice of stabilisation pretreatment and binder depends on the chemistry and structure of the contaminated soil, and is a matter for a specialist subcontractor. Their decision over the most appropriate treatment may rely on previous experiences with similar soils, or on laboratory tests which will be undertaken prior to full-scale treatment. Laboratory tests of this sort should provide details of the stabilisation of the solidified wastes, and on their resistance to physical degradation, chemical reaction and leaching.

Chemical stabilisation

Whilst solidification involves the physical stabilisation of contaminant molecules, other techniques chemically treat contaminated soil without recourse to solidification. These techniques have limited applications to soils containing specific types of contamination.

Likely chemical treatments include oxidising agents, which chemically alter susceptible molecules, producing lower solubility and less toxic oxidised products, or chemical structures within which the original contaminants are bound. Care must be taken with these treatments to ensure that chemical oxidation products conform to the criteria of low toxicity and solubility, and that they are not likely to return to their former chemical state, and so cause pollution in the future.

The use of ozone for the treatment of contaminated soil and water has been widely investigated in recent years, and has been used successfully to treat organic contaminants.[168] Ozone gas is a powerful oxidising agent which will readily destroy hydrocarbon contaminants. Methods have been developed whereby ozone can be pumped into a soil-groundwater system for the treatment of organic contamination. The success of these methods tends to be limited by:

- the nature of the pollutants;
- permeability of the soil;
- organic content of the indigenous soil.

The cost of ozone treatment can be relatively high due to the energy costs incurred in producing the electrical discharge in an ozone generator, and because of the non-specificity of the treatment. The latter means that, as with many oxidising agents, the oxidising activity of the treatment will be consumed not only by the organic contaminant, but also by other organic substances in the material undergoing treatment.

All chemical oxidation methods, which involve the handling of potentially hazardous oxidising agents, are likely to require specialised equipment with which to apply treatment.

11.8.3 Biological treatment

Introduction

The ability of certain groups of aerobic microorganisms to degrade complex organic compounds to simple molecules, such as water and carbon dioxide, has been used to good effect in the treatment of organic contamination at contaminated sites in Europe and North America. The tolerance of microorganisms to high concentrations of toxic organic compounds, and their subsequent detoxification of these chemicals, can be applied to the treatment of pollution both *in situ* and in excavated soil and abstracted groundwater.

The use of microorganisms for the treatment of sewage and waste waters is not a new concept, and the bioremediation of contaminated soil and groundwaters makes use of well-established principles established in this field. Biological treatment thus usually requires the following key parameters to be optimised:

- a microbial population with the ability to degrade the organic contaminants of concern;
- a nutrient supply to suit the nutritional requirements of the microorganisms;
- pH values near neutrality;
- a supply of oxygen;
- water;
- temperatures that will allow microbial and enzyme activity;
- absence of inhibitors of microbial growth.

Whilst these conditions may all be present in an uncontaminated surface soil, such as under tilled agricultural conditions, in a contaminated soil many of these factors may be limiting the potential for microbial growth and concurrent degradation of the wastes. The aim of biological treatment is to optimise these factors in order to stimulate biodegradation.

This process of optimisation will therefore include one or more of the following:

- addition of mineral nutrients to the system;
- aeration;
- implementation of a watering and/or drainage regime;
- inoculation of the system with an appropriate source of microorganisms;
- application of other amendments *e.g.* growth factors, co-metabolites, lime.

The range of organic chemicals amenable to treatment in this way is extensive, but is limited to those compounds which exhibit reasonable aqueous solubilities, as it is soluble organic compounds that tend to be more biologically available (see Box 11.5). Coal carbonisation by-product contamination in soils and groundwaters, including coal tar components, phenols, oils and BTEX, is sometimes amenable to bioremediation, and laboratory studies and full-scale schemes have illustrated the success of microbiological treatments.[119, 252]

Factors affecting the practical application of biological treatment methods can be summarised as follows:[20]

- chemistry of the contaminants and their biological availability;
- temperature and seasonal fluctuations;
- soil conditions for ground engineering and aeration of the system;
- residual concentrations of contaminants remaining after treatment *i.e.* very low concentrations may not be achievable;
- potential formation of soluble and toxic intermediates.

The limitations imposed by climate are most likely to include those of low temperature and extremes of rainfall. Thus, in temperate northern Europe ambient biological treatment will typically be limited to a period of 25-35 weeks, whereas in warmer Mediterranean climates the treatment season will be longer.

Box 11.5: Typical organic contaminants of coal and steel sites and their amenability to biological treatment

Readily treatable contaminants with some aqueous solubility *e.g.* 10^2-10^5 mg/l, include:

phenols, benzene, toluene, xylene, light mineral oils, petroleum hydrocarbons.

Contaminants of lower solubility *e.g.* 10^{-3}-10^2 mg/l, treatable in some cases include:

tar oils, heavier mineral oils, more soluble fractions of coal tar (*e.g.* naphthalene, phenanthrene, acenaphthene).

Contaminants with negligible solubility ($<10^{-3}$), unlikely to be successfully treated by biological means include:

heavy fractions of coal tar (*e.g.* dibenzoanthracene), pitch, coal and coke, ferrocyanides.

In situ treatment

In situ biological treatment uses equipment to stimulate the biological degradation of contaminants present at depth in the soil. This equipment often consists of pipes, pumps, wells and dosing tanks, sometimes partly contained within small transportable units on site, which circulate groundwater through an area of contaminated ground, whilst applying amendments to the system to stimulate microbial degradation.[21] The volumes of contaminated material below the surface may be confined within placed barrier walls, or in areas that are unconfined other than by the careful control of groundwater pumping. A typical scenario is illustrated in Figure 11.7.

A variety of chemical amendments may be applied to the circulating leachate and groundwater within *in situ* biological treatment systems.

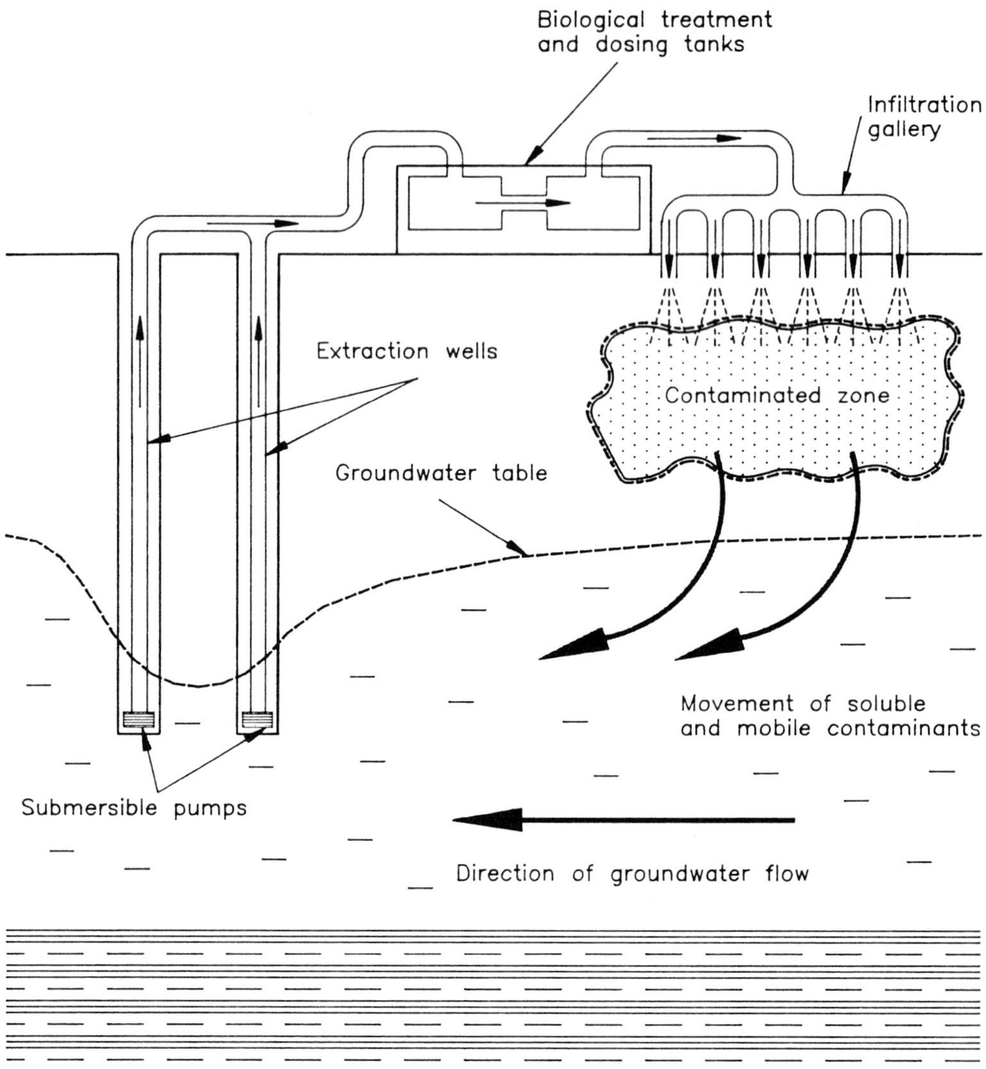

Figure 11.7: Example of a typical recirculating *in situ* biological treatment system

These include nutrient sources (*e.g.* proprietary blends, spent brewers yeast, molasses), solubility enhancers (*e.g.* surfactants) and sources of oxygen (*e.g.* air, hydrogen peroxide). Specific live cell preparations of bacteria are also added in some circumstances, and some such preparations are now available in dried, packaged form. The benefits of using such microbial inocula rather than relying on the stimulation of indigenous soil microorganisms are, however, often difficult to quantify.

There are a number of variations on the general theme of *in situ* treatment. Most rely on the extraction, recirculation and amendment of groundwater over the period of treatment, which may take from a few weeks to a few years depending on the situation. Whilst all schemes should pay due attention to the simultaneous treatment of both soil and groundwater in the saturated soil zone, problems may arise in the treatment of contamination in the vadose zone. In some cases this has been overcome by recirculating groundwater through this zone before it is returned to the groundwater, as illustrated in Figure 11.7. Alternatively, where volatile organic compounds are present, vacuum extraction or bioventing may be incorporated into the clean-up system (see Section 11.8.4).

Ex situ treatment

Where the ground conditions do not allow the *in situ* biological treatment of contamination *e.g.* where permeabilities are too low, it may be possible to excavate contaminated soil and treat it in above-ground, or *ex situ* soil heaps. These heaps are aerated either by conventional agricultural tillage, or by the incorporation of aeration pipework throughout the soil mass. In these cases volatile contaminants may be lost to the atmosphere by air stripping, and initially rapid declines in the concentrations of some contaminants may be observed. However, when compared with *in situ* systems the aeration status of an *ex situ* system tends to be more uniformly controlled and the application of nutrient and microbial amendments can be undertaken with conventional agricultural spraying equipment. To avoid the loss of leachate from the system, soil treatment beds are often built on impermeable bases, incorporating a drainage system. This system may drain to a collection sump from where the leachate can be collected and treated or recycled. In some cases *e.g.* derelict industrial sites, it may be possible to construct soil heaps on areas of hardstanding, and use existing drainage runs to collect leachate. Where this is not feasible, durable plastic membranes and gravel drainage layers can be used in combination, as an impermeable, drained foundation to an *ex situ* treatment system.

In addition to chemical and microbial amendment of excavated soil systems, there is the opportunity to manipulate the physical nature of the soil to aid biodegradation. In this way materials can be added to aid the workability of clay *e.g.* lime, and to improve the structure of the soil to enhance aeration and drainage. The latter may be achieved by the application of uncontaminated organic matter or inert material such as gravel. In an extreme case, organic material may be added to contaminated materials with the intention of creating a composting system. Here, the quantity of organic material will tend to outweigh that of the contaminated soil, and moisture levels (approximately 40%) tend to be higher than for soil treatment heaps (approximately 10-20%), such that the conditions are created for intense aerobic microbial activity which initially produces a rise in temperature within the heap. Suitable organic media for composting include sewage sludge, green wastes, manure, wood chips, bark and straw, and to some extent the use of a composting system to decontaminate soil materials will be limited by the local availability of these organic substrates.

Composting is a complex and dynamic process which relies on a succession of microbial populations during the decomposition of organic materials. Under controlled conditions it can result in the formation of a beneficial medium for soil conditioning. This can be of benefit to the establishment of vegetation during a reclamation scheme (see Section 14.4).

Composting has not yet been widely used for the treatment of contaminated soils, but it is likely to find more applications in the future.

11.8.4 Separation

Introduction

Various techniques are available for the separation of contaminants from soil particles; as treatments for contaminants within the ground and for excavated or extracted contaminants.

Vacuum extraction

Vacuum extraction is a physical technique for the removal of volatile organic compounds from soil and groundwater. It may also be referred to as vapour extraction or soil venting.

Vacuum extraction is an *in situ* treatment, and relies upon the extraction of volatile contaminants from wells within a contaminated area, by the use of vacuum pumps or fans. The application of suction to the wells causes air in the surrounding soil to flow into the well, along the pressure gradient, and out through the extraction system. This air will contain contaminants present in the vapour phase. The decrease in pressure within the soil air will also increase the volatilisation of organic compounds *i.e.* contaminants, and thus aid their removal from the soil. Depending on the nature of the contaminants and their concentrations in the extracted air stream, an emission control system may be used to clean the air prior to discharge to atmosphere. Such control may use condensation, adsorption, biological or thermal treatment.[44]

The application of vacuum extraction is limited mainly by two main factors:

- soil permeability;
- contaminant volatility.

Usually the extraction system is based on a series of wells, the depth and spacing of which is critical to the success of the system. The extraction system configuration is based on an understanding of the contamination profiles and the permeabilities of the soil.[154]

A schematic diagram of a typical system is shown in Figure 11.8.

It is more usual for contamination by volatile and semi-volatile compounds to occur in both the vadose zone and the saturated zone beneath. In this case a dual vacuum extraction system is used. In dual vacuum extraction water is removed from at or just below the groundwater table in the extraction wells, either by application of a vacuum or by a submersible pump. Contaminants in the groundwater are removed with the groundwater, and those which remain in the soil, adhering to soil particles as the groundwater table is lowered, are removed by the vacuum applied to the unsaturated zone (see Section 12.5.3). As organic contaminants tend to be most concentrated at the interface between the groundwater table and the unsaturated zone above, dual vacuum extraction is frequently the most effective treatment method. However, the extracted groundwater will generally require treatment to separate and degrade contaminants (see Section 12.5.3).

The flow of air through the soil induced by vacuum extraction also enhances biological degradation of contaminants. This biological activity can be encouraged by application of soil amendments to the unsaturated zone.[163]

Soil vacuum extraction systems tend to be suited to contamination such as that caused by spillages of organic liquids. The technology has been used successfully to remove light petroleum and chlorinated hydrocarbons, and is of some use where volatile coal carbonisation products contaminate the deeper strata in permeable soils.

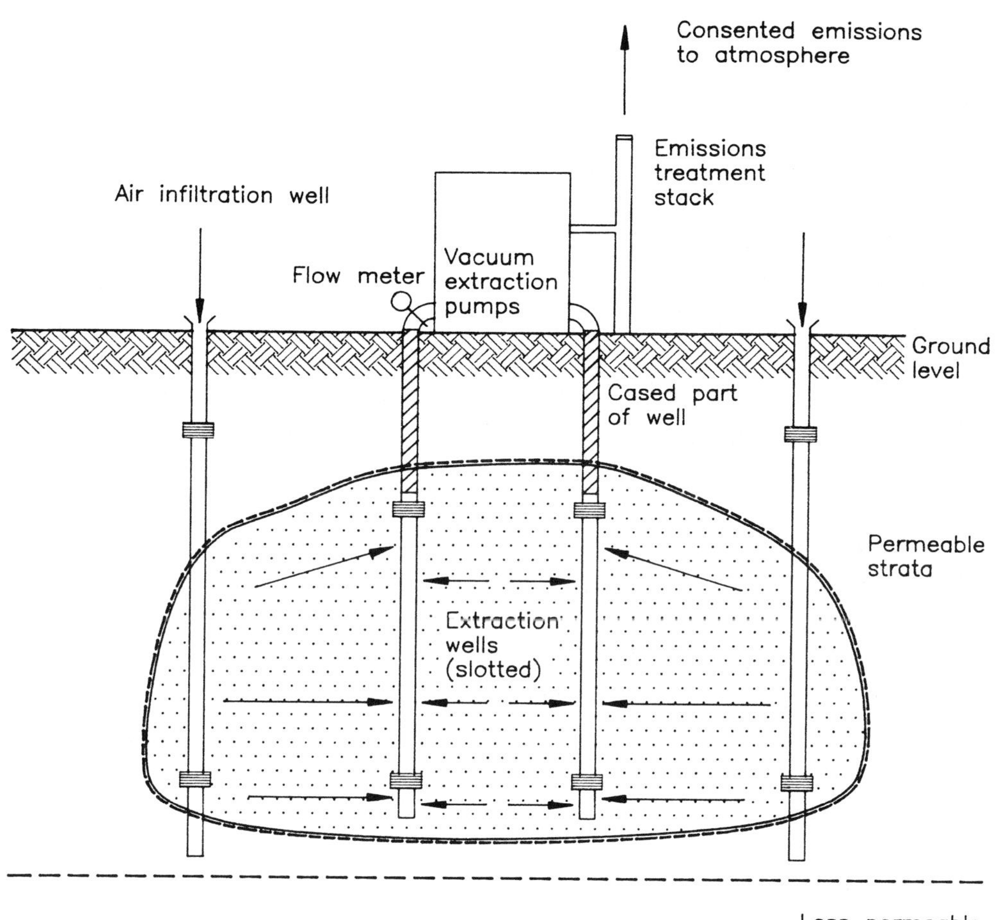

Figure 11.8: Schematic diagram of a vacuum extraction system for volatile organics remediation in soil

Soil washing

Soil washing is a general term applied to a number of water-based processes for the separation of contaminated materials from soils. It is applicable to a wide range of contaminants, from metals to volatile organics, and can achieve efficient decontamination in suitable soils.

The processes generally depend on the observations that either contaminant molecules are water-soluble, or that they are commonly associated with the fines fraction of soils (notably clay and silt minerals, hydrous oxides and organic matter). These contaminated fines can be separated from the soil matrix by various aqueous systems that rely on particle separation, so leaving behind a largely uncontaminated 'soil' residue, which can be replaced on site or reused elsewhere.

A typical process flow diagram for a soil washing facility is shown in Figure 11.9.

The residual contamination arising from a soil washing process can be placed into one or more of the following classes:

- contaminated wastewaters, which then require further treatment to remove contaminants from solution;
- contaminated fines, often separated as sludges, which then require disposal;
- emissions to air, in the case of volatile compounds.

Soil washing is a European technology, developed during the 1980s in the Netherlands and Germany,[178] using well established principles of particle size separation used in the minerals industry. Several operators throughout Europe have this technology at their disposal, and some of the smaller throughput plant is transportable,[9] allowing it to be taken to a site, and avoiding the need to transport toxic materials over long distances. Such equipment, however, requires sources of power and

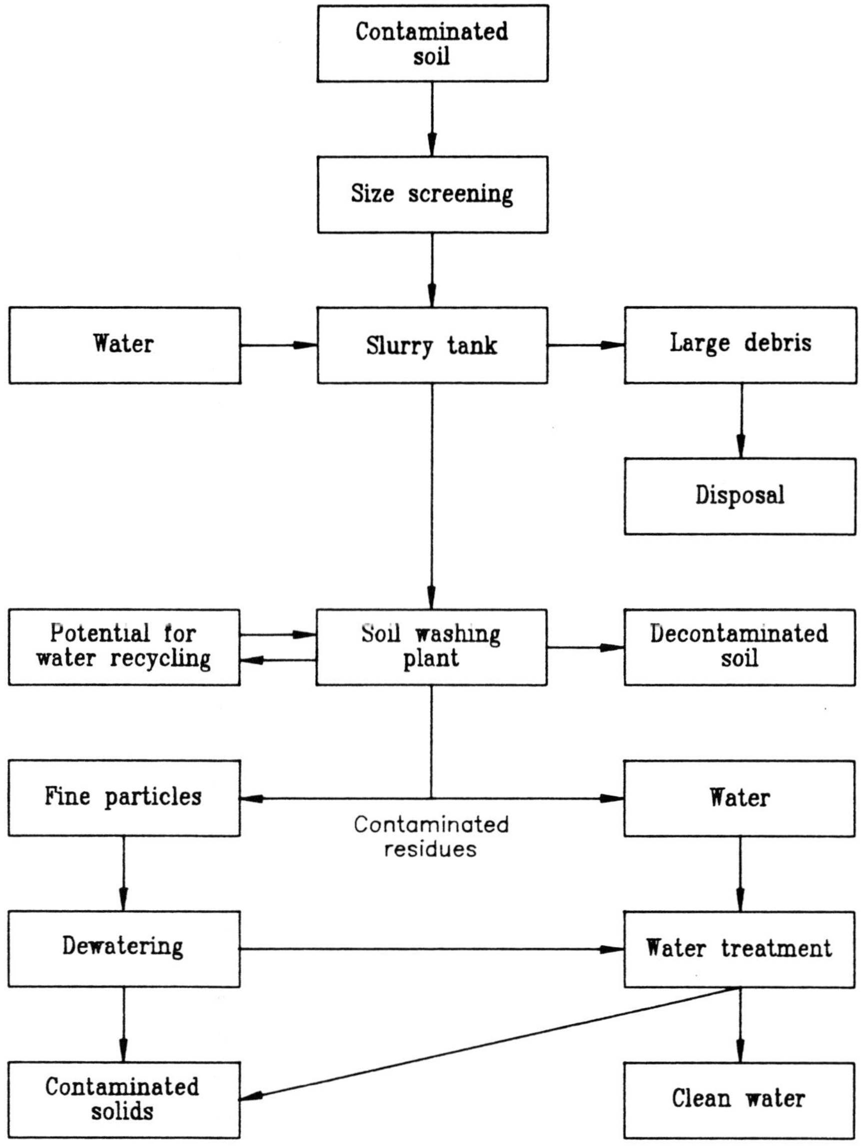

Figure 11.9: Flow chart for principal processes and products of a typical soil washing decontamination treatment

water, as well as on-site space; possibly 1.5ha for a 20t/h soil washing plant.[244]

Generally, soil washing is particularly suited to sandy and gravelly soils, where significant reductions in the volume of contaminated materials can be achieved. In this way soil washing may be able to provide a one step treatment, or may be useful as a pretreatment, prior to, for example, the incineration of contaminated fines material.[208, 244]

Silt and clay soils are less amenable to treatment, and contaminants that are particularly hydrophobic are more difficult to remove. The latter may be overcome to some degree by the use of solvents or surface active compounds (detergents), although these may interfere with the efficiency of washwater treatment.

Variations of soil washing include adaptations of mineral ore processing techniques,[184] high pressure jet grouting technology,[106] and hydrocyclones.

Solvent extraction

Solvent extraction is similar in principle to soil washing, in that contaminants are removed from the soil matrix into a liquid medium. In this case however, the liquid medium is not water, but an organic solvent. Such solvents include non-chlorinated organic solvents *e.g.* triethylamine, carrier oils, and liquified gases *e.g.* propane and/or butane (critical fluid extraction[132]).

The technology is generally applied to the treatment of organic contaminants such as PCBs, volatile organics and petroleum hydrocarbons, and is of less use for the extraction of metals and inorganics. Solvent extraction has shown potential for the treatment of coal tar wastes in soil.

Because of the potentially hazardous nature of solvents used in the extraction process, and the cost of these solvents, solvent extraction treatment plants have the facility to recycle the solvent. Thus the process consists of a soil-solvent contactor vessel, where the soil is washed in the solvent, followed by a solvent recovery facility where the solvent and the contaminants are separated, and the solvent recycled. Separation of solvent and contaminants can be carried out by steam stripping.[111] The process therefore results in the production of concentrated contaminated residues, which require disposal or further treatment.

Generally solvent extraction facilities are transportable, although rates of throughput tend to be low, and the availability of such equipment in Europe is limited.

Leaching, pumping and treating

The leaching of contaminants from soil, and the pumping of contaminated groundwater and subsequent treatment above ground are variations on the themes of soil washing/solvent extraction and *in situ* treatment respectively. Section 12.5.3 considers pump and treat methods for contaminated groundwater.

The types of treatment which can be applied to contaminated groundwater in surface treatment facilities may include some of those discussed earlier, such as air stripping, biological treatment and adsorption, as well as less established techniques such as the biosorption of metals.[19]

The application of leaching methods to the extraction of contaminants from soils is an emerging technology that has potential for the controlled release of contaminants, including metals, from polluted soils.

11.8.5 Application and integration of techniques

On most former industrial sites, and especially those of former coal carbonisation or steel making activities, it is rare that a single treatment technique will meet all the requirements of site clean-up. This is often due to constraints posed by the presence of either:

- complex combinations of contaminants;
- discrete areas of different types of contaminant;
- variations in ground conditions;
- areas required for different end-uses, and so different clean-up standards.

As a consequence, a combination of techniques can provide the most effective solution to site decontamination. On-site decontamination is frequently combined with off-site disposal or physical containment of areas of low treatment potential. For example, in the decontamination of a gasworks site in the UK, *ex situ* biological treatment was used to treat phenolic and coal tar residues, a clay-lined containment cell was constructed on site, and other materials were disposed of to a licensed landfill.[31] Alternative scenarios may use two or three stage treatment to achieve decontamination objectives, and some of these are highlighted in Box 11.6.

The choice of treatment(s) therefore needs careful consideration following site assessment and feasibility studies, and combinations of different techniques can be more reliable than single treatments where contamination at a site involves heterogeneous contaminants and ground conditions.

The application of remedial treatment or a combination of treatments must always take into account the sensitivity of local communities to the on site works. Every effort should be made to avoid inconvenience to local residents during site decontamination by minimising such effects as noise, dust, odours and traffic congestion.[249]

Box 11.6: Some examples of combined treatments used to obtain required standards of decontamination

- *In situ* treatment of volatile contamination by vacuum extraction, and biological treatment of abstracted groundwater.
- Soil washing of organic contaminants followed by the biological treatment of the contaminated fines in an *ex situ* system.
- Treatment of pumped groundwater in an above-ground biological reactor, which subsequently requires polishing to attain required low levels of contaminants, using an activated carbon filter.
- Physical containment of contaminated ground prior to *in situ* bioremediation.

11.9 Standards and objectives for decontamination

11.9.1 Considerations

Whilst the choice of the most appropriate method of decontamination is important, it is also important to adopt appropriate chemical standards for the end result of a decontamination operation. These considerations will be affected by:

- lack of established standards for soil clean-up on a national or regional basis;
- incomplete understanding of the objectives of clean-up in terms of risk reduction;
- incomplete understanding of the exact nature of the contamination present;
- high variability in the distribution of contamination through the soil, both before and after treatment.

Despite the importance of setting decontamination criteria, this has in some instances been undertaken on a piecemeal basis, and has led to problems after the application of the treatment, for example where standards of decontamination have not been agreed by a controlling authority. In some areas of Europe there exist regional or national standards for soil contamination, such as in the United Kingdom,[121] the Netherlands,[165] and various regions of Germany (see Section 2.6.5). These standards form guidance for the practitioner and are advantageous because:

- the target concentrations can be easily defined;
- the choice of remedial treatment will be limited to those methods which are able to achieve the required final concentrations;
- the liability of the site owner or treatment company for future pollution arising from the site is limited providing that the defined standards are met during clean-up.

Where no guidance levels are defined then it is the responsibility of those concerned with the redevelopment of a contaminated site to agree on suitable criteria for decontamination, in liaison with the statutory authorities. Thus, regulation by negotiation aims to bring together all the parties concerned, including the regulator, to gain an agreement on targets for reclamation.[254] Agreement should be made on the basis of:

- the intended long-term use of the site after reclamation;
- the removal of risk associated with the contamination, its effect on health and safety at the site, and the proposed after-use;
- the adequate protection of groundwater in the area, and the environment in general.

Agreement and ratification of target levels between the parties concerned can be a long process, and should therefore start at an early stage. Informed parties should take into account the environmental availability of the soil contaminants in relation to local conditions, and also the background levels of contamination in the region.[95]

Once target concentrations have been set, and a method of remediation chosen, the next stage is to choose the best way to monitor the treatment. Design of a monitoring strategy requires consideration of the following:

- chemical analytical methods to be used;
- sampling regime during the treatment operation;
- strategies to monitor the site after treatment is complete.

11.9.2 Analytical methods

Introduction

The choice of analytical method for the accurate assessment of a specific contaminant concentration is fundamental to the treatment programme. Analysis of samples collected during a site investigation is discussed in Section 2.6.4. For many chemical parameters there are a number of alternative analytical methods available. The following should be taken into consideration when choosing which methods to use:

- a nationally or internationally accepted standard method is generally preferred;
- throughout treatment, monitoring and validation, a single method for a single chemical parameter should be adhered to;
- analysis should be carried out by a recognised analytical laboratory, preferably one that is nationally accredited by a controlling body, and which adheres to strict quality assurance procedures.

Detection limits

With particularly toxic chemicals the concentrations required to ensure adequate risk reduction may be very low; in the region of a few parts per million for soils, or a few parts per trillion for groundwaters. The detection limit of the method chosen must be below the concentrations required after treatment of the contaminated material.

The accurate detection of low concentrations of pollutants is thus an important issue which should be noted at the outset. Mistakes have occurred in some cases where the detection limits of a standard method were too high to allow the adequate determination of concentrations after treatment.

Leaching tests

The importance of the environmental availability of a contaminant should not be underestimated in assessing risk. The solubility and mobility of a contaminant is likely to affect:

- biological availability;
- pollution potential;
- toxicity.

Leaching tests aim to subject a sample of contaminated material to a set of conditions that simulate the rigours of environmental liquid-solid contact, and the chemical and physical changes that this causes. The aims of a leaching test are to extract the available (or potentially available) contamination from a sample into the liquid phase.[246]

The use of agreed leaching tests to accompany standard quantitative procedures for soils can enhance the monitoring and validation of a clean-up scheme, and may offer valuable supplementary data to accompany that of soil chemical concentrations.

Leaching tests are discussed further in Section 12.4.5.

11.9.3 Sampling regime

Introduction

The determination of a sampling pattern for the investigation of contamination at a site is discussed in Section 2.6.2. Similarly, a sampling regime must be agreed for the monitoring and validation of a decontamination operation, even where contaminated materials are to be moved off site. The choice and consistency of sampling pattern is important if meaningful data are required on the clean-up operations.

On-site treatment

Where soils and groundwater are to be treated on the site it is important to undertake sampling and analysis to establish the concentration of contaminant(s):

- prior to the treatment;
- during the treatment;
- at the end of the treatment.

Where statistically meaningful data is required *e.g.* for regulatory compliance, the operator must ensure that the number of samples taken is consistent throughout the treatment, and that enough samples are taken to allow for the variability between results on a single sampling occasion.

Off-site disposal

Where contaminated materials are to be carried off site there also needs to be a suitable sampling regime in place. This should address the contamination status of:

- the materials moving off site;
- materials left on site;
- imported fill materials.

An agreement with a waste transport contractor, or fill materials supplier, concerning a sampling strategy for loads of soil materials moving on and off site should be made at the outset of the reclamation works. This should ensure that only materials of a suitable chemical quality are treated, disposed of, or used during the scheme.

11.9.4 Monitoring

Monitoring of a treatment process is necessary in order to maintain the level of treatment required and to detect any under-performance of the process. In addition long-term monitoring of a site may be required to detect any failures of a treatment at some point in the future. The latter is particularly important where wastes have been retained on site by methods such as containment, solidification or stabilisation.

Monitoring systems commonly relate to gas production *e.g.* from landfills, and to groundwater quality. Both can be successfully monitored by sampling from borehole stand pipes, installed and maintained to a suitable standard on or adjacent to the treated area.

11.9.5 Validation

At the end of a contaminated land treatment programme it is most important to assess the success of the scheme, in terms of contamination decline, and final decontamination levels. This may be particularly so when validating the success of an on-site decontamination treatment that has involved the separation or degradation of pollutants in the soil *e.g.* biological methods.

The validation is likely to conform to the requirements of a regulator, who may stipulate:

- the target concentrations;
- the methods to be used in analysing the contamination levels;
- the minimum number of samples that have to be taken from a given site area;
- a proportion of samples which are required to fall below the target concentration at the end of the treatment.

In the event of a treatment failure, the regulatory body may also require assurance of an alternative treatment or disposal route for materials that remain contaminated.

11.9.6 Costs

The cost of any decontamination scheme should allow for the costs of an appropriate sampling and evaluation programme, according to the factors discussed in Sections 11.9.1 to 11.9.5. In some cases the cost of such a programme may form a significant part of the total cost of the scheme. Any treatment for contaminants which are expensive to analyse accurately *e.g.* dioxins, may incur substantial analytical costs.

It is therefore important to accurately assess the requirements of sampling and analysis at all stages of a decontamination operation, and the costs that accompany these actions. The costs are likely to include:

- time for personnel to carry out sampling;
- allowances for the establishment of monitoring access points and equipment *e.g.* boreholes, piezometers, on-line and automated measurement devices;
- analysis of samples;
- documentation and reporting of results.

12 WATER QUALITY

Chapter contents

12 WATER QUALITY

12.1 Introduction

Industry uses water in various ways. Water in canals and rivers, for example, may be used for transport of goods. Industries have been located to make use of such transport networks and canals built especially to serve them. In recent decades there has been a move to consolidate steel making at large steelworks in coastal locations, and to make use of imported coal and iron ore. Water is also used in a wide variety of processes: as a medium for transport and separation of particles, for example in coal washing and gas cleaning, and for cooling. Industrial sites may contain lagoons where waste process water is trapped to allow the settlement of solids. Modern sites are likely to recycle much of their process water, but historically, discharges of process water may have resulted in severe pollution of adjacent watercourses with long-term contamination of sediments.

There is much European Community legislation relating to water. Legislation on water pollution of relevance to the reclamation of derelict land is described in Section 12.6.

12.2 Water and coal mining

12.2.1 Types of water discharges

Introduction

During coal mining there are three main types of water discharge:

- process water;
- mine drainage;
- surface run-off.

Process water

Water is used in coal preparation plants, in which the coal is separated from non-coal materials. The processes are similar to those described in Chapter 8 for the recovery of coal from spoil heaps. The material from the mine is immersed in water and the lighter coal particles are separated from the heavier shale particles under gravity. Some fine-grained particles of coal are carried over in the process water, and these are removed in a second separation stage which utilises froth flotation. In this process reagents which adhere preferentially to the coal particles are added to the water. The reagents used include oils and polyelectrolytes. The oils used frequently containing phenols. Air is then blown into the mixture and the chemical reagents cause air bubbles to be formed around the coal particles. These particles then float to the surface where they form a froth which can be skimmed off. The water is then removed by filtration to give a coal filter cake. The waste from the froth flotation process consists of water containing fine-grained particles (tailings). The water and tailings are separated by filter presses, or by placement in settlement lagoons, in which the tailings separate out under gravity. The supernatant liquid may then be discharged or returned as process water.[148]

At disused coal preparation facilities the condition of the settlement lagoons will depend on the extent to which they have become filled with tailings and on the water level. Such lagoons may remain as bodies of open water or as areas of unsaturated fine-grained silt.

Mine drainage

During mining operations, many mines require pumping to remove water from the workings. When mining ceases and pumping stops water levels rise and the mine floods. Flooding of the mine alters the hydrological conditions of the area and new discharges of mine water may appear.

Many shallow mine workings, less than 100m deep, are wet and the water level in them is affected by surface storm water which percolates into the mine through cracks, joints and faults in the strata.[181] Deeper mines are more likely to be isolated from surface water by impermeable strata and are likely to be relatively dry. However, sandstones associated with coal in the stratified coal measures of shales, sandstones and coal, are potentially water bearing rocks. Thus, if the sandstones outcrop at the surface they will tend to be recharged by rainwater, resulting in significant inputs of water into the mine.

Coal measures are sometimes overlain by younger rocks. These rocks may be important aquifers, such as the Triassic sandstones which overlie the coal measures over Eastern Britain. Shafts sunk through these aquifers have to be protected by pumping groundwater from the surrounding rock. On abandonment of the mine this pumping will cease and the shafts will act as conduits for movement of water into the mine.

Mines may also be connected underground so that pumping of water from one mine will influence water levels in an adjacent mine. An abandoned mine may not have reached a hydrologically stable condition if it is connected to working mines. Later closure of the working mines, with cessation of associated pumping, could alter the water level in the abandoned mine, causing new issues of water to appear. Changes in water level could also affect the air-flow patterns in the abandoned mine. Such changes have, for example, resulted in the onset of combustion in colliery spoil overlying an old mine entrance. Rising water levels in the mine pushed air out through the spoil and resulted in an underground fire which led to die back of vegetation planted as part of a reclamation scheme.[63]

Surface run-off

Pollution of surface run-off water may arise where spoil, tailings or stockpiled coal are exposed to rainwater. Pollution will be greatest where slopes are steep, and erosion maximised, and the distance to surface watercourses is short, with little opportunity for suspended solids to settle out.

12.2.2 Causes of poor water quality

Introduction

There are three principal types of water pollution associated with coal mining:

- suspended mineral particles;
- salinity (*i.e.* a high concentration of dissolved salts);
- acidity.

Other possible pollutants include mineral oil, exuded by mine strata into the mine water in some coal fields.[103]

In addition to pollution from mining activity, water pollution may result from wastes deposited in mine shafts. For example, at the Ravenscraig steel works in Scotland coke oven wastes and pickling acids were disposed of down disused coal mine shafts at the site. Groundwater contaminated with ammonia, phenols, cyanides and oil was found to be seeping into surface watercourses.[11]

Suspended mineral particles

Suspended mineral particles may be present in surface run-off from colliery spoil or tailings, where these are not protected from water erosion, for example by a cover of vegetation. The particles are generally inert, consisting of clay, quartz or coal.[103] High concentrations of dense suspended solids will blanket the beds of receiving watercourses, killing the benthic (*i.e.* bottom dwelling) organisms, which are a vital part of the aquatic food chain.

Salinity

The major salts in mine drainage waters are chlorides and sulphates. The salinity of mine waters is dependent upon the soluble ion content of the coal and surrounding rocks and tends to increase with depth. The more saline waters tend to contain significant concentrations of barium, strontium, ammonium and manganese as well as chloride and sulphate.[103] Some freshly exposed colliery spoils are also saline (see Section 5.3.2) and may cause saline surface run-off.

Acidity

Acid mine drainage is the principal cause of poor water quality arising from coal mines and is dealt with in detail in Section 12.2.3.

12.2.3 Acid mine drainage

Introduction

Acid mine drainage (AMD) results from the exposure of sulphide-containing minerals to water and oxygen. Sulphides are oxidised to sulphate and acid is produced. At the resulting low pH values many metal salts become more soluble, so the drainage water is therefore both acidic and metal-rich.

The predominant sulphide mineral associated with coal measures is pyrite. Pyrite is a widely distributed mineral, present in sediments formed under anaerobic conditions. Pure pyrite is iron (II) sulphide, FeS_2, although other metals may be present as impurities, including arsenic, aluminium, gold and, notably, manganese. AMD resulting from pure pyrite oxidation contains iron as Fe^{3+}, sulphate (SO_4^{2-}) and acidity, as H^+. The iron forms a precipitate of hydrated ferric oxide (ochre) which coats the beds of watercourses, rendering them unsuitable as habitats for benthic organisms. The high concentrations of other metals, notably aluminium, may also be toxic to aquatic life, particularly fish. AMD-affected waters are thus usually without fish and sometimes devoid of any aquatic life.

Pyrite may be present both in colliery spoil and *in situ* coal measures underground, AMD may therefore be produced both by the interaction of surface water with colliery spoil, and by the interaction of groundwater in the mine with pyrite-containing rocks. The latter produces polluted adit discharges and is the major source of AMD pollution.

There are four stages in the generation of AMD:

- oxidation of pyrite;
- mobilisation of the products of oxidation;
- neutralisation;
- secondary mobilisation.

Each stage is described in turn below.

The typical composition of AMD resulting from coal mining is shown in Table 12.1. The presence of aluminium and manganese ions in the AMD results from the secondary mobilisation of metal impurities present in the pyrite and in surrounding minerals.

Table 12.1: Typical composition of AMD from coal mining[259]

Constituent	Concentration (mg/l) except pH
SiO_2	90
Mg	80
Ca	200
Al	50
Fe	50-300
Mn	20-300
Sulphate	20-2000
pH	3.0-5.5

Oxidation of pyrite

The oxidation of pyrite has been discussed in Chapter 5, and the chemical reactions involved shown in Box 5.1. Oxidation of pyrite may be by atmospheric oxygen, which is a slow process, or by ferric (Fe^{3+}) ions, which is rapid. Ferric ions are produced by the oxidation of ferrous (Fe^{2+}) ions, released by the oxidation of pyrite. This reaction is slow except when it is catalysed by iron-oxidising bacteria such as *Thiobacillus ferrooxidans*. These bacteria are only active under acidic conditions (pH 2-4), so once the pH has dropped to these values pyrite oxidation is increased considerably and the pH falls rapidly.

Mobilisation

In the absence of water, the products of pyrite oxidation accumulate on the mineral surface and thereby inhibit further reaction. AMD is formed when these products dissolve and are transported away by the water. Alternate wetting and drying of pyritic spoil, or minerals in underground strata, enhances pyrite oxidation by washing away the oxidation products leaving a fresh reactive surface.

Neutralisation

Acid-consuming minerals, such as the carbonates ankerite and siderite, react with the acidity, increasing the pH and thereby reducing the solubility of metals. The metals are then precipitated as metal salts such as hydroxides, carbonates or hydrogen carbonates.

Table 12.2 gives examples of acid-consuming minerals and their neutralising capacities.

In any situation, both sulphide minerals and acid-consuming minerals are likely to be present. The relative amounts of the two types of minerals will determine whether long-term production of AMD will result.

Table 12.2: Summary of some acid-consuming minerals and their neutralising characteristics (from BCAMD Task Force, 1989[25])

Mineral	Composition	Acid-consuming potential †	Buffer pH
Calcite, Aragonite	$CaCO_3$	100	5.5-6.9
Siderite	$FeCO_3$	116	5.1-6.0
Ankerite	$CaFe(CO_3)_2$	108	-
Dolomite	$MgCa(CO_3)_2$	92	-
Gibbsite	$Al(OH)_3$	26	4.3-3.7
Limonite/Goethite	FeOOH	89	3.0-3.7

†The acid-consuming potential is given as the weight (grammes) of the mineral required to provide the same neutralising effect as 100g of calcite.

Secondary mobilisation

Due to its acid nature, AMD will cause weathering of other non-pyritic, non-acid-consuming minerals. This weathering can lead to the leaching of further metals into the AMD. In particular aluminium, which is toxic to fish at low pH, may be leached from clay minerals. In this way AMD can lead to the mobilisation of toxic materials from otherwise inert materials.

The extent and characteristics of this secondary mobilisation will depend on the type of minerals encountered by the AMD, in addition to chemical factors (pH of AMD, redox potential, adsorption phenomena, chemical composition of AMD), biological factors (presence of iron-oxidising and other bacteria) and, to a lesser extent, physical factors (particle size and shape, temperature, pressure of pore gases).

12.3 Water and steelmaking

Large volumes of water are used in the steel industry, mainly for the cooling of heat generating plant such as furnaces, in the processing of steel products, and in the cleaning of flue gases. The processes involved in steel making have been described in Section 9.1. Settlement lagoons for waste water may be present.

Water pollution may arise from the presence of water-mobile contaminants on the site. The wastes and contamination from iron and steel making have been described in Section 9.2, and those from coal carbonisation (used at integrated steelworks to produce coke) in Section 10.2.

Slags, the major waste, or by-product, of iron and steel production, generally do not give rise to water pollution. Occasional problems include the leaching of sulphur compounds from blast furnace slags and high alkalinity from steel slags.[83, 225] Flue dusts from iron and steel making furnaces, particularly electric arc furnaces, may contain metals in a water-soluble form.

The greatest water pollution is, however, likely to be found in areas where coal carbonisation was carried out, and in the associated by-products facilities. Polluting substances associated with these areas include volatile organic compounds such as benzene, toluene and xylenes, more complex organic compounds such as naphthalenes and polyaromatic hydrocarbons, and inorganic substances such as sulphates, cyanides and ammonium.

12.4 Assessment

12.4.1 Introduction

This section deals with assessment of disused coal and steel sites from the water pollution perspective. It is through the pollution of water that contaminated sites are most likely to have an impact outside of their boundaries. Water pollution is therefore an important issue which should receive close attention during site assessment and reclamation design. The principles of site assessment have been outlined in Chapter 2.

The first step in assessment of whether a site is, or has the potential to be, a source of water pollution, is to find out from archive studies and soil investigations whether there are substances on the site which are potentially polluting. Assessment could then proceed as follows:

- identification of water bodies which could be polluted by substances on the site *i.e.* potential targets;
- collection of information on these targets to assess the impact of any pollution;
- characterisation of existing water quality;
- determination of water pollution potential of contaminated materials.

12.4.2 Identification of target water bodies

Both ground and surface waters may be vulnerable to pollution.

Surface waters include streams, rivers and lakes within or adjacent to the site. Pollution may occur through erosion of particulate matter or seepages of dissolved or liquid contaminants into these waters.

Groundwaters vary from perched, near-surface, groundwater tables to deep regional aquifers.

The continuity between different groundwater units and between ground and surface water can be inferred from geological information obtained from published data and from borehole drilling. Pollution of groundwater units may lead to pollution of surface waters into which they feed. Such surface water may be remote from the site and, due to the slow movement of pollutants in groundwater, it may be some years before such pollution becomes apparent. Similarly, surface watercourses may flow across outcrops of aquifers downstream of a contaminated site. For example, parts of the chalk groundwater system of Kent, southern England, an important aquifer for water supply, have become polluted with chlorides from streams contaminated by coal mine drainage.[265] Ground investigations, in particular the drilling of boreholes (see Section 2.5.2), and the disturbance of contaminated materials associated with reclamation works, may connect previously separate groundwater units. If one of these units was uncontaminated prior to disturbance it may become contaminated by the other.

The ways in which ground and surface waters are interconnected is greatly influenced by mining activity. Mine passages may have intercepted flows of groundwater, with the result that groundwater emerges as adit discharges, rather than from the pre-mining springs. The effects of pumping of mines and flooding which follows the cessation of pumping after mine closure have been discussed in Section 12.2.1.

The direction and rate of groundwater movement are important factors to determine in order to assess groundwater pollution. Direction of flow is perpendicular to the groundwater contours, obtained by measurement of standing water levels in wells or boreholes. The rate of groundwater flow can be calculated from the hydraulic gradient and hydraulic conductivity (see Box 12.1).

Characteristics of surface and groundwaters may vary considerably under different weather conditions and at different times of year. Surface watercourses may only flow after high rainfall events, or there may be a steady flow throughout the year. Similarly groundwater bodies vary in

Box 12.1: Calculation of rate of groundwater flow[52]

Groundwater flow can be described using Darcy's Law:

$$V = -K \frac{dh}{dl}$$

where:

V	=	specific discharge *i.e.* the rate of flow through unit cross sectional area.
K	=	hydraulic conductivity, a measure of the permeability of the material.
$\frac{dh}{dl}$	=	hydraulic gradient, difference between the groundwater level at any two points divided by the distance between them.

The unit cross sectional area to which V relates includes soils and voids. Groundwater flow is only through the voids so the actual velocity is the product of the specific discharge and the fraction of the unit cross sectional area occupied by voids *i.e.* the porosity of the material. If the porosity is 20%, the velocity of groundwater flow will be five times the specific discharge.

their response to rainfall. In some aquifers the amount of water within the matrix of the rock is small but there are many fissures through which groundwater flow is rapid. In such 'fissure-flow' aquifers there are large variations in groundwater level, with a rapid response to rainfall. The result of such variations in groundwater level can be that contaminated ground, which in the summer may be many metres above the groundwater table, may be inundated in the winter. By contrast, in aquifers in which flow is predominantly intergranular *i.e.* between particles within the matrix of the rock, storage capacity is far greater but flow slower. The response to rainfall events is therefore muted and the groundwater level does not show such large fluctuations. Mining activity is likely to increase the fissure-flow characteristics of an aquifer, increasing the variation in groundwater level.

12.4.3 Sensitivity of target water bodies

The impact of pollution from a particular source is influenced by the nature of the receiving water. In a heavily industrialised area, where extensive pollution of ground or surface water is already prevalent, water pollution from a contaminated site will have less impact on water quality than where the site in question is the major source of water pollution in the area. Evaluation of the potential benefits of preventing water pollution from the site requires information on the water quality upstream of the site and on inputs of pollution from other sources. The policy of the regulatory authority is also important. For example, there may be a long-term objective to improve water quality which will lead to a reduction in other pollution inputs. Such objectives are frequently the result of European Community legislation on water pollution (see Section 12.6). Reductions in inputs of pollution from other sources in the vicinity of a site will increase the relative importance of the pollution arising from that site, and thus the benefits to be gained by the prevention or treatment of that pollution.

Demands made on target water bodies will also influence the impact of water pollution. Surface water may support flora and fauna of ecological value. Some water habitats, particularly estuaries, can be important habitats for birds, despite historic pollution. Surface and groundwater is frequently abstracted for a variety of uses, and the contamination of water is most likely to be the subject of legal proceedings when pollution from a site causes water to be unsuitable for its intended use. The outcome of legal action is often to require the owners of the site causing pollution to remediate the pollution at their expense, *i.e.* the polluter pays.

12.4.4 Existing water quality

The effect of a site on ground and surface waters can be determined by collecting information on the existing quality of such waters. Any past water quality monitoring data should be consulted and sampling of surface and groundwaters carried out. Samples should be analysed for potential pollutants. Repeated sampling gives more reliable data than single samples as water quality can vary greatly at different times of year and under different weather conditions (see Section 12.4.2).

12.4.5 Water pollution potential of contaminated materials

Results of water monitoring may be inconclusive as to whether or not particular materials are sources of water pollution. An additional approach is to carry out leaching tests on the materials in question to determine the mobility of the potential pollutants they contain. Such tests involve placing a sample of the material in contact with a solvent, usually water, for a set period of time. The sample and water mixture may be shaken, or water may be allowed to percolate through the sample, simulating contact with rainwater. On completion of the contact period, the water is filtered and the filtrate analysed for the parameters of interest. Several standard leaching tests have been developed, for example DIN 38414-S4 in Germany.[74] Leaching tests provide a simple means of estimating the potential mobility of contaminants from a solid phase into a liquid phase.

Reclamation may reduce the particle size of a material or alter its environment. Leaching tests may be used to help predict the effects these changes will have on the leachability of substances from the material, and thus the likely water pollution impacts. They may also be necessary to allow decisions to be made on the disposal or reuse of materials which are to be excavated.

For pyritic wastes it may be necessary to evaluate the acid generation potential of the wastes (see Section 12.2.3). There are several methods by which this can be done.[25, 49, 144] The basis of the methods is to calculate the total possible acid production and then subtract the acid-consuming potential of the material to give a figure for net acid production (see also Box 14.7). The total possible acid production is that produced by oxidation of all the pyrite present in the material. Pyrite content may be either measured directly or inferred from the analysis of sulphur species. The measurement of total sulphur is the simplest procedure, though subtraction of sulphate sulphur gives a more accurate estimation of pyrite content, but one which still includes organic sulphur. Methods for the determination of pyrite in colliery spoils have been reviewed.[66] The method recommended by this review involves sequential oxidation with hydrochloric and then nitric acid followed by determination of the dissolved iron content. Further tests may be carried out on samples of materials found to have the potential for acid generation. These tests, known as kinetic tests, involve subjecting samples to weathering under laboratory-controlled or on-site conditions to determine the rates of acid generation.

12.5 Control of water pollution

12.5.1 Surface water

Surface water quality generally improves rapidly once pollution inputs cease. Therefore the strategy should always be to prevent inputs of pollution. Pollution arising from contaminated ground can be prevented by isolating the ground from water. Such isolation may be achieved by the installation of impermeable covering systems, cut-off drains to divert surface water before it reaches the contaminated material (see Section 13.6.3), and impermeable vertical barriers to prevent lateral movement of groundwater into or out of contaminated ground (see Section 11.3). Removal of contaminated materials from adjacent to watercourses may be necessary to facilitate this isolation from water. Treatment of contaminated ground to remove contaminants or to immobilise them (see Chapter 11) will also have the effect of reducing inputs to vulnerable surface watercourses, provided these treatments reduce the concentration of water-soluble contaminants.

Erosion of particulate matter directly into watercourses can be an important pollution input. Reclamation should minimise erosion through landform design and, where appropriate, the establishment of a vegetation cover.

Reclamation works constitute a particularly vulnerable period as there is inevitably some disturbance of materials on site which can lead to increased pollution of surface waters. In particular there may be 'wash-out' of suspended matter into watercourses at times of high rainfall. On vulnerable sites the input of surface run-off to working areas should be minimised by the installation of cut-off drainage prior to work commencing. Rainwater incident on the working areas should be drained to a treatment area and such treatment should, as a minimum, provide for the removal of suspended solids. Oil interceptors may also be required and further treatment by biological or chemical means may be necessary depending on the nature of the pollution and on the receiving water. It may be possible to discharge water to a sewage works or industrial water treatment works and so reduce the requirement for on-site treatment.

In some situations the pollution of surface water will arise even after reclamation. This is particularly so with acid mine drainage where pollution is generated by an ongoing natural process. In these situations long-term treatment of the water may be required to meet discharge standards set by regulatory authorities. The control and treatment of acid mine drainage is discussed in Section 12.5.3.

12.5.2 Groundwater

Treatment of polluted groundwater may be necessary for the following reasons:

- groundwater pollution is causing pollution of nearby surface waters;
- the groundwater is abstracted for some purpose which is affected by the pollution;
- to enable development of land;
- to prevent further dispersal of pollution;
- to improve and protect water resources which may be used at some time in the future.

Movement of groundwater is generally slow, with long migration times for pollutants. Therefore, unlike surface waters, reduction in pollution does not rapidly follow the cessation of pollution inputs. Clean-up of groundwater can be difficult and expensive, and is generally achieved by extraction of groundwater via pumping from wells, followed by treatment of the extracted water. Treated water may be re-injected into the ground or discharged to surface water.

Such pump and treat operations require extraction of a volume of groundwater several times greater than the volume of contaminated groundwater. This is because movement of contaminants is retarded, in comparison with movement of groundwater, by sorption onto soil particles. An understanding of the sorption and desorption characteristics of the contaminants in question is thus vital to enable prediction of the volume of groundwater requiring extraction. Further complications arise from the non-uniformity of contaminant distribution and the heterogeneity of the aquifer.[152] This will result in the dispersal of contaminants during groundwater extraction, increasing the effective retardation factor (the ratio of the velocity of groundwater to the apparent velocity of the contaminant). Insufficient consideration of this factor has lead to groundwater clean-up operations taking far longer than originally predicted.[142]

The volume of contaminated water extracted during groundwater pumping can be reduced by measures which minimise the extraction of groundwater from unpolluted strata, or which keep contaminated and uncontaminated groundwater separate. One such method is the dual pumping system developed in Denmark.[18] In this system groundwater is removed simultaneously from the contaminant plume and from underlying uncontaminated groundwater. Simultaneous extraction from pumps placed at the top and the bottom of a well which penetrates through the contaminant plume to uncontaminated groundwater, creates a water divide within the well, as shown in Figure 12.1. Water above the divide migrates towards the top pump and water below the divide towards the bottom pump. Variations in the rates of pumping can be used to adjust the position of the divide. In this way uncontaminated groundwater can be extracted separately and dilution of the contaminated groundwater avoided. If there is uncontaminated water above as well as below the contaminant plume, three pumps will be necessary.

Systems using two pumps in one well have also been used to remove free oil product from the surface of the water table. A deep pump removes groundwater to create a cone of depression at the well into which oil migrates from the surrounding ground, see Figure 12.2. This oil is then collected by a skimmer pump.[26] A disadvantage of this method is that oil migrating into the cone of depression may become immobilised by sorption onto soil particles. This is frequently referred to as 'smearing', and creates a reservoir of hydrocarbon contamination which can be a source of long-term groundwater pollution. A simpler system pumping oil and water from the surface of the groundwater only may be more effective. Extracted oil and water can then be separated or treated by a range of techniques.

A method which overcomes the problem of smearing is vacuum extraction; a vacuum is used to extract volatile contaminants from the unsaturated, vadose zone of the soil and to remove water from the saturated zone (dual vacuum extraction). Reduction in the groundwater level forms a cone of depression but volatilisation of hydrocarbons by the vacuum applied to the vadose zone reduces their adherence to the soil particles, giving more effective removal[154] This technique is discussed in Section 11.8.4.

Extracted groundwater is generally treated by the physio-chemical methods used for the treatment of industrial waste waters, though concentrations in groundwater are typically far less than in industrial effluents. Treatment systems used commonly include oil separators, carbon filters, coagulation/flocculation, chemical oxidation and aeration. Biological treatment systems are also used. Aeration systems, such as air stripping,[45] to remove volatile compounds, may require some treatment of exhaust gases to control air pollution. The major problem in design

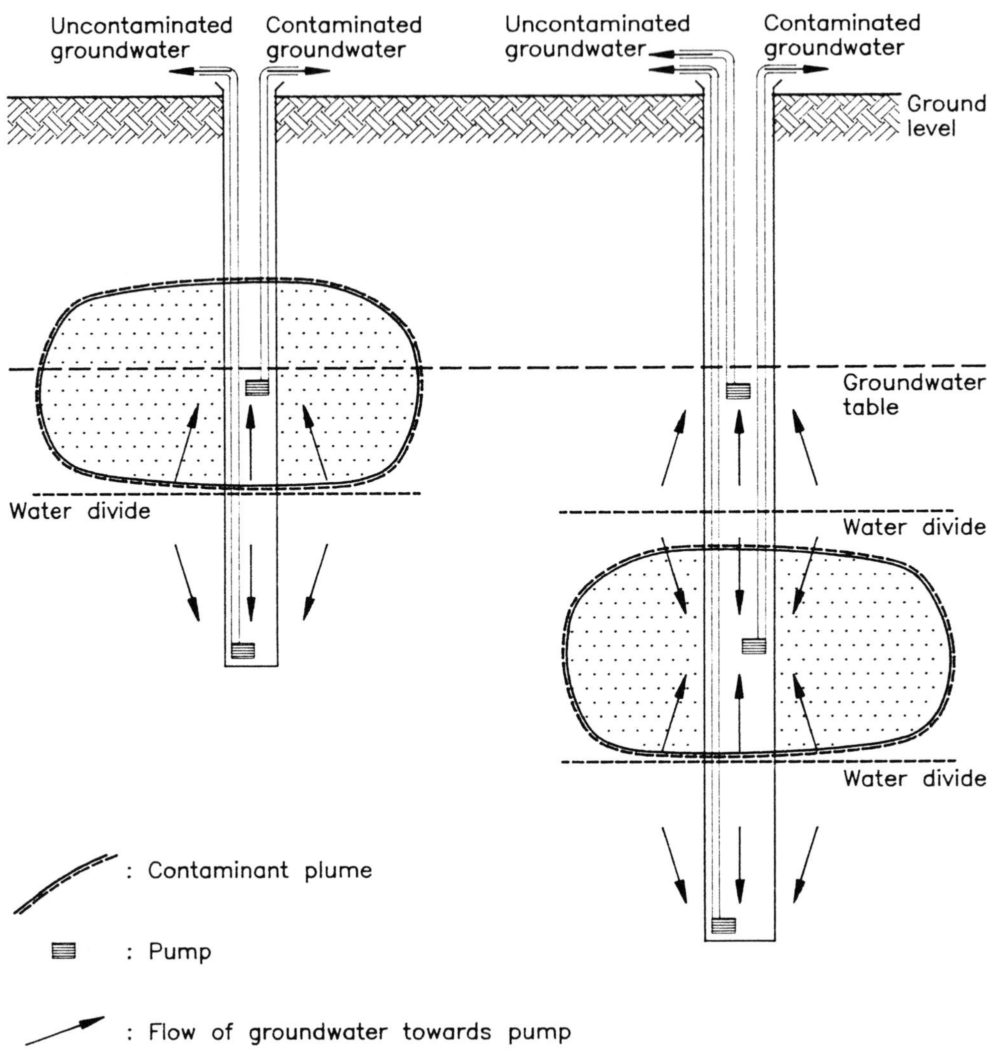

Figure 12.1: Dual pumping system

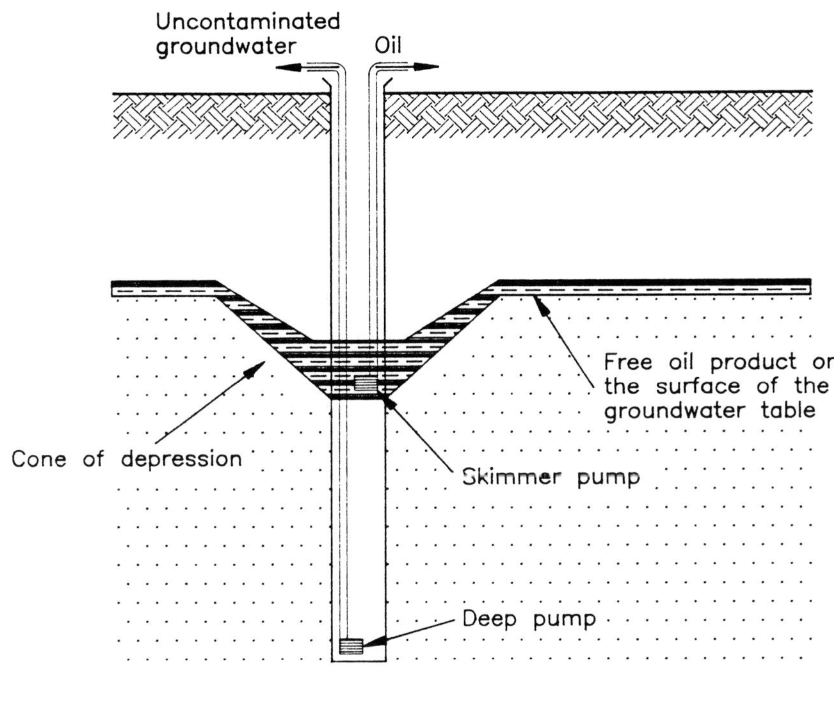

Figure 12.2: Removal of free oil product from the surface of a groundwater table

of treatment systems for extracted groundwater arises from the difficulties in predicting the quality and quantity of the water.[150]

12.5.3 Control and treatment of AMD

Introduction

There are a wide variety of potential control and treatment methods for AMD. Some of these methods are outlined in Table 12.3. Methods act by:

- preventing the oxidation of pyrite;
- preventing the formation of AMD;
- collecting and treating the AMD.

Prevention of pyritic oxidation

These methods act by:

- restricting the supply of oxygen;
- restricting the supply of water;
- inhibiting the iron-oxidising bacteria that catalyse the oxidation of pyrite.

Oxygen and water are both reactants in the oxidation of pyrite, so restricting their supply will reduce the rate of oxidation.

Phosphate-rich minerals, such as apatite, have been used in laboratory and small-scale field trials.[266] These minerals coat pyrite particles with insoluble iron phosphates and thereby isolate the pyrite from water and oxygen. Large doses of phosphates are generally required. In one study, however, acid production was reduced by 96% at a dose of less than 3 parts per thousand.[137]

Table 12.3: Acid mine drainage control measures

	Principle
Prevention of pyrite oxidation	
Bactericides	Inhibit the activity of *Thiobacillus ferrooxidans* and other iron-oxidising bacteria
Phosphates*	Coat pyrite particles with relatively insoluble iron phosphates thus isolating them from water and oxygen.
Soil or bog covers	Isolate waste from oxygen
Submergence of waste	Isolate waste from oxygen
Inundation of mines	Isolate waste from oxygen
Alkaline treatment	Raise pH resulting in metal precipitation and armouring of reactive surface of waste
Prevention of AMD formation	
'High and dry' waste placement	Isolate waste from ground and surface water and rainfall
Capping of waste	Prevent infiltration of rainfall
Stream/river sealing	Prevent infiltration of surface water into mines
Other civil engineering measures e.g. groundwater diversion, cut-off drainage	Prevent access of water to waste
Collection and treatment	
Alkaline addition	Raise pH resulting in metal precipitation
In-line system	Alkaline addition
Electrolysis *	Metal ions are separated from the water by the passage of an electric current through the AMD
Wetlands *	Cause metals to precipitate as oxides or sulphides and to be retained within the substrate material
Anoxic limestone drains	Raise pH and may cause metal precipitation

* Experimental or pilot-scale only

Submergence of spoil in water can be an effective means of reducing contact of the spoil with oxygen since the diffusion rate of oxygen in water is low.[25] Similarly, flooding of disused mines can significantly reduce pyrite oxidation within the mine. However, the cost of constructing impoundment structures to enable submergence of spoil is high and the potential environmental impact of placing mining waste in water bodies is great. Old wastes, which may contain large amounts of the products of pyrite oxidation, may liberate these products when first inundated, producing extremely acid water.

Sealing of mine entrances and allowing old workings to flood is common practice in many mining areas. When a disused coal mine in Pennsylvania was allowed to flood, acidity at the outfalls fell by some 75% and sulphate concentrations were halved, once the acidic mine water in the mine prior to flooding had been flushed out.[136] Use of this method should, however, be employed with caution and only after a detailed hydrological and geological survey to ensure that flooding the mine will not give rise to new issues of acidic water.

Covering of colliery spoil with organic matter, such as peat substitutes, then flooding to form a wetland, can reduce oxygen ingress to spoil since the covers themselves have a high oxygen demand.

Treatment of spoil with alkaline materials, such as lime, can be used to raise pH and thereby reduce pyrite oxidation through:

- precipitation of ferric ions (Fe^{3+}), which oxidise pyrite, removing them from solution;
- formation of an 'armour' of precipitated metals, principally ferric hydroxide, on the reactive surfaces of the pyrite preventing further reaction;
- inhibition of iron-oxidising bacteria, such as *Thiobacillus ferrooxidans*, thus preventing production of ferric (Fe^{3+}) ions from ferrous (Fe^{2+}) ions.

Large quantities of lime are generally required to treat pyritic spoils (see Section 14.4.5).

Organic materials, such as sewage sludge, are also reported to inhibit pyrite oxidation in colliery spoil, enabling a vegetation cover to become established.[162] Such organic materials may react by chelating ferrous and ferric ions, so they are unavailable for the reactions involved in producing acidity, and by providing buffering capacity against falls in pH.[195]

Bactericides can be used to inhibit *Thiobacillus ferrooxidans*, the bacterium primarily responsible for catalysing the oxidation of iron in pyritic ores. *Thiobacillus ferrooxidans* increase the reaction rate by up to 1,000,000 fold.[226] Anionic surfactants such as sodium lauryl sulphate (SLS), which disrupt the bacterial cell wall, have been much used, and food preservatives such as benzoic acid and sorbic acid have been investigated as their release into the environment may be more acceptable than that of surfactants.[195]

In full-scale trials, a single application of 0.25% SLS at about 5,000 litres per hectare on to mine waste was found to greatly reduce acidity and sulphate and iron concentrations in surface water for a period of 3-6 months.[181] Other laboratory and pilot-scale tests of SLS, sodium benzoate and potassium sorbate at concentrations of 15-40 ppm achieved reductions of iron and sulphate concentrations of above 70%.[218]

Slow-release forms of bactericide compounds are commercially available. Anticipated release lifetimes greater than 5 years are proposed for some products.[137] In the absence of these slow release forms, bactericides may need to be applied several times a year.[136] Although primarily applied to solid waste, commercial slow release bactericides are available for use in treatment ponds.[69] Removal figures of greater than 90% for iron, manganese and aluminium have been reported.

Prevention of AMD formation

Methods aiming to prevent formation of acidic drainage do so by isolating the pyrite-containing material from water. As water and the oxygen dissolved in it are necessary for the oxidation of pyrite, these methods will also inhibit pyrite oxidation. However, their main effect is to prevent the dissolution of the products of oxidation in flows of water, *i.e.* the formation of acidic drainage water, allowing instead the accumulation of the products of oxidation on the surface of pyritic materials.

An integrated program of control measures involves:

- diversion of surface water;
- interception of groundwater;
- prevention of rainwater infiltration;
- waste placement methods.

Surface water diversion can be achieved through the use of ditches or berms. These structures can be effective in the long-term, given sufficient maintenance. Surface water diversion measures, coupled with revegetation of the waste can reduce acid production by approximately 50%.[137]

'High and dry' placement of waste to isolate it from groundwater can be very successful in reducing AMD. For example, spoil can be placed on a porous pad with drains on the uphill side of the waste, to allow groundwater to bypass the waste.[161]

Capping of waste, to reduce or prevent rainwater infiltration, can be very effective, especially when combined with drainage methods to divert surface and groundwater (see Section 11.4). Cover systems use materials such as top soil, clays (including bentonite), tertiary-treated sewage sludge or compost. Synthetic materials are also used, including asphalts and tars, concretes, cements and plastic liners. Materials such as plastic films or liners act as impermeable barriers to water movement, but are easily damaged, can deteriorate and are expensive. Swelling clays, such as bentonite, are prone to cracking under dry conditions, allowing water to percolate into the underlying spoil through the cracks.[195]

All types of capping systems require long-term maintenance and the need for future replacement must be considered as part of a longer-term strategy.

Stream and river sealing can be used if hydrological investigation shows that infiltration of surface water from watercourses to an underground mine is a significant source of acid mine water. Watercourses can be redirected, culverted, piped or the stream bed sealed, and any deep faults can be grouted. An experimental grouting procedure has been used which involves injecting expanded polyurethane grout into the stream bed.[1] Disturbance to the stream bed is far less than would have been involved if the bed had been concreted. Flow losses into mine workings were successfully reduced in this way.

Collection and treatment of AMD

Principal methods of AMD treatment include:

- alkaline addition;
- constructed wetlands;
- anoxic limestone drains.

Alkaline addition involves the addition of chemicals such as sodium hydroxide, sodium carbonate, calcium hydroxide (lime), calcium carbonate or ammonia to AMD in a suitable impoundment. At the raised pH ferric ions precipitate out of solution as hydrated ferric oxide (ochre). Any ferrous ions dissolved in the AMD must be oxidised to ferric ions before this precipitation can occur, so oxidation, by mechanical aeration or injection of air or oxygen is often necessary.

Alkaline addition requires a long-term expenditure on chemicals and day-to-day supervision. Periodic collection of the precipitated metals is necessary and the resulting metal-rich sludge must be disposed of in an approved manner.

Alkaline addition has been used extensively at both active and disused mines but may often be a prohibitively expensive form of treatment. A pilot-scale 'in line system' (ILS) for treating mine drainage has been developed in the USA.[2] This system injects alkali (sodium hydroxide) into the flow of AMD, which is then passed through a jet pump and a static mixer. The jet pump entrains air and the static mixer provides effective oxidation of the AMD by forcing the air bubbles into solution. The degree of aeration achieved has been found to reduce alkali requirements.

A chemical treatment using barium sulphide to remove sulphate from AMD and so raise the pH has been developed in South Africa.[37] Electrolytic treatments are also being developed as an alternative to chemical neutralisation.[266] In the electrolysis cell hydrogen ions are reduced to form hydrogen gas at one electrode, thus raising pH, whilst iron and manganese are precipitated at the other electrode.

Constructed wetlands are increasingly being used to treat AMD from coal mines. There are over 400 such systems operating in the USA[136] and a few such systems operating in the UK. They have also proved effective in treating municipal sewage, metal mine drainage[202] and may be suitable to treat landfill leachate.[230]

In a constructed wetland, water passes through an area planted with wetland plants. These plants are typically emergent species such as reedmace (*Typha latifolia*) or common reed (*Phragmites australis*). The plants are rooted in a substrate which may be soil, gravel or organic-rich material such as spent mushroom compost, whilst their leaves and stalks are projected above the surface of the water which covers the substrate. Water flow may occur through the substrate, over the surface, or both.

Figure 12.3 shows some of the metal removal processes thought to occur in a wetland treating AMD. It is generally considered that the most important of these are metal oxidation/hydrolysis and sulphate-reduction. The first of these captures metals as oxides/hydroxides on and within the substrate, the second, as sulphides within the substrate.

Constructed wetlands offer several important advantages over other treatment systems, including:

- relatively low capital costs;
- little day-to-day supervision required;
- the provision of additional environmental and wildlife benefits.

However, substrate lifespan and disposal options have not yet been fully researched and design parameters are still under development. Also, a large area may be required to treat the flows of AMD commonly encountered.

A small wetland area for the treatment of AMD from a reclaimed colliery spoil heap has been constructed in Lothian Region, Scotland[62] and is shown in Photograph 12.1. The system consists of three lagoons, lined with clay, with top soil as the substrate and planted with *Phragmites australis*. Drainage water from the spoil is pretreated by anoxic limestone drains to raise the pH from 2.5 to 3.5. The wetland has been successful at reducing pollution of the adjacent watercourse.

Anoxic limestone drains (ALDs) are buried trenches of limestone. Under aerobic *i.e.* aerated, conditions the passage of AMD across limestone results in the 'armouring' of the reactive surfaces of the limestone. This armouring means that only a limited amount of the limestone can dissolve. Provided anoxic conditions within the ALD can be maintained, and AMD can be intercepted whilst still anoxic, the limestone of the drain can react with the AMD to generate alkalinity and raise pH without such armouring.[220]

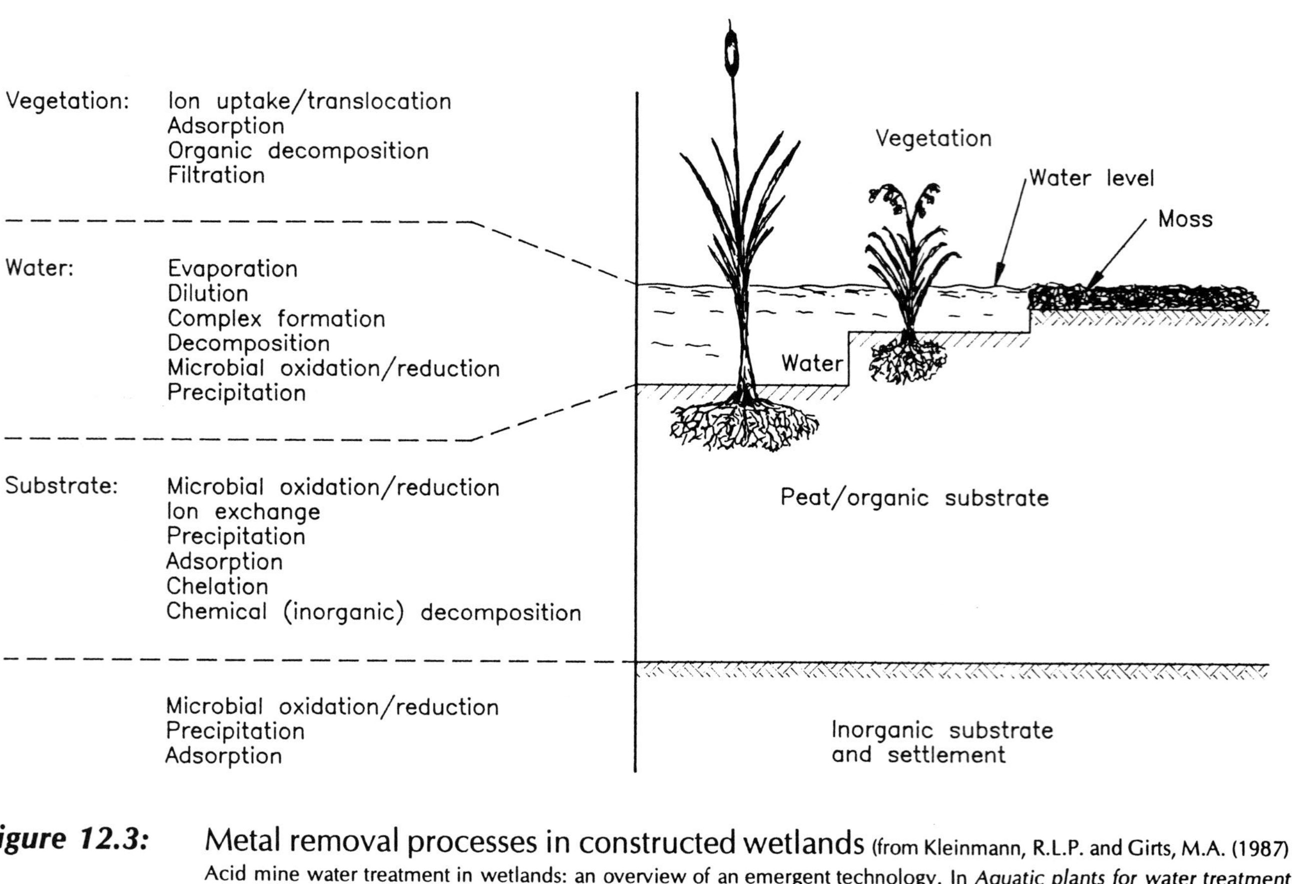

Figure 12.3: Metal removal processes in constructed wetlands (from Kleinmann, R.L.P. and Girts, M.A. (1987) Acid mine water treatment in wetlands: an overview of an emergent technology. In *Aquatic plants for water treatment and resource recovery* (Orlando, FL. 1986), edited by K.R. Reddy and W.H. Smith, 255-261, Orlando, FL.:Magnolia Publishing Inc.)

Photograph 12.1: The wetland treatment area at Gilmerton Bing, Lothian, UK. Water draining the colliery spoil enters the lagoon in the foreground from the anoxic limestone drains. *Phragmites australis* can be seen in the third lagoon of the system (source: Landscape Development).

Increased alkalinity favours metal oxidation, hydrolysis reactions and metal removal by precipitation, and therefore ALDs should discharge into a settling lagoon or constructed wetland in which the metal precipitates can be removed by settlement processes.

ALDs were originally developed as a pre-treatment for AMD entering constructed wetlands, but certain flows of AMD may lend themselves to treatment using only a combination of ALDs and settling lagoons.

12.6 Legislation and standards

12.6.1 Introduction

This section considers legislation on water at the European Community level. Such legislation is in the form of directives, which must be implemented in each Member State through its own laws and regulations. Directives concerning water pollution of relevance to land reclamation fall into two groups: directives concerned with pollution by particularly dangerous substances, and directives which lay down standards for water put to certain uses.[85, 89, 98]

12.6.2 Dangerous substances directives

These directives identify substances, or types of substances, which are particularly hazardous in the water environment and require Member States to control discharges of those substances with the objective of reducing or eliminating the pollution associated with them. Many of the substances covered will be found at steel and coal sites, so these directives have relevance to the control of pollution from these sites.

The primary directive concerning the discharge of "certain dangerous substances" is Council Directive 76/464/EEC of 4 May 1976 on pollution caused by certain dangerous substances discharged into the aquatic environment of the Community. This directive, commonly referred to as the 'dangerous substances directive', is a framework directive, which sets out general principles of control. Various daughter directives then set out emission limit values, quality objectives and reference methods for measurement and monitoring procedures for particular substances. 'Emission limit values' stipulate the maximum amounts of the substance which can be released from particular industrial processes, per unit of production. 'Quality objectives' define maximum concentrations within receiving waters. Quality objectives may be different for fresh and saline waters, but they do not vary according to the use of the water. The protection of groundwater is dealt with in the Groundwater Directive;

Council Directive 80/68/EEC of 17 December 1979 on the protection of groundwater against pollution caused by certain dangerous substances.

Both directives contain two lists of families and groups of substances in an Annex. The lists in the two directives are similar, but not identical (see Boxes 12.2 to 12.5). Directive 76/464/EEC requires Member States to take steps to eliminate pollution by List I substances and reduce pollution by List II substances. In the groundwater directive Member States are required to prevent introduction of List I substances into groundwater and limit introduction of List II substances so as to prevent pollution. In both directives the definition of pollution takes into account the harm caused by a substance, rather than simply the presence of a substance.

Box 12.2: Dangerous Substances Directive 76/464/EEC. List I

List I of families and groups of substances

List I contains certain individual substances which belong to the following families and groups of substances, selected mainly on the basis of their toxicity, persistence and bioaccumulation, with the exception of those which are biologically harmless or which are rapidly converted into substances which are biologically harmless

1. organohalogen compounds and substances which may form such compounds in the aquatic environment,
2. organophosphorus compounds,
3. organotin compounds,
4. substances which have been proven to possess carcinogenic properties in or via the aquatic environment,
5. mercury and its compounds,
6. cadmium and its compounds,
7. persistent mineral oils and hydrocarbons of petroleum origin, and for the purposes of implementing Articles 2, 8, 9, 14 of this Directive:
8. persistent synthetic substances which may float, remain in suspension or sink and which may interfere with any use of the waters.

Box 12.3: Dangerous Substances Directive 76/464/EEC. List II

List II of families and groups of substances. List II contains:

- substances belonging to the families and groups of substances in List I for which the limit values referred to in Article 6 of the Directive have not been determined,
- certain individual substances and categories of substances belonging to the families and groups of substances listed below, and which have a deleterious effect on the aquatic environment, which can, however, be confined to a given area and which depend on the characteristics and location of the water into which they are discharged.

Families and groups of substances referred to in the second indent

1. The following metalloids and metals and their compounds:

1. zinc	6. selenium	11. tin	16. vanadium
2. copper	7. arsenic	12. barium	17. cobalt
3. nickel	8. antimony	13. beryllium	18. thallium
4. chromium	9. molybdenum	14. boron	19. tellurium
5. lead	10. titanium	15. uranium	20. silver

2. Biocides and their derivatives not appearing in List 1.

3. Substances which have a deleterious effect on the taste and/or smell of the products for humans consumption derived from the aquatic environment, and compounds liable to give rise to such substances in water.

4. Toxic or persistent organic compounds of silicon, and substances which may give rise to such compounds in water, excluding those which are biologically harmless or are rapidly converted in water into harmless substances.

5. Inorganic compounds of phosphorus and elemental phosphorus.

6. Non-persistent mineral oils and hydrocarbons of petroleum origin.

7. Cyanides, fluorides.

8. Substances which have an adverse effect on the oxygen balance, particularly: ammonia, nitrites.

Box 12.4: Groundwater Directive 80/68/EEC. List I

List I contains the individual substances which belong to the families and groups of substances enumerated below, with the exception of those which are considered inappropriate to List I on the basis of a low risk of toxicity, persistence and bioaccumulation.

Such substances which with regard to toxicity, persistence and bioaccumulation are appropriate to List II are to be classed in List II.

1. Organohalogen compounds and substances which may form such compounds in the aquatic environment.

2. Organophosphorus compounds.

3. Organotin compounds.

4. Substances which possess carcinogenic, mutagenic, or teratogenic properties in or via the aquatic environment.

5. Mercury and its compounds.

6. Cadmium and its compounds.

7. Mineral oils and hydrocarbons

8. Cyanides.

An important difference in the approach taken by the two directives is the relationship between List I and List II. In the groundwater directive substances belonging to the groups in List I are to be treated as List I substances unless they pose a low risk of toxicity, persistence and bioaccumulation. In the dangerous substances directive substances are to be selected for List I status, from the families or groups of substances given in List I, on the basis of their toxicity, persistence and bioaccumulation. Substances do not have List I status unless emission limit values and quality objectives have been laid down by daughter

Box 12.5: Groundwater Directive 80/68/EEC. List II

List II contains the individual substances and the categories of substances belonging to the families and groups of substances listed below which could have a harmful effect on groundwater.

1. The following metalloids and metals and their compounds:

1. Zinc	11. Tin
2. Copper	12. Barium
3. Nickel	13. Beryllium
4. Chromium	14. Boron
5. Lead	15. Uranium
6. Selenium	16. Vanadium
7. Arsenic	17. Cobalt
8. Antimony	18. Thallium
9. Molybdenum	19. Tellurium
10. Titanium	20. Silver.

2. Biocides and their derivatives not appearing in List I.

3. Substances which have a deleterious effect on the taste and/or odour of groundwater, and compounds liable to cause the formation of such substances in such water and to render it unfit for human consumption.

4. Toxic or persistent organic compounds of silicon, and substances which may cause the formation of such compounds in water, excluding those which are biologically harmless or are rapidly converted in water into harmless substances.

5. Inorganic compounds of phosphorus and elemental phosphorus.

6. Fluorides.

7. Ammonia and nitrites.

directives. Until this happens substances belonging to the families and groups of substances given in List I of the Annex are to be treated as having List II status.

Daughter directives to 76/464/EEC are shown in Box 12.6 and the substances with List I status as a result of those directives in Box 12.7.

Box 12.6: Daughter Directives to 76/464/EEC

82/176/EEC	On limit values and quality objectives for mercury discharges by the chlor-alkali electrolysis industry.
83/513/EEC	On limit values and quality objectives for cadmium discharges.
84/156/EEC	On limit values and quality objectives for mercury discharges by sectors other than the chlor-alkali electrolysis industry.
84/491/EEC	On limit values and quality objectives for discharges of hexachlorocyclohexane.
86/280/EEC	On limit values and quality objectives for discharges of certain dangerous substances included in List I of the Annex to Directive 76/464/EEC.
88/347/EEC	Amending Annex II to Directive 86/280/EEC on limit values and quality objectives for discharges of certain dangerous substances included in List I of the Annex to Directive 76/464/EEC.
90/415/EEC	Amending Annex II to Directive 86/280/EEC on limit values and quality objectives for discharges of certain dangerous substances included in List I of the Annex to Directive 76/464/EEC.

Box 12.7: List I substances

Substances with List I status under Directive 76/464/EEC as a result of the daughter directives shown in Box 12.6.

	Directive Number	Date of entry into force
Aldrin, Dieldrin, Endrin and Isodrin	88/347	1/1/89
Cadmium and its compounds	83/513	1/4/86
Carbon tetrachloride	86/280	1/1/88
Chloroform	88/347	1/1/90
DDT (all isomers)	86/280	1/1/88
Hexachlorobenzene	88/347	1/1/90
Hexachlorobutadiene	88/347	1/1/90
Hexachlorocyclohexane (all isomers)	84/491	1/4/86
Mercury and its compounds:		
Chlor-alkali	82/176	1/7/83
Other sectors	84/156	12/3/86
Pentachlorophenol	86/280	1/1/88
1,2-Dichloroethane	90/415	1/1/93
Trichloroethylene	90/415	1/1/93
Tetrachloroethylene	90/415	1/1/93
Trichlorobenzene (1,2,4 & technical blends.)	90/415	1/1/93

There are around 4,500 potential List I substances used within the European Community. In 1982 the Commission selected 129 of these substances for priority inclusion in List I. This priority list now contains 109 substances, including organochlorines, PCBs, aromatic solvents and several pesticides, though it is still known as the 'List of 129'.[85]

For List I substances emission limit values and quality objectives are set on a Community-wide basis by the daughter directives. Implementation by Member States is through a system of prior authorisations for discharges. In contrast, control of List II substances is at a national level. Member States are required to implement pollution reduction programmes for substances belonging to the families and groups of substances in List II which have a deleterious effect on the environment. The selection of such substances, and setting of appropriate quality objectives is a matter for Member States.

Under the Groundwater Directive procedures must be established for prior investigation followed by authorisation or prohibition of all discharges, disposal or tipping of List II substances and indirect discharges of List I substances. Indirect discharges involve percolation through ground or subsoil before introduction into groundwater. Direct discharges of List I substances are prohibited unless investigation has shown that the receiving groundwater is permanently unsuitable for other uses, as may be the case in long established industrial areas. Discharges of trace amounts of List I and List II substances, which will not cause any deterioration in groundwater quality, may be permitted. The Groundwater Directive seeks to conserve groundwater resources and applies to all groundwater, irrespective of current use.

12.6.3 Directives specifying water quality standards

There are five directives which specify standards for water for various uses, or for the support of various types of aquatic life. These directives are listed in Box 12.8. The relevance of these directives to reclamation is that they provide standards for evaluating whether or not water is polluted, and thus whether it requires remedial treatment on site.

All directives except 80/778/EEC, on water for human consumption, specify two types of values: mandatory, 'imperative' (I) values, and non-binding, 'guide' G values. The directives specify sampling and analytical procedures and the percentage of samples which must conform to each

Box 12.8: Directives specifying standards for water.

75/440/EEC concerning the quality required of surface water intended for the abstraction of drinking water in the Member States.

76/160/EEC concerning the quality of bathing water.

78/659/EEC on the quality of fresh waters needing protection or improvement in order to support fish life.

79/923/EEC on the quality required of shellfish waters.

80/778/EEC relating to the quality of water intended for human consumption.

value. The Surface Water for Drinking Directive, 75/440/EEC specifies G and I values for three categories of surface water (A1, A2, A3) and the type of treatment required for each category. Directive 78/659/EEC on freshwater for fish applies to surface waters designated by Member States as needing protection or improvement in order to support fish life. There are separate G and I values for waters supporting salmonid and cyprinid fish. Member states are required to implement pollution reduction programmes to ensure that waters conform to I values within five years of their designation. The G and I values given in these directives, which are relevant to land reclamation, are shown in Table 12.4 and 12.5.

In contrast to the other directives the standards in directive 80/778/EEC, on water for human consumption, are referred to as guide levels (GL) and maximum admissible concentrations (MAC). These have been defined for a wide range of parameters and are often used when interpreting results of groundwater analyses. GL and MAC values for parameters of relevance to land reclamation are shown in Table 12.7.

12.6.4 Other standards

In addition to Community-wide standards laid down by EC directives individual Member States may have set quality objectives for List II substances under the Dangerous Substances Directive, 76/464/EEC.

Standards for contaminated land may also include standards for water quality. The Dutch list (see Section 2.6.5) gives the most complete coverage in terms of range of parameters for groundwater quality. A study for the Department of the Environment in the UK on the redevelopment of gas works and similar sites recommends threshold trigger values in water samples for phenols, free cyanide, thiocyanate and sulphide.[86] These trigger values are shown in Table 12.8.

Table 12.4: G and I values specified in Directive 75/440/EEC on Surface water for drinking.

Parameter	Unit	A1 G	A1 I	A2 G	A2 I	A3 G	A3 I
pH		6.5-8.5		5.5-9		5.5-9	
Total suspended solids	mg/l SS	25					
Conductivity	μS/cm^{-1} at 20°C	1000		1000		1000	
Nitrates	mg/l NO_3	25	50*		50*		50*
Fluorides	mg/l F	0.7-1	1.5	0.7-1.7		0.7-1.7	
Dissolved iron	mg/l Fe	0.1	0.3	1	2	1	
Manganese	mg/l Mn	0.05		0.1		1	
Copper	mg/l Cu	0.02	0.05	0.05		1	
Zinc	mg/l Zn	0.5	3	1	5	1	5
Boron	mg/l B	1		1		1	
Arsenic	mg/l As	0.01	0.05		0.05	0.05	0.1
Cadmium	mg/l Cd	0.001	0.005	0.001	0.005	0.001	0.005
Total chromium	mg/l Cr		0.05		0.05		0.05
Lead	mg/l Pb		0.05		0.05		0.05
Selenium	mg/l Se		0.01		0.01		0.01
Mercury	mg/l Hg	0.0005	0.001	0.0005	0.001	0.0005	0.001
Barium	mg/l Ba		0.1		1		1
Cyanide	mg/l CN		0.05		0.05		0.05
Sulphates	mg/l SO_4	150	250	150	250*	150	250*

Categories of treatment :

A1 simple physical treatment and disinfection
A2 physical treatment, chemical treatment and disinfection
A3 intensive physical and chemical treatment, extended treatment and disinfection

Table 12.4 continued:

Parameter	Unit	A1 G	A1 I	A2 G	A2 I	A3 G	A3 I
Chlorides	mg/l Cl	200		200		200	
Phosphates	mg/l P_2O_5	0.4		0.7		0.7	
Phenols (phenol index)	mg/l C_6H_5OH		0.001	0.001	0.005	0.01	0.1
Dissolved or emulsified hydrocarbons (after extraction by petroleum ether)	mg/l		0.05		0.2	0.5	1
PAH	mg/l		0.0002		0.0002		1.0001
Total pesticides †	mg/l		0.001		0.0025		0.005
COD	mg/l O_2					30	
Dissolved oxygen saturation rate	% O_2	>70		<50		<30	
BOD_5	mg/l O_2	<3		<5		<7	
Nitrogen by Kjeldahl method (except NO_3)	mg/l N	1		2		3	
Ammonia	mg/l NH_4	0.05		1	1.5	2	4 ‡
Substances extractable with chloroform	mg/l SEC	0.1		0.2		0.5	

PAH Polycyclic aromatic hydrocarbons
† Parathion, BHC, dieldrin
‡ exceptional climatic or geographical conditions
COD Chemical oxygen demand
BOD_5 Biochemical oxygen demand (at 20°C without nitrification)

Table 12.5: G and I values specified in Directive 78/659/EEC on freshwater for fish

Parameter	Salmonid waters		Cyprinid waters		Comments
	G	I	G	I	
pH		6 to 9		6 to 9	
Suspended solids (mg/l)	525		525		Values do not apply to suspended solids with harmful properties.
BOD_5 (mg/l)	53		56		
Nitrites (mg/l NO_2)	≤0.01		≤0.03		
Phenolic compounds (mg/l C_6H_5OH)		2		3	Inspection by taste
Petroleum hydrocarbons		3		3	Inspection by visual appearance and taste
Non-ionised ammonia (mg/l NH_4)	≤0.005	≤0.005	≤0.005	≤0.005	
Total ammonium (mg/l NH_4)	≤0.04	≤1	≤0.2	≤1	
Total residual chlorine (mg/l HOCl)		≤0.005		≤0.005	Higher values can be accepted if pH>6
Total zinc (mg/l Zn)		≤0.3		≤1.0	See Table 12.6
Dissolved copper (mg/l Cu)	≤0.04		≤0.04		See Table 12.6

Table 12.6: Variation of copper and zinc limit values with water hardness.

		Water hardness (mg/l $CaCO_3$)				
		10	50	100	300	500
Zinc I values:	salmonid waters (mg/l total Zn)	0.03	0.2	0.3		0.5
	cyprinid waters (mg/l total Zn)	0.3	0.7	1.0		2.0
Copper G values:	(mg/l dissolved Cu)	0.005	0.022	0.04	0.112	

Table 12.7: GL and MAC values specified in Directive 80/778/EEC on water for human consumption.

Parameters	Units	Guide level (GL)	MAC	Comments
Hydrogen ion concentration	pH unit	6.5 ≤ pH ≤ 8.5		- The water should not be aggressive. - The pH values do not apply to water in closed containers. - Maximum admissible value: 9.5.
Conductivity	$\mu S\ cm^{-1}$ at 20°C	400		- Corresponding to the mineralisation of the water. - Corresponding relativity values in ohms/cm: 2500.
Chlorides	mg/l Cl	25		- Approximation concentration above which effects might occur: 200mg/l
Sulphates	mg/l SO_4	25	250	
Calcium	mg/l Ca	100		
Magnesium	mg/l Mg	30	50	
Sodium	mg/l Na	20	175 150	- As from 1984 and with a percentile of 90. - As from 1987 and with a percentile of 80. These percentiles should be calculated over a reference period of three years.
Potassium	mg/l K	10	12	
Aluminium	mg/l Al	0.05	0.2	
Free carbon dioxide	mg/l CO_2			- The water should not be aggressive.
Nitrates	mg/l NO_3	25	50	
Nitrites	mg/l NO_2		0.1	
Ammonium	mg/l NH_4	0.05	0.5	
Kjeldahl Nitrogen †	mg/l N		1	
($K\ Mn\ O_4$) Oxidisability	mg/l O_2	2	5	- Measured when heated in acid medium.
Substances extractable in chloroform	mg/l dry residue	0.1		
Dissolved or emulsified hydro-carbons ‡	mg/l	10		
Phenols (phenol index)	C_6H_5OH mg/l		0.5	- Excluding natural phenols which do not react with chlorine.

Table 12.7 continued

Parameters	Units	Guide level (GL)	MAC	Comments
Boron	mg/l B		1000	
Iron	mg/l Fe	50	200	
Manganese	mg/l Mn	20	50	
Copper	mg/l Cu	100		- Above 3000 μg/l astringent taste, discolouration + corrosion may occur.
Zinc	mg/l Zn	100		
Phosphorus	mg/l P_2O_5	400	5000	
Fluoride	mg/l F			
	8-12 °C		1500	- MAC varies according to average temperature in geographical area concerned.
	25-30 °C		700	
Arsenic	mg/l As		50	
Cadmium	mg/l Cd		5	
Cyanides	mg/l CN		50	
Chromium	mg/l Cr		50	
Mercury	mg/l Hg		1	
Nickel	mg/l Ni		50	
Lead	mg/l Pb		50	
Antimony	mg/l Sb		10	
Selenium	mg/l Se		10	
Pesticides	mg/l			- Includes PCBs and PCTs
- individual compounds			0.1	
- total			0.5	
Other organochlorine	mg/l	1		- Haloform concentrations must be as low as possible.
compounds	mg/l			
PAH			0.2	

† excluding N in NO_2 and NO_3
‡ after extraction by petroleum ether.
PAH polyaromatic hydrocarbons

Table 12.8: Tentative trigger values for substances in samples of water from former gas works sites (from Environmental Resources Limited, 1987[86])

Parameter	Units	Threshold trigger value	Action trigger value
Phenols	mg/l	1	20
Free cyanide	mg/l	5	50
Thiocyanate	mg/l	5*	NL
Sulphide	mg/l	5	50

* If any water sample has a red coloration the threshold value is exceeded

NL No limit set

These trigger values apply to all site end-uses.

13 LANDFORM AND EARTHWORKS

Chapter contents

continued...

13 LANDFORM AND EARTHWORKS

13.1 Introduction

For the purpose of this book, the term 'landform' refers to any topographical feature created by Man. Such features will be either superimposed on or cut out of the general topography of the site.

'Earthworks' refers to all operations carried out on site which relate to excavation, removal and deposition of ground materials.

The landforms that are to be found in association with the coal and steel industries are many and various, and existing landforms may be a hindrance to reclamation and after-use or may present opportunities for re-use.

The types of landforms found on coal and steel sites can be divided into two groups:

- landforms associated with the functioning of a site as a colliery or as a steelworks *e.g.* embankments or cuttings for roads and railways, canals, hardstandings, stockpile areas and levelled ground occupied by the works;
- landforms associated with the disposal of wastes *e.g.* spoil tips and lagoons.

The first group tends to create fewer problems for reclamation and is more likely to have the potential for re-use although this depends on the requirements of the final landform design.

The landforms to be found on coal and steel sites, and their effect on reclamation design, are influenced by the general topography as follows:

- in narrow valleys, spoil heaps will be found occupying the valley bottom, with a profound effect on the character of the area (see Photograph 13.1);
- on flat land, spoil heaps are likely to be very prominent (see Photograph 13.2);
- in hilly areas, spoil may be tipped on a hill top or valley-side and stability may be the most important issue, but in many cases the tips are likely to be a dominant feature in the landscape (see Photograph 13.3);
- areas such as those of buildings, hardstanding and car parks, when located on sloping land, will have required terracing of the original ground;
- steelworks are normally found on flat land, near the coast or along a valley bottom, as they require a supply of water and because of the large size of the associated buildings and structures (see Photograph 13.4);
- for both collieries and steelworks, valley bottoms may be raised to give a flat working area, and for this purpose and also to provide space for tipping of waste, the original river or stream is often culverted (see Photograph 13.1);
- in low lying locations, typically estuaries and coastal plains, large areas of land have been raised using imported fill material to provide sites for steelworks.

A range of other factors which affect the types of landform found on abandoned coal and steel sites are given in Box 13.1.

Photograph 13.1: Former steelworks and colliery, Ebbw Vale, South Wales. The large tips to the left of centre comprise slags, whilst to the right of centre is colliery spoil. The river running along this valley was culverted to provide extra tipping space. This site was reclaimed and used for the 1992 Garden Festival, Wales, UK (source: Welsh Development Agency)

Photograph 13.2: Spoil heaps can be prominent features in the landscape, especially in flat areas. Their height can be lowered by half by removal of just one eighth of the volume of a conical tip (source: Uwe Ferber)

Photograph 13.3: Tipping of spoil in an elevated position. A tip breaking the skyline is especially conspicuous (source: Welsh Development Agency)

Photograph 13.4: Steelworks typically require a large area of level or terraced land, because of their size, and a good supply of water. This steelworks is already partially demolished (source: Richards, Moorehead and Laing Ltd)

Box 13.1: Additional factors affecting landform

Factors affecting the types of landform found on former coal and steel sites include the following:

- subsidence (see Chapter 3). Subsidence is not limited to the area at the surface where colliery buildings and spoil heaps are located. In flat areas, subsidence can lead to the formation of new water bodies (see Photograph 13.5) which it may be appropriate to incorporate into the after-use scheme (see Section 17.4.5). If not, ground levels will need to be raised and this may provide an opportunity for disposal of spoil materials;
- steel slag, which is often tipped in a molten state, tends to form large, solid masses, with very steep, even cliff-like landforms (see Photograph 13.6);
- underground mining for iron ore tends to produce spoil heaps and subsidence problems similar to those to be seen at coal mining sites;
- the basic shape of spoil heaps is determined by the method of tipping (see Figure 13.1);
- the form of spoil heaps is often dictated by the area of land available at the time of tipping and the volume to be placed on that land (see Photograph 13.7);
- tipping of colliery waste was often more wasteful of space in the past than in modern operations;
- modern tipping often results in a general raising of the ground level, rather than the creation of a distinct spoil heap;
- older tips are loose-tipped and therefore less consolidated than modern 'constructed' tips (see Chapter 6).

Photograph 13.5: Subsidence of the ground surface following deep mining can lead to the formation of new water bodies. The lake in the right foreground is used for fishing and other recreational activities (source: EPF)

Photograph 13.6: These 'bluffs' formed of steel slag tipped in a molten state, were retained and utilised as a dramatic feature in the 1992 Garden Festival, Wales, UK (source: Richards, Moorehead and Laing Ltd)

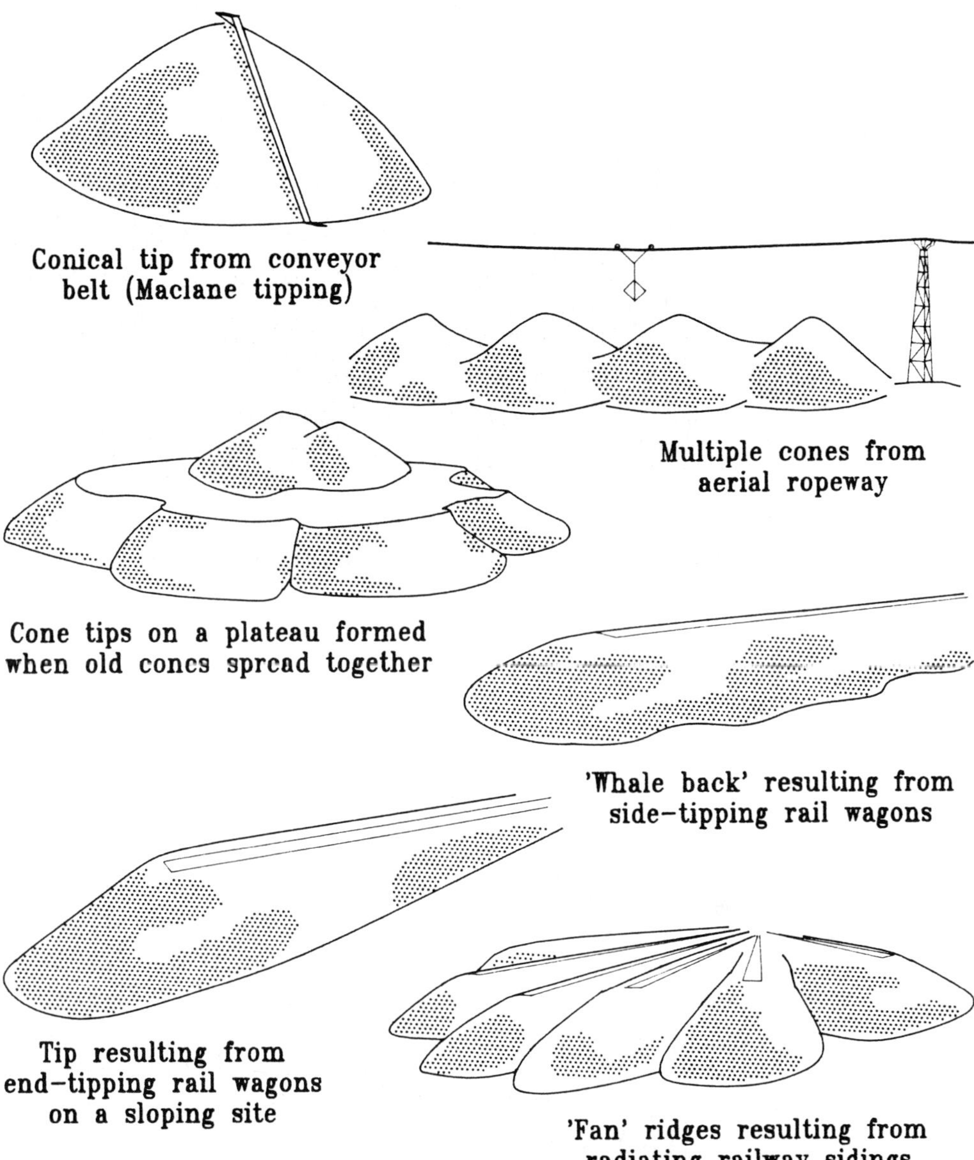

In recent years, tip construction by lorries, box–scrapers, bulldozers, *etc.* has frequently occured, creating further variety in tip shapes.

Figure 13.1: Shapes of tips resulting from the method of tipping (after Tandy, 1975[234])

Photograph 13.7: Shortage of tipping space can determine the form of spoil heaps and can be a severe constraint on reclamation options (source: Welsh Development Agency)

The issues arising from the various landforms found on derelict coal and steel sites can be summarised as follows:

- scale with respect to surrounding areas of land;
- actual size;
- location within the site;
- constraints on after-use caused by landform ('hard' after-uses are more specific in their landform requirements than 'soft' uses and therefore tips of waste material often provide fewer constraints to 'soft' after-uses);
- opportunities for re-use provided by the landform as it exists;
- re-use potential of waste materials;
- stability of waste heaps (see Chapter 6);
- materials with difficult handling and chemical characteristics (see Chapter 5 and 9);
- subsidence of land and possible flooding (see Chapter 3);

- visual aspects such as shape, prominence, colour and texture;
- vegetation cover;
- relationship with the surrounding landscape;
- waste heaps may have become an accepted part of the local scene and may even be regarded as an amenity by local people;
- undisturbed tips may be an important element in the history of the site, illustrating techniques used at various times;
- undisturbed tips can be of mineralogical and ecological value;
- expense - changing unsafe or inappropriate landforms usually involves considerable engineering works.

For any of the above reasons, there can be a need either to amend and remodel or to retain the landforms on a site. This chapter discusses various aspects of landform remodelling, covering the need for remodelling of landforms, the various approaches to and constraints on remodelling, the design process and technical aspects of remodelling.

13.2 The need for remodelling of landforms

Reshaping landforms is often necessary and the reasons for remodelling of landforms fall within the following categories:

- instability;
- presence of hazardous materials;
- visual considerations;
- land-use needs or opportunities;
- re-use potential of materials.

The reasons for landform remodelling are further discussed in Box 13.2.

Whilst there can be many good reasons why landforms should be changed, it is important that the design team critically examine any ideas for such changes. On abandoned sites, landforms often mellow with age, and if they are safe, can sometimes be retained as benign reminders of

Box 13.2: Reasons for remodelling of landforms

The following questions should be asked by the design team to determine whether remodelling of landforms is necessary:

- is there a hazard, or potential hazard, present on the site, such as unstable tips or abandoned lagoons?
- is the site visually intrusive? Can detrimental impact be reduced or eliminated by remodelling?
- is there a need to clear away or modify inappropriate landforms, or create new ones, to cater for the proposed after-use and the long-term management of the site?
- would remodelling add to the prospects for economic regeneration in the area, through improvement of the environment or provision of developable land?
- would there be benefit from creation of variety of landform on a site which is currently very bland?
- even if the shape of the landform is usable, does the ground offer adequate bearing capacity for the proposed after-use?
- could remodelling serve to reduce or eliminate erosion, by reducing gradients and assisting vegetation establishment?
- would remodelling offer the opportunity to create useful barriers between different land uses e.g. for screening, noise baffling or dust control?
- is the existing landform causing problems with drainage, perhaps leading to flooding, which could be resolved by remodelling?
- would remodelling offer the opportunity to remove, treat or encapsulate noxious materials, or even produce a disposal site for imported waste?
- is there saleable material on the site which requires tips to be reworked, and can this be used to generate income e.g. coal content of old colliery spoil tips, use of spoil as inert fill for road construction, as cover material on domestic waste sites or to create new land elsewhere; or seams of coal which are now economically workable by opencast mining but are buried under tips?

the industrial past. Indeed, there is an increasing recognition of the merit of retaining historical features on old industrial sites and, more significantly, that the site as a whole forms an historical landscape which may be of some considerable value.

Methods of assessing the historical value of a site are explained in Section 2.7. Box 13.3 expands on some of the arguments for retaining spoil heaps.

Abandoned tips may be perceived as aesthetically unacceptable features for three reasons:

- their historical connotations, being symbolic perhaps of a low standard of living and environmental degradation;
- their location amongst the surrounding scenery *e.g.* pleasant countryside or areas of housing;
- their visual characteristics *i.e.* bulk, height, colour and shape.

The first reason for objecting to the continuing presence of waste tips *i.e.* their historical connotations, is dependent on local and individual attitudes. Acceptance of the historical significance of industrial relics can reduce or even eliminate any objections based on the second and third reasons.

When considering the visual impact of abandoned industrial sites, it should be recognised that whilst waste tips may be the most noticeable features of the landscape, due to their sheer size and bulk, the associated derelict buildings, redundant equipment and other artifacts may be at least as significant in an assessment of the visual quality of a given area. These features can collectively form an untidy mixture of unwanted and irrelevant materials, with harsh, unnatural shapes and colours.

Box 13.3: Some arguments for the conservation of spoil heaps

There are a growing number of examples where removal or reclamation of spoil heaps has been resisted, either by specialist interest groups or the general public. The following examples illustrate this trend, relating to the conservation of heritage, wildlife and fossils.

1. In North-East England, a public inquiry was held in 1990 to examine proposals to reclaim the site of the former Kilton iron ore mine. After hearing evidence from two in favour and twenty four against reclamation, the Inspector wrote in his report that he considered that the tip was not an eyesore nor a disincentive to local economic development. It had a role to play in the development of tourism in the area as an historical feature in the landscape and he considered that the proposed scheme would be detrimental to the very wildlife it was claimed would benefit from it. In fact the Inspector found evidence that the shale tip at Kilton was now a uniquely recognisable industrial archaeological relic in East Cleveland, and concluded that the advantages of its retention far outweighed the benefits of its removal.

   The Inspector's recommendation that the scheme should not go ahead was accepted by the Government.

2. In the coal basin of Nord-Pas de Calais, in the North-East of France, over 250 colliery spoil heaps have been formed. Some have developed a vegetation cover of significant ecological value. Alpine species, which are adapted to poor, stoney soils, and species from the south of France which tolerate hot conditions, have colonised some of the tips.

   This ecological value has been recognised in some instances, for example on tip no. 36 at Noeux-les-Mines. There, a stairway was constructed to the summit, in association with the local naturalists society, to safeguard the rare species on the site. On the Pinchouvalle tip, one of the most imposing in the region, reclamation proposals were orientated towards exclusive use of the site for nature conservation and related activities, including habitat conservation, paths and observation points.

3. In an article entitled "Britain's tip heaps - an unlikely subject for conservation?"[6] the author highlights the fact that colliery spoil heaps can be "store-houses of irreplaceable scientific specimens of fossil plants, fish, amphibia and insects". The soft rocks of the coal measures are rarely seen in natural exposures and the mines are normally inaccessible, so the tips themselves can provide a unique resource for study. Examples are quoted where the fossil content of tips are of international importance and cases are cited where material was salvaged from tips, for subsequent study, before reclamation.

In summary, the overall objectives of landform remodelling are to ensure that:

- the land is safe;
- the land is usable;
- the land can be managed economically;
- the quality of the local environment is enhanced;
- economic regeneration is accelerated by creating new landforms which can accommodate a particular new development (see Photograph 13.8).

13.3 Approaches to remodelling landform

13.3.1 Introduction

Where the landform is to be remodelled well prepared reclamation proposals will address the constraints resulting from the past use of the sitc and will look to thc futurc, by preparing the way for the preferred after-use.

The relationship between landform and after-use is fundamental. The landforms that exist on a site may be suitable only for some uses and may therefore influence decisions on after-use. Alternatively, landforms may have to be amended to suit the required after-use (see Section 13.6.5 and Box 13.12). If final after-use is uncertain at the time that reclamation works are being designed, a landform which provides flexibility of use is then advantageous. Generally, however, a better solution will be produced if the landform can be designed for a specific use, which is then implemented.

A considerable amount of time may elapse between the completion of reclamation and final development of the after-use. The full benefits of remodelling will not be felt until development is complete and it is important in the achievement of local acceptance and continued support that local people realise why reclamation was carried out in a particular way.

Photograph 13.8: The location of this site close to housing and in a flat, open landscape suggests a need for reclamation, including remodelling of the tip. In locations like this, there is no obvious need for creation of new development land, but the environmental improvements that can be brought about by reclamation are important in the promotion of economic regeneration (source: EPF)

Both administrators and designers need to be aware that, because of changing economic, social and technical factors, justification can arise for the remodelling of land which has already been reclaimed.

In the remodelling of landforms, there are three broad options to be considered. These are:

- the retention of the landform;
- minimal interference;
- major earthworks.

The rationale behind the first option, where existing landforms are retained for historical or other reasons, was discussed in Section 13.2.

The possibility of minimal interference is an important option, since if satisfactory results can be achieved at lower cost than by major earthmoving operations, a greater area of reclamation is facilitated for the same expenditure. Relatively minor changes to a landform can substantially reduce its visual impact. Similarly an unsatisfactory landform can be disguised by planting with trees and shrubs rather than being reshaped. Where a suitable after-use can be identified which requires no reshaping of the landform, consideration needs to be given to whether remodelling to suit another more demanding after-use is really justified. This approach to reclamation is particularly applicable to amenity after-uses and is discussed in Box 13.4.

It is important always to consider a combination of options within one site. Coal and steel sites can be very large and very variable in both their problems and their potentials. An over-simplistic approach may be wasteful of resources or produce an inferior result. For example, major earthworks may be required in one part of a site but another part could be retained as it is, conserving mature vegetation, thus reducing overall costs and producing a more sensitive and interesting scheme. On the other hand, one must guard against over-complication which can make the implementation of reclamation proposals difficult and expensive.

13.3.2 Major earthworks

Major alterations to a landform may be undertaken:

- reworking of tips for saleable material which, in conjunction with good planning, can produce an improved landform as part of a reprocessing operation;
- removal of waste materials for use off site, the feasibility of which will depend on transport costs compared with the value of alternative materials;

- reshaping on site, with no export of material, by spreading out materials, creating new landforms or filling holes and depressions;
- reshaping on site in conjunction with opencast coal mining;
- reshaping in conjunction with the treatment of contaminated land;
- importation of waste materials from elsewhere, to be incorporated into a new landform on site.

Box 13.4: Reclamation without remodelling

Reclamation schemes involving large scale regrading of waste heaps or extensive civil engineering works inevitably involve considerable expenditure. This expenditure may be justified in order to remove hazards or to prepare land for development, but such expenditure is harder to justify for the provision of less intensive uses such as informal recreation, nature conservation and landscape improvement.

Revegetation techniques (see Chapter 14) have been developed which allow many derelict sites to be returned to the use of the community with a minimum of expenditure on earthworks. These techniques widen the choice of practical land reclamation approaches, and widen the range of objectives which are economically feasible.

These revegetation techniques form part of a 'low-cost philosophy' which may be summarised as five simple points:

- involve the public in the choice of after-use;
- match the after-use with the existing site features;
- match the intended vegetation types with the characteristics of the site;
- work towards the long-term development of the vegetation;
- work with nature rather than against it.

This low-cost approach is not a substitute for conventional civil engineering in schemes involving construction or land remodelling for urban or industrial uses, but it can play a valuable part in the treatment of non-development areas within such schemes.

Within these possibilities, there are two fundamentally different approaches, namely to produce either a 'naturalistic' or an 'artificial' landform. Circumstances will dictate which is the more applicable of the two end results.

The naturalistic approach seeks to emulate natural landforms and is particularly applicable to rural sites, and with after-uses such as agriculture, forestry, nature conservation and informal recreation. Figure 13.2 shows a selection of naturally occurring landforms which can provide inspiration for the remodelling of tips. Removal of harsh lines and angles, as well as reducing the height and softening the bulk by planting, can produce results which blend very satisfactorily with the surroundings (contrast Photographs 13.9 and 13.10). Such work must be designed and implemented with sensitivity and be in keeping with the locality.

The artificial approach, in contrast, is likely to be more applicable on urban sites and for after-uses such as built development and formal recreation (see Photograph 13.11). In these circumstances, emulating natural topographical features would not produce the required landforms. The artificial approach also includes the possibility of innovative landform design by artists or sculptors, sometimes known as 'earth art'. Examples are shown in Figure 13.3.

Both the natural and artificial approaches encompass the following:

- the opportunity to use the sculptural effect of landform in an aesthetic way, to provide a sense of enclosure, separation, contrast, or drama and in the creation of watercourses and water bodies;
- the opportunity to shape the land to suit the specific requirements of certain after-uses which would not otherwise be possible;
- the opportunity to recreate previous landforms, be they relics of the industrial past which have been disrupted by subsequent activity, or the pre-industrial topography itself.

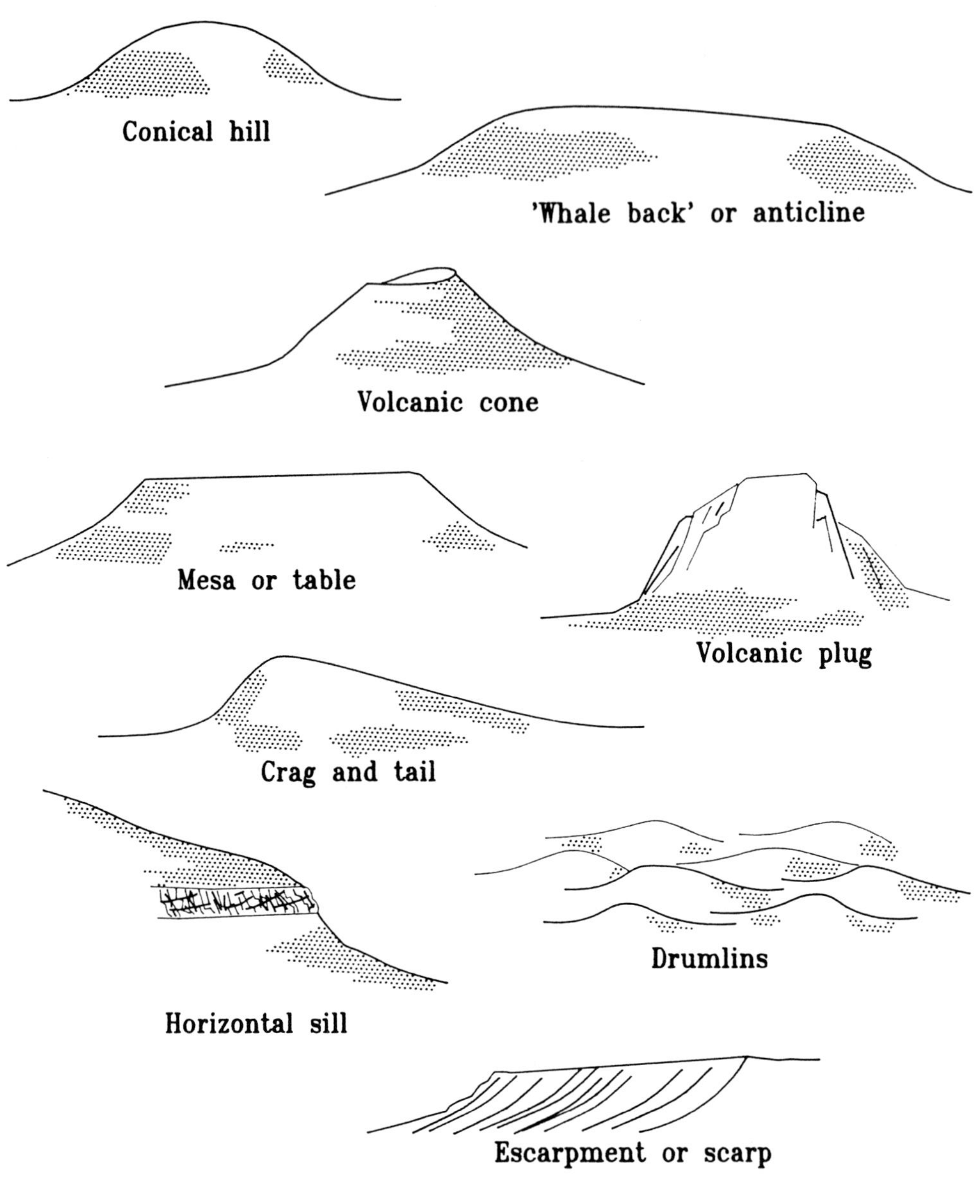

Figure 13.2: Some naturally occurring landforms (after Tandy, 1975[234])

Photograph 13.9:
Reducing the height and bulk of this tip, removing the prominent straight line of the incline and greening the spoil all contributed to the blending of this site with its surroundings. With time, tree planting would further enhance the effect (source: Welsh Development Agency)

Photograph 13.10: In contrast with the scheme shown in Photograph 13.9, treatment of this site concentrated on producing a stable grassed landform but failed to integrate the site with its surroundings (source: Richards, Moorehead and Laing Ltd)

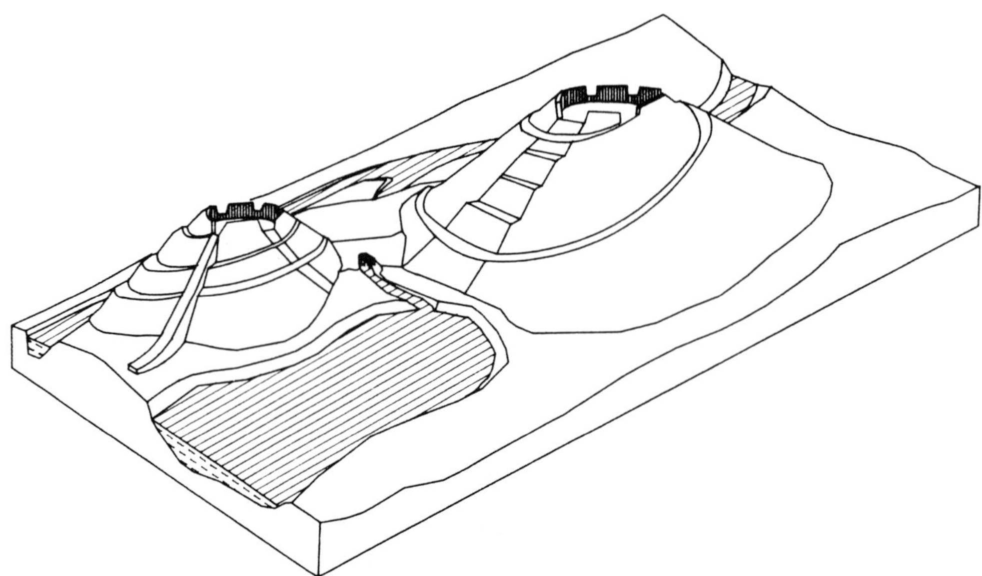

The smaller cone was proposed as an alternative to covering the foreground and adjoining farmland with the last output of waste from the colliery. The lake is converted from old sludge lagoons (after Tandy, 1975).

(source: Mike Petts)

Figure 13.3: Examples of innovative landform design

Photograph 13.11: Reclamation schemes which prepare land for industry, housing, sports pitches, etc. require the formation of level, terraced or gently sloping ground. Such schemes are often designed to maximise developable space, by using steep and regular batters, with the consequence that the new landform has an artificial appearance. A landform of this type can be softened or disguised by plentiful tree planting (source: Richards, Moorehead and Laing Ltd)

13.3.3 Utilising existing site features and characteristics

Having carried out their site assessment, the design team should be fully aware of the opportunities presented by the site. For example:

- the location, size and shape of a site may lend itself to a particular purpose;
- materials found on site can have useful engineering qualities such as strength and durability;
- there may be a good view from an elevated landform;

- sites may have the benefit of substantial infrastructure and utilities;
- existing watercourses may provide a convenient system for draining surface water from the redeveloped site;
- water flowing through the site could be utilised creatively to form an attractive feature;
- existing vegetation can be retained to provide a sense of maturity to the scheme, or to maintain the stability of a steep slope;
- older tips may be well colonised by vegetation of ecological value;
- there may be materials on site which are more suitable as a growing medium than others, and/or would encourage ecological diversity if used appropriately (see Box 13.5, Section 13.3.4). These materials should be identified and saved for future use;
- terraces formed for buildings and the buildings themselves can be re-used;
- existing features of historical value can be retained and incorporated into the scheme;
- existing landforms may be useful for specific after-uses *e.g.* long, steep gradients for dry-ski slopes, undulating areas for off-road motorcycling, or large, evenly graded areas for industrial development, housing or playing fields;
- it may be desirable to retain tips, or parts of tips, for their mineralogical or fossil value (see Box 13.3) and for study of the processes which produced the waste;
- intricacy of landform can be retained to ensure variety, both visually and ecologically.

13.3.4 Other general principles

There are other general principles which influence the approach to earthworks design:

- after-uses such as heavy engineering, light industry, housing and sportsfields have specific requirements as regards minimum area, shape and surface gradients which will determine the scale of the earthworks required when these uses have to be accommodated;
- in order to reduce long-term maintenance liabilities it is useful to minimise the reliance on constructed features, by integrating the design of landform, drainage and vegetation (see Photograph 13.12);
- there is a need, in most cases, to minimise maintenance liabilities, therefore a stable landform, healthy vegetation and robust drainage are desirable qualities;
- incorporating variety of slope, drainage, and surface materials in the scheme will enrich a site in visual and ecological terms (see Box 13.5). This should be considered both at the macro (whole site) scale, where the site is seen in the context of its surroundings, and also at the micro scale, where either restoration of the local 'grain' of the landscape is sought, or new local interest is created around buildings;
- modification of microclimate can be very beneficial, to create shelter, make best use of aspect and sunshine, avoid frost hollows, or create noise baffles and security bunds;
- vegetation and buildings on adjacent land can change and these changes can lead to masking or exposure of the reclamation site.

When considering the possible reshaping of the land the design team should assume that the landform will be permanent, and so should design for the best possible result.

Photograph 13.12: Integration of landform, drainage and vegetation. Vegetation softens the appearance of a ditch provided for occasional storm-water. Gentle side slopes also reduce visual impact (source: Welsh Development Agency)

13.4 Constraints on remodelling

13.4.1 Introduction

There is increasing potential for interesting and diverse reclamation schemes, because of:

- increasing knowledge and experience;
- improving technical skills;
- improved machinery;
- computer assistance;
- an increasing awareness of the benefits of multi-disciplinary design teams;
- availability of funding for post-reclamation development.

Box 13.5: Landform modelling for ecological diversity

Diversity of habitat and diversity of species are valuable characteristics of sites where wildlife conservation is an objective. These characteristics can be encouraged by the construction of a diverse landform, containing a range of:

- slope angles;
- aspects;
- exposure and shelter;
- drainage conditions;
- substrate textures;
- substrate pH;
- substrate nutrient concentrations.

Diversity of plant species is encouraged by neutral or alkaline pH and low nutrient concentrations which prevent a few vigorous, competitive species outgrowing all others. The wastes found on abandoned coal mine sites and iron and steel works are inherently infertile, and have a range of textures and pH values. Landforms and drainage conditions can be manipulated by civil engineering works. The different substrates can be mixed to produce gradations of physical and chemical characteristics.

Landform diversity can be designed to produce or favour specific habitats and vegetation types, either in a scheme which is to be fully vegetated at completion, or where the new landform is left to revegetate by natural colonisation (see Box 14.4) The vegetation and substrates recorded during pre-reclamation surveys can provide a guide to the range of habitats which could be designed into the scheme. Working together, the ecologist, civil engineer and landscape architect can produce a landform which will support a range of habitats and species.

There remain, however, factors which will limit the options in relation to landform design and creation. These are described in the rest of this section, under the following headings:

- land;
- materials;
- reclamation techniques;
- timing and season;
- after-use and management;
- costs.

13.4.2 Land

The extent of the site may be defined by ownership and/or by surrounding land-uses (see Photograph 13.7). Such boundaries may limit the scope for spreading out waste materials in order to reduce the height of tips.

If major reshaping of tips is anticipated, the design team should be given the opportunity to consider the amount of land needed in advance of land acquisition, so the best reclamation options may be produced.

One objective of reclamation may be to reduce the visual impact of a tip by the blending of landform and land-uses with the surroundings. The higher the tip, the flatter the landscape and the more confined the site, the more difficult this will be to achieve. Photograph 13.2, showing tips in northern France, illustrates this point. The same problem occurs in mine tips in the Ruhr area of Germany:[33]

> "Adapting the slag-heaps to fit with the mainly flat landscape of the coal mining area is problematical. Large masses of material to be dumped over a small space and the often immediate proximity of settlements make the problem even more difficult or quite impossible. Dumps 40 to 90m above ground level are bound to break up the landscape. Whether they are regarded as disturbing elements mainly depends on their shaping, landscaping and possibilities for later use."

Removal of tips will usually benefit neighbouring properties, but there are circumstances where it can be to their detriment. For example, the sheltering effect of an old tip may be lost, or a new but unattractive view may be opened up.

13.4.3 Materials

The materials to be found on site may constrain the reclamation process in a number of ways:

- the chemical and physical characteristics of a material will affect its compatibility with the proposed after-use (see Chapters 5 and 9);
- the cost of moving material about a site depends on the quantity and also the distance that it has to be moved;
- the natural angle of repose of a material cannot be exceeded without special measures to retain or stabilise the slope;
- regraded material may require special treatment before it will support the proposed after-use;
- the handling characteristics of the materials will determine the ease of moving them;
- the nature of the materials on site will affect drainage needed during the works;
- there is likely to be a need for a volume balance of materials within the site;
- the degree of compaction before and after moving affects the volume of material and therefore the final levels. Also, compaction after moving makes vegetation establishment more difficult, and uneven compaction can lead to differential ground settlement;
- the production of excessive dust in dry weather should be prevented;
- the potential for ignition of combustible materials should be anticipated;
- disturbance of an old tip can expose fresh spoil and lead to the loss of weathered material which is more amenable to plant growth.

13.4.4 Reclamation techniques

The current best practice will always set the limits on what can be done and at what cost. Design of a scheme such that available techniques allow reasonably economical implementation is a normal requirement. For example, an understanding of the abilities and limitations of earthmoving equipment is important in this respect. Without this understanding, proposals may be produced which are impractical or expensive to implement. Relevant to this issue would be a wish to design a varied and interesting landform. The right balance has to be found between variety in the result and the practicality of its creation.

13.4.5 Timing and season

The timing of implementation is significant for several reasons:

- weather and ground conditions significantly affect earthworks operations. For example, in the winter heavier rainfall can require temporary drainage during the works and the ground may be so wet as to prevent earthmoving operations continuing. Conversely, dry conditions can lead to problems of the raising of dust by vehicles. Dust production can be a particular nuisance at burning colliery tips (Box 13.6);
- the timing of earthworks should be planned so that they coincide with the seasonal requirements of vegetation. For example, a grass sward may be needed to prevent surface erosion, but grass can only be established in the right season, dictated by local climatic conditions;
- social and economic factors can both affect the progress and timing of reclamation schemes, possibly overriding the plans of the design team. Rapid progress may be required to meet the demands of the local community, and funding may be linked to strictly controlled financial periods of the year. Such factors will usually be outside the control of the design team.

Box 13.6: Weather conditions and their effects on earthworks.

Problems with earthworks can arise as a result of either wet or dry ground conditions.

Wet ground conditions

Only a relatively small proportion of rainwater will penetrate to a significant depth in a well-established and undisturbed colliery spoil tip. However, if the surface is disturbed, water penetrates much more readily resulting in saturated conditions in the layers near the surface. Surface erosion will be encouraged under these conditions and a previously stable tip can, subject to the amount of rainfall, become unstable.

Excavation of an old tip therefore requires consideration of how water penetration or its effects can be reduced, with provision of temporary civil engineering measures, such as cut-off drainage and embankments.

Dry ground conditions

Dry conditions can lead to problems with dust. This is especially so with sites subject to burning. However, reclamation contractors often prefer to work during the winter because, compared with other types of site, the ground conditions can be less saturated, thus allowing fuller use of earthmoving equipment. Winter working will reduce the dust problem.

The earthworks contract should include clauses which restrict working in conditions where dust blow would be a problem. Mobile water sprayers can be used to control dust generation, especially on haul roads.

In addition to these precautions, dust monitoring equipment can be installed on and around a site. Even if dust control precautions are successful, monitoring can be useful for public relations purposes *i.e.* to demonstrate to local residents that the potential dust problem is well controlled.

Restrictions on earthmoving

Large scale earthmoving will typically involve spoil and slags. Spoil derived from coal measures tends to be soft, breaking down rapidly when subject to weathering and also when moved in a very wet or saturated condition. Slags derived from iron and steel making are generally hard, brittle and not degradable. When moved in wet conditions, colliery spoil tends to become impermeable and slags retain their permeability.

Specifications for earthworks should restrict the moving of colliery spoil by reference to moisture content which should then be measured on a regular basis. Alternatively, based on experience, earthmoving can be controlled by the engineer using rainfall records maintained on site and judgement on the day by day condition of the spoil. In prolonged periods of wet weather, earthmoving is commonly brought to a complete stop.

Movement of slags in wet weather does not raise any serious problems and conventional earthmoving techniques and practices will be found to produce acceptable results.

It is advisable to ensure that potential contractors are aware of the expected timing of reclamation works and are provided with comprehensive site investigation data on ground and climatic conditions.

13.4.6 After-use and management

A scheme will be judged by the end-result of reclamation, not the process of reclamation, and not by comparisons of the site before and after treatment. Therefore:

- topography must relate to the gradient limits of the proposed after-uses and also to the management of the site (see Boxes 13.12 and 13.13);
- the right balance between flexibility of after-use and design for a specific purpose has to be achieved;
- the results of reclamation should be both aesthetically good and technically sound. This will place significant limits on the options for landform design and creation.

13.4.7 Costs

The financial implications of proposals to remodel the topography of a site are likely to be very significant. Because of transport costs or waste disposal limitations, earthworks will be frequently limited to a balance of cut and fill within the site. Where the need for developable land is a priority, it may be necessary to consider spreading material on to neighbouring land, but this is likely to introduce an extra cost through the need to acquire land which may not be derelict. Where some income can be generated by the reclamation work, this will be of great help in promoting the scheme, as will creation of land with development potential (see Section 13.2).

13.5 The design process

13.5.1 Introduction

Competence in design is learned through training and experience, and the following aspects are discussed in this section:

- team work;
- the holistic approach;
- design for after-use;
- earthworks design;
- implementation.

13.5.2 Team work

Good reclamation design is the result of team work and this general principle applies equally to the more specific topic of landform design. Successful team work depends on the establishment of clear objectives, the application of discipline and ultimately the performance of team members. The multidisciplinary nature of reclamation work demands project leadership and good project management if the work is to be successful.

The main professional disciplines relevant to landform design are civil engineering and landscape architecture. Others which have a bearing on the subject include:

- geotechnical engineering;
- geology and hydrogeology;
- soil science;
- geomorphology;
- ecology;
- other environmental sciences;

- planning;
- forestry;
- horticulture;
- bioengineering;
- industrial archaeology.

13.5.3 The holistic approach

All relevant issues should be identified and considered and, through an analytical approach, solutions sought which satisfy all reclamation objectives. This holistic approach should develop naturally if the team includes the necessary disciplines and works together effectively. The following paragraphs provide a few examples of this approach.

The surroundings of a site provide the setting for the new landform and land-uses and provide clues to assist in the design of the scheme. Both the surroundings and the site itself should therefore be studied to achieve the best design.

In the early stages of the design process, the team should be searching for both the positive attributes of the site and any problems to be dealt with. There may be features or characteristics which merit incorporation into the scheme. Section 13.3.3 of this chapter lists many of the possibilities. A temptation to clear a site and start again in all respects is usually wasteful and should be avoided.

It is best not to become involved in detail too soon in the design process, and to consider the options and agree the general principles at an early stage.

The practical constraints and opportunities identified at the site assessment stage, together with the objectives for reclamation, will form the basis for new proposals. Imagination and vision are also needed to produce the best solutions.

13.5.4 Design for after-use

Determination of the after-use of a site is fundamental to good reclamation, and this applies as much to landform design as to any other aspect. Considerations of after-use need to operate at two levels; land-use planning and site planning.

Land-use planning provides the overview which determines the general viability of proposed after-uses and ensures that zoning of after-use is compatible with the surroundings.

Site planning is the subsequent process which fits the proposed after-uses to the site, and *vice versa*. Site planning will thus aim to define the following:

- the location, size and disposition of all after-use features;
- appropriate communication links with surrounding land;
- an achievable and cost-effective layout;
- integration of landform, drainage and vegetation with each other and with the after-use(s);
- suitable microclimate and ground conditions for the after-use(s);
- a site which will be physically stable and manageable in the long-term.

The site planning process should, with all but the very simplest site, lead to the production of a masterplan (see Section 2.2).

It is important that a masterplan is realistic in what it shows, but there is no need to design a scheme with great precision at the preliminary stage of the design process.

A masterplan which shows the proposed after-use of a site is very useful to the reclamation team, who can use it to ensure that they design a landform that will be well suited to the after-use.

13.5.5 Earthworks design

The starting point for the design of earthworks is a plan of proposed after-uses as described in Section 13.5.4. This plan will allow the design team to draft landform proposals, using rough sketches and a topographical survey, contour plans and cross-sections. Alternative schemes may well be examined at this stage and rough costings will be needed to enable comparison and to ensure cost-effectiveness.

Various techniques are available to assist comprehension of the proposals and to communicate ideas:

- a general plan to provide an overview of proposals;
- cross-sections showing existing and proposed topography;
- perspective sketches, often referred to as 'artists impressions';
- computer constructed perspectives, which can be invaluable in creating reliable visualisations;
- photomontage, where perspective artwork is combined with a photograph to give an impression of realism;
- physical models, which offer a full exploration of the three dimensional implications of a scheme;
- video films can be created to simulate the experience of passing through a site, either using a model and a special camera or by using computer generated images.

As well as exploring the appearance of the proposals, it is also important to consider where the proposed landform could be seen from and, conversely, what will be visible from the new landform.

Once draft proposals are agreed which satisfy the objectives of reclamation, the proposed after-uses and aesthetic criteria, it is necessary to calculate the volumes of cut and fill material. With experience, the design team will have produced a draft scheme which is already close to providing the necessary balance of cut and fill.

There are a number of ways to calculate earthworks volumes:

- formulae;
- the grid method;
- the contour method;
- the cross-section method;
- physical models;
- computer models.

The principles, advantages and disadvantages of these methods are described in Box 13.7.

All methods of volume calculation need to take into account the degree of compaction of materials as existing and as required, using information on the likely factors of increase or reduction in volume.

Volume calculations will enable rough costings to be refined. Subsequently, if the scheme is being designed to be within a fixed budget, modifications to landform volumes can be undertaken.

There should be full integration of the proposals for landform, drainage and vegetation, to ensure an efficient and effective scheme.

13.5.6 Implementation

Once a feasible scheme has been formulated, communication of the design intentions to those responsible for implementation is necessary to obtain approvals from authorising bodies and to inform firms at tender stage and contractors during site work. Effective communication is achieved through a combination of drawings, written information and person to person contact.

Box 13.7: Methods for calculating earthworks volumes

Earthworks design cannot proceed very far before volume calculations are needed. At first, rough approximations are adequate, so methods of calculation can be relatively simple. As the design progresses, increasingly precise calculations are needed, especially if a balance of cut and fill is required.

1. Formulae - for slopes that approximate to simple geometric forms, volumes can be estimated by mathematical formulae.

2. Grid method - uses the change of level between existing and proposed surfaces, at a network of points over the site. More suitable for simple landforms. Useful for rough comparison of alternatives.

3. Contour method - convenient for simple landforms, this uses the volume of each 'slice' of ground represented by a contour.

4. Cross-section method - using measurement of the area of cross-sections through the landform, drawn at regular or selected intervals, this method can produce reliable figures. Frequently the favoured method, but is time consuming if done thoroughly enough to give accurate figures.

5. Physical models - have been used for volume calculation, using photogrammetry. Useful for large and complex schemes. A volume balance is achieved early on in the design process. Provides an indication of how material will have to be moved around the site. Has other benefits - visualisation of proposals and wind-tunnel tests.

6. Computer models - using digitised data, the method allows convenient comparisons of options, rapid and accurate calculation of volumes and production of cross-sections and also accurate visualisation. Often too costly for small schemes but very cost-effective for large or complex ones.

Communication should be as clear and precise as possible and should include:

- information on the nature of materials;
- results of special surveys/investigations;
- plans and cross-sections showing the existing situation and proposals;
- volumes to be moved;
- haul routes to be used;
- requirements as to placement of materials;
- health and safety requirements, including dust control;
- any other restrictions on working methods;
- specifications of either methods of working or performance required.

13.6 Technical aspects of landform remodelling

13.6.1 Introduction

The techniques for remodelling derelict land are described here with emphasis on those elements of particular relevance to the reclamation of coal and steel sites.

13.6.2 Topographical survey

A fundamental tool in the remodelling of landform is a good quality survey of the existing topography and site features (see Section 2.4.2). It should preferably be extended into the surrounding areas of land.

The survey can be produced either from aerial photographs by the process of photogrammetry, or by ground survey. Modern electronic survey equipment has revolutionised ground survey techniques and productivity. Box 13.8 provides a comparison of the two techniques.

Box 13.8: Comparison of aerial and ground survey methods

Despite the major improvements in ground survey equipment, aerial survey still gives a more realistic (though less precise) representation of the land surface, and is more economical than a ground survey for a complex site. Aerial survey also avoids possible difficulties with access to private land.

The main disadvantage with aerial survey is that, on vegetated areas, the actual surface of the ground is obscured, so height information is inaccurate or absent altogether. This problem is worst when the existing vegetation is tall and dense, comprising thick scrub or woodland.

Aerial photography for mapping purposes is best carried out when deciduous trees are not in leaf. Evergreen species are particularly troublesome because they obscure the ground throughout the year.

The acquisition of aerial photographs will assist the design team, even if they are not used to produce the survey, since they provide a detailed record of surface features.

The use of computer programmes for the modelling of land surfaces has provided a new method of earthworks design to the design team. The topographical survey in digital form is compared to a proposed landform and the volumes of cut and fill are calculated with greater speed and accuracy than manual methods (see Box 13.7). A more interactive approach to landform design is facilitated whereby more options can be examined and compared, because the volume calculation process is so rapid. Furthermore, as the model is in three-dimensional form, it can be viewed from any angle, with true perspective, providing views to assist with the visualisation of proposals.

Computerised surface modelling systems can also be very beneficial when setting-out proposals on site. See Box 13.9 for details.

Box 13.9: Computer surface modelling as an aid to setting-out on site

A surface modelling system can be used to produce quickly the cross-sections through the existing and proposed landforms that are traditionally used on a site to set-out the earthworks.

Alternatively, the system can be used to produce a plan of "isopachytes". These are lines which connect points of equal depth of cut or fill. They therefore identify the location and depth of cut and fill areas, with the zero isopachyte indicating the boundary between cut and fill.

On site, a plan showing isopachytes is extremely useful as an aid to understanding the extent and amount of the required earthworks. The isopachytes show, at any one point, how much material is to be removed or added and have been found to greatly assist the control of earthmoving on a site.

13.6.3 Drainage

Introduction

The installation of an effective drainage system is a crucial element in the remodelling of a landform and is needed to:

- prevent erosion of spoil and/or capping materials;
- reduce the amount of water penetrating areas of mine spoil;
- prevent unwanted ingress of water into mine workings;
- prevent the build-up of water within spoil heaps;
- prevent the migration of silt and contaminants to adjacent land and watercourses both during and after the works;
- aid vegetation establishment in areas susceptible to waterlogging;
- prevent flooding.

The appropriate measures for a site will depend primarily on the local climate, the water regime of the surroundings, the materials on site and the proposed topography, surface finish and after-use.

It is important that drainage measures are designed not only according to technical criteria but also with their visual impact in mind, since they can be very intrusive. For example, lined concrete channels are extremely prominent, especially in hilly terrain. Open channels of large dimensions can also present physical obstructions to movement by people and animals.

In designing drainage the opportunity can also be taken to incorporate wetlands to increase ecological diversity and also to assist in pollution control (see Section 12.5.3).

The following sections describe the types of drainage works usually necessary and discuss design considerations.

Cut-off drainage

Cut-off drainage is provided to intercept surface water arriving above areas of spoil, waste or contaminated materials, and to convey the water safely to the outlet drainage system.

Temporary ditches

Temporary ditches are provided prior to earthworks to intercept surface water and sediment and, if necessary, to convey the water to a treatment unit prior to discharge to the outlet drainage system.

Field drainage

Field drainage may be required in restored areas to assist in the removal of surplus groundwater so as to prevent waterlogging.

Permanent drainage

The function of a permanent drainage system is to intercept surface water run-off or groundwater and convey it to an outfall. The system may also convey existing watercourses across the site. Where permanent drainage is to be installed, the following should be taken into account:

- the effect of a reclaimed site on the downstream watercourses to which the site contributes - in most cases, there will be an increase in peak flow rates immediately downstream of the reclaimed site due to a more efficient drainage system, which has a more rapid response time to short duration, high intensity storms. The effect may be negligible where the site is part of a much larger catchment. However, for a smaller catchment, the capacity of bridges and culverts to receive these additional flows must be evaluated. In some cases, flood attenuation measures may be required, for example soakaways, balancing ponds or flood plains;
- flood flow calculations - various methods have been developed to calculate flood flows resulting from rain falling on a catchment. The application of one method, the Rational Method of Flood Estimation to a reclamation scheme is described in Box 13.10;
- drainage layout - will depend on the proposed landform, the future use of the site and pollution considerations. Both pipes and ditched systems may be employed at the same site. The principal characteristics of the two systems are compared in Box 13.11;
- construction material - care should be taken to ensure that the materials used in the construction of drainage systems blend in with the reclamation scheme finish. Grassed channels or stone faced channels are the least obtrusive, however the former require more maintenance than lined channels and the latter can be costly;

Box 13.10: Factors used in calculating peak run-off in the Rational Method of Flood Estimation[173] and their application to a reclaimed site

When calculating run-off for a reclaimed site and for small catchments of which the site forms a significant part, the Rational Method of Flood Estimation can be used. This gives the peak rate of run-off directly from the catchment area, rainfall intensity and run-off coefficient.

Catchment area

Catchment areas may be obtained from contoured maps. Care must be taken to ensure that leats and underground connections between catchments are taken into account in calculating catchment areas.

Rainfall intensity

For a particular site, the rainfall intensity depends on the storm duration and return period. For a low-risk rural site, a two year design return period would be reasonable. Where there is a risk of flooding to properties, a higher standard of protection would be required.

The consequences of a flood event more extreme than the design event should be considered. This is particularly so in remote areas where meteorological and other data for the site in question may not be reliable. Measures to deal with greater than design event flows include techniques such as reinforcing grassed areas adjacent to channels with geotextile to prevent erosion.

Run-off coefficient

The run-off coefficient is the percentage of rainfall on the catchment which appears as surface run-off in the storm drainage system.

The permeability of the surface material is the main factor determining the run-off coefficient and may vary from 60% for rapidly draining soils to 90% for heavy soils or 100% for surfaced areas.

In choosing a run-off coefficient, it must be remembered that maximum run-off may occur following earthworks and before vegetation establishment. This is the most vulnerable period in terms of soil erosion. The longer vegetation is likely to take to establish, the more weight must be given to this consideration.

Where a membrane or impermeable layer is used in a capping treatment a higher run-off coefficient than usual for the covering material would be appropriate.

- existing drainage systems - historically, mining engineers sometimes constructed elaborate systems of water supply and drainage to service their mines. Some of these still operate and some are redundant. These systems should not be ignored during reclamation schemes and the opportunity should be taken to use them if appropriate.

Box 13.11: Comparison of ditched and piped drainage systems

Ditched systems

Advantages

- economical;
- if constructed along a site boundary they also provide site security;
- provide storage for flood water - this is particularly useful if the outlet to the system is of limited capacity;
- flexible - can be easily enlarged or converted to a piped system;
- can be laid to slacker gradients than piped systems;
- if unlined can act as soakaways, so reducing flow at the outlet.

Disadvantages

- require regular maintenance and are prone to erosion;
- large ditches can be dangerous and restrict mobility around a site;
- need to be laid at slack gradients so if steep falls are required cascades or weirs have to be provided;
- can be visually intrusive.

Piped systems

Advantages

- access around the site is not impeded;
- low maintenance;
- long life;
- multifunctional.

Disadvantages

- limited capacity;
- high cost;
- provide limited storage for flood water;
- not as flexible as ditched systems;
- may become blocked;
- inlets and catchpits require regular maintenance.

Reservoirs and lagoons

Both coal and steel sites may contain ponds, lagoons or larger bodies of open water such as reservoirs. A body of water provided for the coal or steel processing operations may be considered a significant part of the industrial heritage of the site and, if appropriate to the proposed after-use, may be retained.

However an open body of water is potentially very dangerous if raised above the general ground level. Unless it has been regularly maintained, which is most unlikely on abandoned sites, embankments and hydraulic structures may well be in a poor state of repair.

The defects of a dam, for instance, may be such that the cost of repair cannot be justified. However there may be considerably more work and expense involved in abandoning a reservoir than merely taking off the outlet valves or cutting a slot through the dam wall. Sediment is a major hazard when exposed in the reservoir basin and its erosion into a downstream watercourse can cause serious water quality problems.

A reservoir may have provided considerable attenuation to flood flows. A loss of attenuation after breaching may require the enlargement of culverts and watercourses downstream.

Water treatment

During reclamation works polluted water may arise after rainfall and this will have to be dealt with. Drainage works prior to the commencement of earthmoving should aim to minimise the production of polluted water, and temporary treatment facilities may be needed.

Where consents are required to discharge water from the site, stringent water quality standards must be met and treatment systems may need to be devised in order to meet these standards. Further details of such treatments are given in Section 12.5.

13.6.4 Erosion control and stability

Newly created landforms are especially sensitive to erosion and Section 13.6.3 explains the need for temporary drainage measures to counteract this problem. Once a slope is vegetated, surface stability is normally secured. There is therefore some urgency to establish vegetation. However, there is a complex interaction between vegetation, soil, climate and the activities of humans, animals or machines, as illustrated in Figure 13.4.

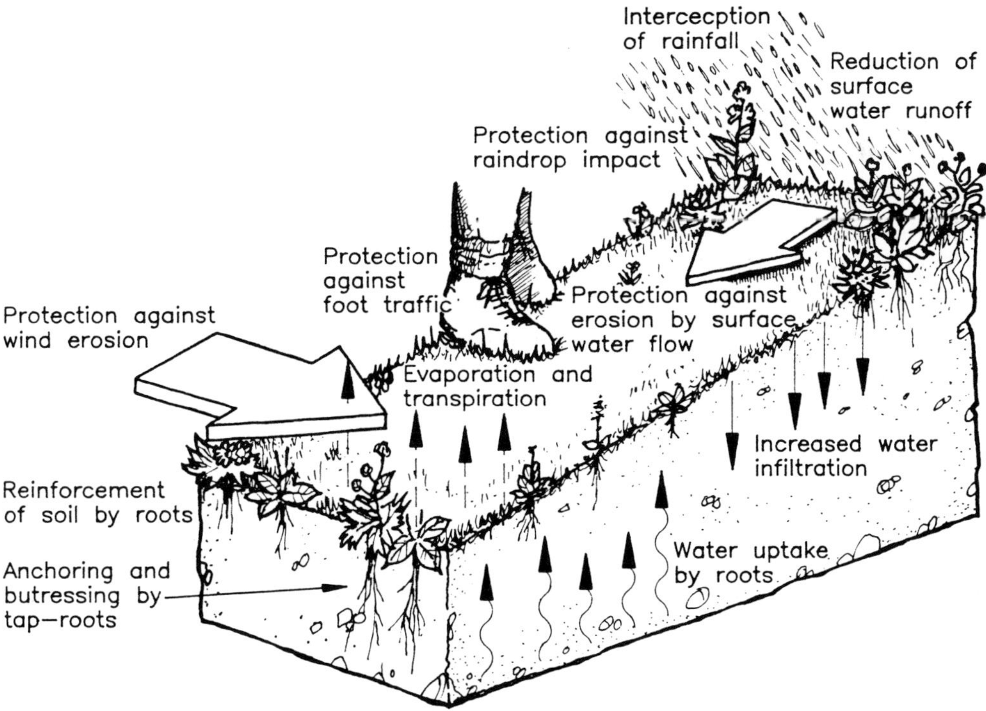

Figure 13.4: Some influences of vegetation on the soil (from CIRIA Book 10, Use of vegetation in civil engineering, 1990[57])

13.6.5 Slope gradients

The gradient of any sloping area is important in relation to:

- how it is created;
- drainage provision and erosion control;
- stability;
- after-use;
- maintenance.

The design team should be aware of the capabilities and limitations of available earthmoving equipment *e.g.* maximum operating angles, turning circles.

For drainage and erosion control, the theoretical optimum gradient of a newly created slope is that at which it will drain naturally without erosion of soil particles. This slope is frequently in the order of 1 in 40 or 1 in 50 for soils and colliery spoil. However, the necessary gradient is dictated in practice by the after-use of the site or by limitations on land-take for the purpose of spreading out spoil materials. These parameters can lead to a need for artificial drainage measures on sites with very shallow slopes, or a requirement for drainage devices to minimise erosion on steeper slopes. Gradients of more than 1 in 3 should be avoided if possible in highly erodible materials. If avoidance of such gradients is not possible the surface should be treated by special stabilisation techniques, such as the use of geotextiles to resist erosion until vegetation is established. A more economical solution is the application of contoured or angled ripping across slopes.

Box 13.6 has referred to the different engineering characteristics of colliery spoil and slags. Colliery spoil should be regarded as being a highly erodible material and slags as being a much lower risk in terms of their susceptibility to erosion.

The slope limitations of various after-uses are described in Box 13.12 and for maintenance procedures in Box 13.13.

13.6.6 Aesthetics

The design team should strive to produce a result which looks right as well as one that functions well. The aesthetics of design is a specialist area which requires training and experience to be fully appreciated.

Spoil heaps are typically very prominent features in the landscape, but this does not necessarily mean they have a negative visual impact. Neither are they necessarily the only or the most obtrusive elements of a derelict site (see Section 13.2). If remodelling of spoil heaps is proposed for visual reasons only, it is important that careful consideration is given to the rationale behind the proposals.

In remodelling the landform of a derelict site, attention should be paid to the characteristics of its surroundings. Blending of landform, surface finish and materials with those of the surrounding area will help considerably to minimise the visual impact.

With a naturalistic landform design (see Section 13.3.2), blending with the surroundings may be more achievable than with an artificial design. In the latter case, screening or disguising the new landform with tree and shrub planting may be the only way to ensure minimum impact. Vegetation can also be used to reduce the impact of tips which are not being remodelled (see Box 13.4). In all cases, new vegetation schemes must be designed with care to ensure that they reduce impact rather than accentuating or adding undesirable characteristics.

The same care is needed with other features of a new landform, especially drainage and boundaries, which if insensitively designed, will draw attention to the site and can be seriously intrusive themselves.

Box 13.12: Slope design for after-use

Angles of slope are critical in all landform design. Existing slopes may be standing at 1 in 2, 1 in 1½, or in extreme cases at 1:1. The gradients to which they must be lowered will depend upon the use to which the land is to be put.

Industrial areas call for very careful consideration since whilst large flat areas require minimal earthworks under a building, flat areas are extremely difficult to drain because of the minimum gradients called for in drainage design. A general slope of 1 in 100 in one direction, or in two directions at right angles to each other, is a reasonable compromise on sites extending to about 5 ha. when the final layout of the development is unknown. For larger sites, gradients of 1 in 150 to 1 in 100 in one direction ease the problems of drainage.

This gradient allows the site to be drained easily by means of land drains or open channels and this is an important element in providing an immediate cover of grass which serves as a control to soil erosion and presents a reasonably attractive site to developers and local residents.

If a layout can be determined for the development, then areas for buildings, car parks, highways and landscaping can be provided with appropriate ground profiles even if development will not take place until some time in the future.

Sports pitches should be steeper than 1 in 150 otherwise they will not drain properly, and must not be so steep as to interfere with play. 1 in 80 is acceptable for association football. **Casual walking** is pleasant on slopes up to about 1 in 8, but people tend to take a winding route to ease the slope on anything steeper than 1 in 10.

If an area is to be planted with **woodland**, a minimum slope of 1 in 10 is advised. Thus, if a site is flatter than this overall, a 'ridge and furrow' landform is recommended with the sides of the ridges sloping at 1 in 10.

Typical gradients for some after-uses with specific requirements are as follows:[143]

Sports pitches	1 in 40 (1·5°) across line of play 1 in 80 (0·7°) along line of play.
Dry ski-slope	c 1 in 4 (14°-17°)
Grass skiing	c 1 in 5 (11°)
Watersports (launching)	1 in 8 (7°)
Amenity woodland	1 in 10 (6°) to 1 in 2 (27°)
Water bodies	(margin) up to 1 in 10 (6°) - for safety reasons and for plants/wildfowl

Box 13.13: Slope design for maintenance and management

If land is to be maintained by grazing, or used for forestry a wide range of slopes are possible. Maintenance of grassed areas by machinery, however, requires care to ensure that working gradients are not exceeded. Safe limits for tractors vary according to:

- the type of tractor;
- the equipment attached to the tractor;
- whether the ground is bare or vegetated (grassed slopes can be much more dangerous than bare soil);
- direction of travel;
- weather conditions;
- the condition of the grass.

As a result of the above, it is possible to use a tractor on slopes of around 1 in 2 in the right circumstances, but at other times 1 in 7 can be hazardous.

The maximum slope for the safe use of a pedestrian grass mower is 1 in 3.

Whilst shallow slopes are easier and safer to maintain, landform design should not aim to avoid steep gradients altogether. In some cases such gradients will enhance the scheme's visual interest. Maintenance problems on steep slopes can be reduced by creating areas which require less management input e.g. woodland, grazed grassland.

Further factors to consider in order to minimise the visual impact of remodelled landforms are given in Box 13.14.

13.6.7 Microclimate

Proposals for remodelling of landforms should aim to improve or at least maintain the micro-climatic conditions in and around the site. Care should be taken to avoid inadvertent effects such as wind funnelling. The design team should also be aware of possibilities such as reducing the sheltering effect of a tip when it is remodelled. Some aspects of microclimate are described in Box 13.15.

Box 13.14: Reducing the visual impact of spoil heaps

In the remodelling of a tip, the following additional factors should be considered so as to maximise the reduction in visual impact:

- a landform seen in silhouette is more prominent than if seen against a landscape background;
- the height of a tip is more noticeable than the bulk, so reducing height is likely to reduce visual impact. This need not involve a large proportion of the volume of a tip. A long, low bulk is likely to be more visually acceptable than a tall, severe landform;
- removal of harsh angles and straight lines in a tip should be a priority;
- straight lines, even slopes and flat surfaces are very prominent, providing strong visual lines to form and bulk. Curved lines and varying gradients are less noticeable and less obtrusive;
- spreading spoil away from an observer reduces the apparent bulk by foreshortening;
- large volumes can be placed unseen behind a false summit, where a landform is viewed from below.

13.6.8 Materials handling

Tipped materials on abandoned coal or steel sites can vary from burning colliery spoil, to fused masses of smelter waste, to saturated sludges. The nature of the waste materials has a major effect on the cost of moving them and the options for treating them.

As a general principle, for reasons of economy, double handling should be avoided and haul routes should be as short as possible.

Excavation into a burning tip creates particular difficulties and is dealt with in Section 7.5.

Box 13.15: Some aspects of microclimate

Landform influences micro-climate by affecting air-flow and by receiving varying amounts of solar radiation, due to variation in aspect and surface reflectivity. Waste tips influence the wind and temperature conditions of their surroundings and can affect air quality in terms of quantity and distribution of gas emissions and dust. They can also provide a baffle against noise emissions.

In a largely flat landscape, a tip, depending upon its shape and size, forms an obstacle for the airstream and weaker air currents circulate round it. When the wind speed increases on its windward side and on its flanks, moderate and strong winds flow over it. On the leeward side, there is first formed a zone of weaker air movements and this is followed further to the lee side by a zone of stronger, more squally winds blowing in irregular directions. These overflow winds, known as the lee effect, cause increased air movement in the lee of the tip.

A planted tip, as a self-contained wooded area, causes a vertical exchange of air masses and also the surface of the canopy causes air turbulence.

Aspect, slope, colour and vegetation cover influence the temperature of a tip surface and the air above it. South-facing slopes of unvegetated, dark shales can reach 60°C or higher in sunny, windless conditions. Such extremes of temperature are gradually reduced as vegetation cover becomes established.[33]

An interesting consequence of the conditions that can be found on colliery spoil heaps is that they can be suitable for growing vines. The suitability of tips for viticulture is partly due to the dry, stoney conditions which vines like, but also because the dark material is a good absorber of solar energy and this is given off again, warming the air above the surface. Thus, in Belgium, the temperature at the summit of a spoil heap has been found, at all times of year, to be 2-5°C higher than at the foot.[193] Slopes of 25-45°, at the latitudes found in Belgium, are optimal for reception of solar energy.

Another particular problem arises from lagoons, used to settle out waste from coal washing plants. These lagoons can be found within spoil heaps, having been tipped over after falling into disuse. It may be necessary to use a drag-line excavator to deal with lagoon sediments, which tend to remain in a saturated or super-saturated condition (see Section 8.3.4).

Where material is being placed in an area which will be vegetated, care should be taken to avoid handling the material in very wet conditions, and trafficking over the area by earthworks plant should be minimised. If trafficking is unavoidable, the placed material should be cultivated before placement of the next layer, to reduce compaction (see Tables 5.6 and 5.7).

If topsoil or sub-soil exist on site these materials should be conserved for re-use. Natural soils are a valuable resource which may not be available in large quantities. It is especially important to take care with the handling and storage of topsoil, since irreparable damage can be done to the soil, in terms of potential for plant growth.

Box 13.16 gives more information on soil handling practices and Figure 13.5 illustrates techniques for soil spreading using heavy machinery, which avoids excessive surface compaction. Section 14.4.3 discusses the advantages and disadvantages of using natural soils in reclamation work.

13.6.9 Compaction

When waste materials are moved they can be placed loose or they can be compacted. If placed in layers of suitable thickness and compacted to the appropriate degree, stable slopes can be created where they would otherwise not be stable, and ground can be created which has adequate bearing capacity and is consistent enough to carry buildings. Information on alternative techniques of ground improvement is provided in Section 13.6.10.

Invariably in the past, spoil heaps were created by loose tipping. In order to create a stable landform it may therefore be necessary to move, place and compact most or all of the material in a tip.

Where vegetation is to be established, compaction should be avoided, but where it is unavoidable or already existing, on haul routes for example, compaction must be relieved by ripping as part of the soil cultivation process (see Section 14.4.2).

Box 13.16: Soil handling, soil quality and vegetation establishment[57]

The selection, handling and treatment of soils which are to be used within the potential root zone of the vegetation, say within 1m of the final ground surface, should take account of:

- their potential as a medium for plant growth;
- the construction of a soil profile;
- the relationship between soil layers.

Irreparable damage can be done, in terms of potential for plant growth, if a soil is handled incorrectly. The problems to avoid are:

- stockpiling in such a way that the natural aerobic soil organisms are killed;
- destroying the existing soil structure;
- compacting soil to excessive densities that reduce water infiltration and inhibit root growth.

When moving large quantities of soil it is usual to employ the most cost-effective, and sometimes the largest, machinery available. For engineering purposes, the compaction effects of large machinery are beneficial, but tracking over surface soils with machinery which imposes a high ground pressure destroys soil structure and produces conditions which are very inhospitable to plant growth. The cost of precautions to avoid damage to soil may be offset against the benefits of improved plant establishment and growth. However the conflict between soil compaction for stability and looseness to permit plant growth is one which has to be considered carefully in each situation.

Damaged soil may never be fully restored by cultivation, but there are a number of practices which, if followed, will reduce the extent of soil damage:

- where possible, excavators and dumpers rather than scrapers should be used to move soil;
- double handling should be avoided as much as possible;
- stockpiles, where they are necessary, should be low and not heavily compacted but graded to shed rainfall. Long-term stockpiles should be seeded to avoid erosion;
- handling of soil should be strictly controlled according to soil moisture conditions. Similar restrictions should apply to tracking over existing or respread soils, when wheel damage can be extensive. Tracking should only be allowed over a soil the strength of which, as measured with a pocket penetrometer, is greater than the ground pressure of the machinery involved;
- indiscriminate tracking by heavy earthmoving machinery over surfaces of existing or spread soil should be avoided. Vehicles should keep to the same wheel-tracks as much as possible, so confining the damage. These wheel tracks can be specially treated later by deep cultivation;
- the temptation to travel repeatedly over an area of spread soil to grade off the surface should be resisted, since this will only produce a smooth, compacted soil surface which will make vegetation establishment difficult.

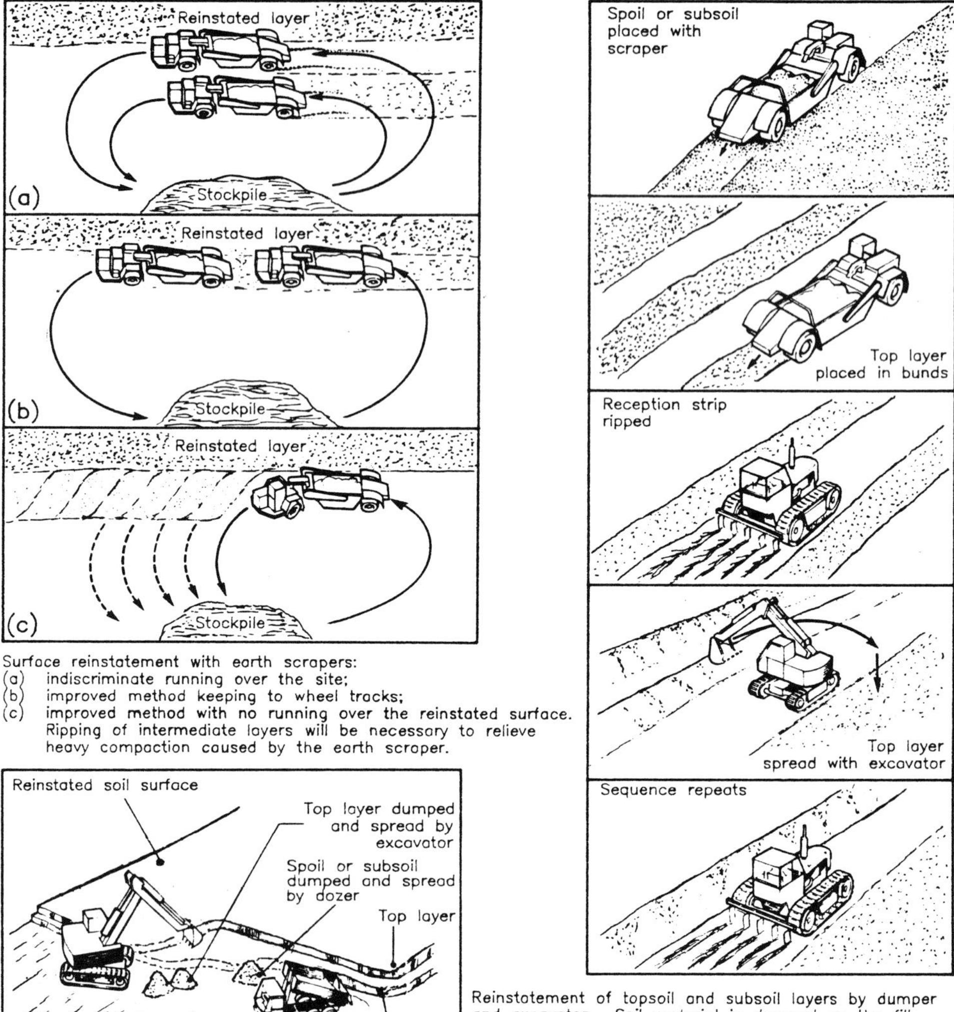

Figure 13.5: Soil spreading techniques which avoid excessive compaction (from CIRIA Book 10, Use of vegetation in civil engineering, 1990[57])

Loosening of surface layers to enable vegetation to establish and thrive does however encourage penetration by water. Water penetration can be a problem on very flat areas, where natural drainage is impeded, and therefore designs should avoid such circumstances by introducing minimum gradients appropriate for the intended development. For reasons of stability, penetration of water into some tips is also to be discouraged (see Section 6.4).

13.6.10 Ground improvement

Derelict coal and steel sites may contain areas of filled ground which are very variable in nature and consistency and may be unsuitable for building on without special treatment. Details of ground improvement techniques are contained in Box 13.17. However existing fill may sometimes be sufficiently compact for building construction without the need for further treatment.

Consideration should be given to the constraints caused by the presence of existing buried foundations, flues and tanks. The proposed layout of roads, utility services and building plots should be designed to minimise the potential problems from underground obstructions, and in some cases old foundations and other buried obstructions will need to be removed. In order to reduce costs, certain parts of the site may need to be designated as landscaped areas or hard-standing, requiring minimal or no disturbance of buried obstructions.

A solution which is sometimes adopted on reclamation sites involving poor ground conditions is to provide a compacted mattress of suitable material at least 1.5m thick over the whole of the development area, using inert slag, colliery spoil or crushed demolition materials. This approach provides flexibility in the planning of the proposed development and overcomes problems of hard ground due to old foundations or areas of hard tipped materials. Any old basements or other underground voids should be backfilled using suitable granular material compacted in layers.

Box 13.17: Ground improvement techniques[53]

Dynamic compaction

Deep compaction of the fill is effected by the repeated dropping of a heavy weight using a crane e.g. a 15 tonne weight dropped from a height of up to 20m, on a grid pattern.

Treatment may be carried out using high energy impacts for the primary and secondary grids, the latter being offset from the former. Craters formed by the impacts are filled using earthmoving equipment and a second stage of more uniform treatment is carried out using a reduced drop height.

An alternative method of treatment for shallower fills is to use a rapid impact compactor, which involves dropping a 7 tonne weight from a height of 1.2m onto a circular plate.

Vibro techniques

Vibro techniques involve the compaction of granular soils or the formation of stone columns using a vibrating cylindrical poker suspended from a crane. The poker may be 300mm to 450mm diameter and can weigh up to 4 tonnes. Treatment depths of around 6m are typical, but depths up to 30m have been achieved. The poker penetrates the ground as a result of the vibratory action assisted by flushing jets in the nose cone and sides. Compressed air is generally used for flushing, although water may sometimes be employed.

Vibro treatment is carried out on a grid pattern, using closer grid spacings where higher bearing capacities are required e.g. beneath pad foundations or edge beams.

'Vibrocompaction' is the term generally used to describe the densification of granular soils using vibro techniques. Additional material may be introduced from the top of the hole as part of the compaction process. An alternative in unstable ground is to use the bottom feed process whereby stone is introduced directly from the tip of the poker via a feed pipe.

'Vibrated stone columns' is the term used for the stone columns introduced and compacted by the vibro process in order to improve the bearing capacity.

Preloading

Preloading involves over-consolidation of the fill by temporary surcharging to improve the bearing capacity prior to construction taking place. Surcharging is usually carried out using several metres of fill material placed using earthmoving equipment and left *in situ* over a period of two to three months.

Excavation and recompaction

This method involves the treatment of loose granular fill material by excavation and recompaction in thin layers under controlled conditions using conventional earthmoving equipment.

Areas of biodegradable fill materials may need to be removed and replaced with suitable inert fill prior to development, in order to overcome possible landfill gas migration and ground subsidence.

Upon completion of the reclamation works, geotechnical investigations will be necessary to provide information for the design of building foundations and road formations. Special foundations may be required to overcome poor or variable ground conditions, possible subsidence or differential settlement.

13.6.11 Volumes

Earthworks design necessarily involves calculation of volumes of material. The principles, advantages and disadvantages of various methods are outlined in Box 13.7 in Section 13.5.

When an old colliery spoil heap of loosely tipped material is moved and then heavily compacted to provide land which will support buildings, and highways for example, there is a decrease in volume by a factor of about 10%. This reduction factor has to be taken into account when designing reclamation works that involve moving spoil material.

Where a balance of cut and fill is not achieved in the design, it will be necessary either to import material on to the site to complete the designed landform, or alternatively to export the surplus off the site.

Despite calculating a balance of cut and fill at the design stage, there is nevertheless likely to be an imbalance in the earthworks when the scheme is implemented. This imbalance can be due to a number of reasons, such as the actual reduction factor not being quite the same as that predicted, and amendments to the landform design during reclamation work. It is good practice therefore to include in the scheme a balancing area where the final levels can be varied without any detrimental consequence for the overall aims of the project.

13.6.12 Timing

The timing of the implementation of a reclamation scheme may be important because of:

- seasonal differences in the weather;
- the affect of weather on ground conditions;
- the need to prepare a site for vegetation establishment;
- community needs;
- financial arrangements.

This was discussed in Section 13.4.5 and in Box 13.6.

13.6.13 Environmental impact

The reclamation of a coal or steel site should reduce its environmental impact, in terms of, for example, air and water pollution and visual intrusion. However, the reclamation process can cause temporary impacts, which should be anticipated and planned for in the design stage. These include:

- noise from earthmoving plant;
- dust raised by machinery and blown onto adjacent areas;
- pollution of watercourses through disturbance of materials;
- increased traffic in the area, especially if bulky materials are imported or exported;
- disposal of contaminated materials, if taken off site.

13.6.14 Records

It is in the nature of reclamation work that the results produced on site are frequently not exactly the same as the proposals shown on working drawings. After reclamation has been completed, experience has demonstrated that it is essential to have thorough and precise records and plans of what was actually done in the reclamation scheme. These will be invaluable:

- if any problems arise, such as accidental exposure of encapsulated contaminated material;
- if there is a requirement in the future to change the landform;
- in the post-reclamation use, development and management of the site.

It is therefore advisable to ensure that an 'as-built' survey is carried out on completion of reclamation and that an archive is set up to record the details of what was done. More information on site records is given in Section 15.4.3.

14 THE ESTABLISHMENT AND CARE OF VEGETATION

Chapter contents

continued...

14 THE ESTABLISHMENT AND CARE OF VEGETATION

14.1 Introduction

The most successful revegetation schemes are those where the establishment of vegetation matches the needs of the intended land use.

The vegetation found on any area of land will either contribute to or detract from the use of that land. For vegetation to contribute to land use, clear definition of the proposed uses is needed. This definition of uses will enable the appropriate vegetation to be selected, established and maintained. In a reclamation scheme where no particular land use is intended the objectives of reclamation will define the vegetation types needed (see Box 14.1).

Figure 14.1 gives examples of the wide range of uses to which land may be put and the vegetation types which are appropriate to these uses. The figure clearly shows why it is important that the vegetation process is guided by well-defined objectives. Vegetation performs an important engineering function through its influence on the soil and soil moisture regime. These 'bioengineering' functions are summarised in Figure 13.4.

A temporary vegetation cover has been established on some reclaimed sites to enhance their appearance to potential developers (see Box 14.2).

The establishment and care of vegetation is a long-term operation. Almost all vegetation needs some continuing care and on reclamation sites this care may be intensive over five, ten or more years. It is therefore important that the requirements for vegetation care are understood and accepted when objectives are agreed, or that the objectives are set according to the resources which will be available in the future. Many reclamation schemes have deteriorated because the objectives and resources for long-term care were not properly matched.

New use \ Vegetation type	Individual trees	Forest tree species	Native tree species	Mixed woodland	Shrubs	Pasture grasses	Mown grass	Rough grass	Wildflower mixture	Wetland species	Aquatic species
Productive grazing						⊠					
Marginal grazing						○		○			
Commercial forestry		⊠									
Marginal forestry		○	○	○							
Sport							⊠	○			
Caravan and campsites	○		○	○	○			○			
Car parks	○				○			○			
Picnic sites	○		○	○	○		○	○	○		
Walking	○	○	○	○	○	○	○	○	○		
Ball games							○	○			
Childrens play	○			○	○		○	○			
Wildlife	○	○	○	○	○	○	○	○	○	○	○
Landscape improvement	○			○	○				○		

⊠ : Essential ○ : Possible

Figure 14.1: Vegetation types for new uses of reclaimed land (from Robinson Jones Partnership, 1987[209])

Box 14.1: Vegetation in reclamation

The nature of the vegetation required will be determined by the objectives of reclamation.

Land use objectives

Vegetation may be fundamental to the intended use of land which has been reclaimed; e.g. agriculture, forestry, sport, wildlife conservation. In these cases it is essential that the intended use of the land is decided before reclamation begins so that ground preparation, fertility, species selection and plant establishment can be designed correctly.

Environmental improvement objectives

In some circumstances there may be no particular use to which land will be put after reclamation; e.g. where the objectives of reclamation are to improve the environment by controlling erosion, absorbing dust and noise, reducing wind speeds and improving the landscape. In these circumstances it is essential that the nature of the required improvement is known so that this objective may guide the revegetation process.

The design of a scheme of vegetation establishment is therefore a process of balance between what is desirable to fulfil the objectives set for the site, and what is practical within the limitations presented by the site, its substrate and the resources available. This chapter describes the approaches and techniques which can enable desirable vegetation to be established and developed on sites where coal mining and iron and steel-making have ceased.

Box 14.2: The benefits of temporary revegetation

Many former coal mining or steel making sites have the potential for new development uses, but lie temporarily derelict awaiting the economic or other circumstances which will trigger development. Other sites are reclaimed and prepared for development but remain empty for many years before being built on. These sites can revert to a semi-derelict state which makes them unattractive to developers.

Simple, economical techniques for revegetation can be applied to these sites to create a temporary vegetation cover which will improve their attractiveness and possibly facilitate a temporary use. Vegetation types which can be used for this purpose include:

- grass/clover swards;
- flowering legumes;
- fast-growing, flowering shrubs;
- low maintenance grass/flower swards;
- fast-growing trees such as Alder and Willow.

Where substrate conditions permit areas of a site may also be left to vegetate naturally and increase in ecological value.

The benefits of temporary revegetation are that it:

- gives rapid visual improvement, ahead of major long-term work programmes;
- demonstrates that land is owned and cared for, discouraging vandalism or dumping;
- allows temporary uses at minimal initial cost;
- improves marketability of development land without impeding that development;
- is inexpensive and can be abandoned for development;
- provides maturity if some vegetation is retained in the ultimate development;
- provides ecological value.

14.2 Approaches to vegetation design

14.2.1 Integration with landform design

The design of revegetation works is a key step in the process of reclamation scheme design, and should be integrated with the design of the landform since together these aspects govern the future appearance and use of the site. Box 14.3 shows how the process of vegetation design should include much more than simply the selection of species.

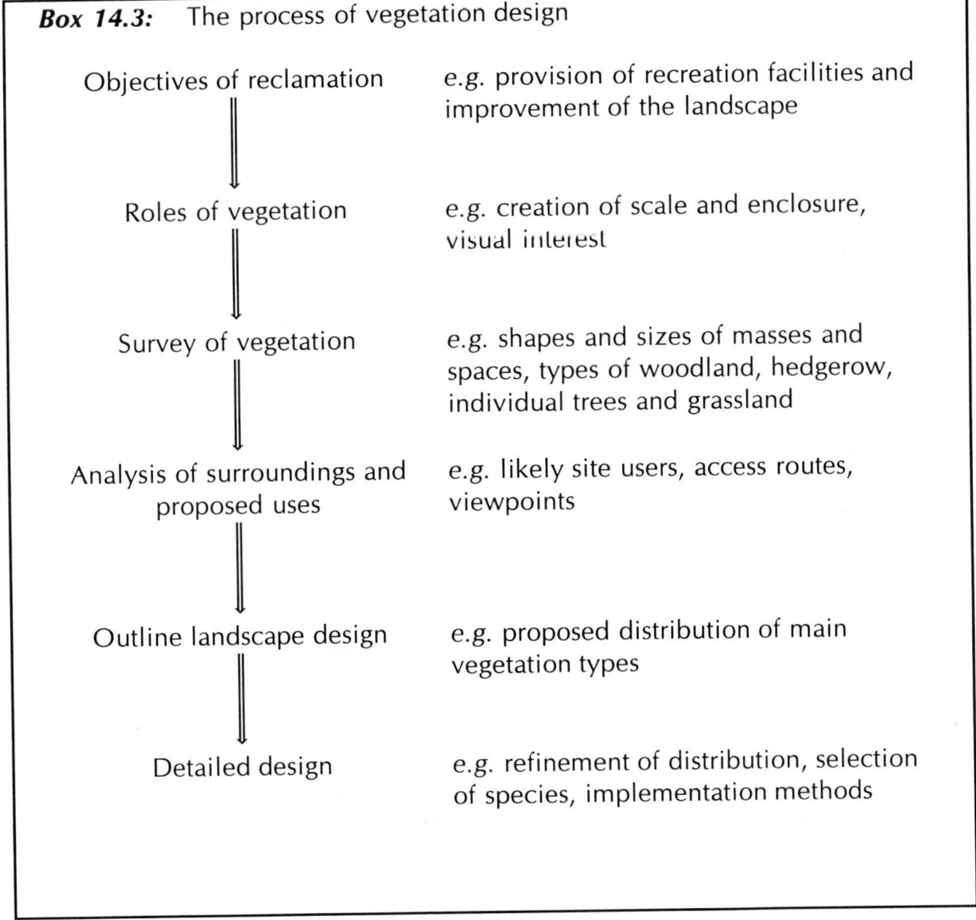

14.2.2 Natural approach to revegetation

Some sites which are to be reclaimed may already have a substantial vegetation cover which has enabled informal new uses to be developed by local residents. These sites and uses can be formalised, for example by additional vegetation works and new accesses. This natural vegetation development provides a model for a natural approach to vegetation design (see Box 14.4 and Photographs 14.1 and 14.2).

Box 14.4: Natural colonisation : a model for vegetation design

Nature and time have combined to produce a vegetation cover on some derelict sites which are now used by the community for recreation. These sites have begun to blend into their surroundings. By studying and copying these natural processes it is possible to achieve similar results in a shorter time scale at relatively little cost. Vegetation types and species which are able to grow on derelict sites are those best suited to the harsh conditions. By working mainly with these vegetation types, the amount of site improvement required for success can be reduced, thus reducing costs. The long-term maintenance requirements of vegetation selected and established in this way will also be reduced since the objective will be to maintain a semi-natural state, not a highly artificial one.

The natural approach to revegetation seeks to establish a vegetation cover appropriate to the selected after-use using simple methods. By a careful process of surveying and conserving naturally developed vegetation, assessing the characteristics of the site and selecting plants which will grow satisfactorily with the minimum of site amendment, a naturalistic vegetation can be developed with limited expenditure. The natural approach makes use of on-site materials, readily available inexpensive amendments and natural colonisation to achieve a diverse and robust vegetation cover which has the added attraction of being relatively low in cost.

Natural colonisation can be used and manipulated to produce a diverse plant cover on sites, or parts of sites, where the substrate has been suitably prepared and a rapid cover is not essential for surface stabilisation. The process of colonisation and succession can be monitored and manipulated by carefully planned inputs of seed, plants and nutrients.

Photograph 14.1:
This former mineral railway has developed naturally to become a well-used path (source: Richards, Moorehead and Laing Ltd)

Photograph 14.2: An attractive mixture of mountain grasses and heather has developed naturally on this old colliery spoil heap (source: Welsh Development Agency)

14.3 The implications of substrate characteristics for revegetation

14.3.1 The principal characteristics

The waste materials from the coal and steel industries present a range of physical and chemical characteristics which may restrict plant growth. The principal characteristics which have major implications for revegetation are:

- extremes of pH;
- toxicity;
- lack of plant nutrients;
- low organic matter content;
- coarse-grained material;
- compaction or consolidation.

14.3.2 Extremes of pH

Substrate pH influences plant growth mainly through its effect on the solubility of chemical elements, including those which are directly toxic to plants and those which are required as nutrients. Figure 14.2 shows the relationship between pH and the availability of the major plant nutrients. Most productive agriculture requires a soil pH of between 5.5 and 7.5 for satisfactory crop growth, and a pH of 6.5 is often quoted as optimal.[153] At this pH nutrient availability to plants is at a maximum and toxicity at a minimum. Soils of lower pH are commonly found supporting commercial forestry, extensive grazing or semi-natural vegetation, and a limited range of plant species will colonise very acidic substrates of pH 3.5 or lower. Few leguminous plants are well-adapted to acidic soils and nutrient availability may be limited by immobilisation in such soils. The production of self-sustaining swards therefore requires the correction of excessive acidity, or the selection of vegetation tolerant of the acid conditions which exist.

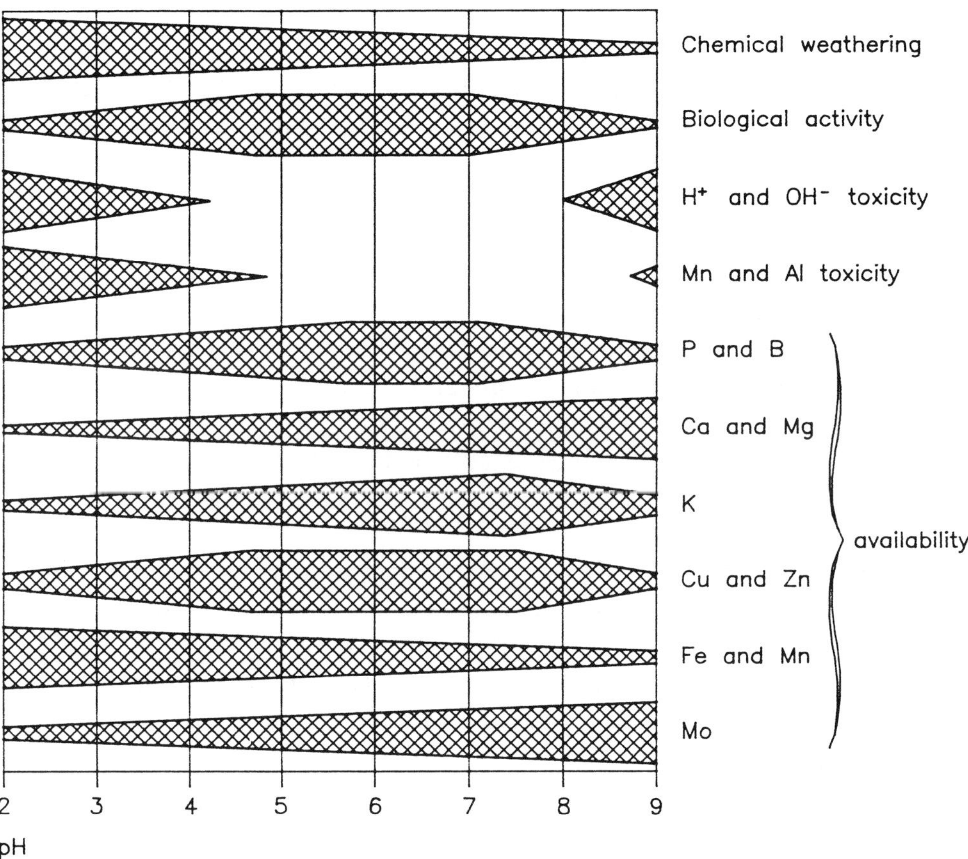

Figure 14.2: The influence of pH on nutrient availability

Highly alkaline substrates such as blast furnace slag are also inhospitable to productive agricultural species, but swards of *Festuca rubra*, *Dactylis glomerata*, *Phleum pratense* and *Lolium perenne* have been established and maintained experimentally on weathered slag to which sewage sludge and annual NPK fertiliser dressings were applied.[99] Weathered slag has a pH of 7.5-8.5, which is similar to that of naturally calcareous soils and chalks. The species of these nutrient-poor, calcareous habitats sometimes colonise weathered slags to produce a diverse grassland flora (see Box 9.5). Diverse grasslands have also been established experimentally on weathered slag substrates using controlled fertiliser additions.[14]

14.3.3 Low nutrient status

Nitrogen and phosphorus are available to plants at extremely low concentrations in most colliery spoils and steel slags. Both nutrients are essential for the growth of plants and when supplies are inadequate young plants will fail to establish whilst established plants will become moribund

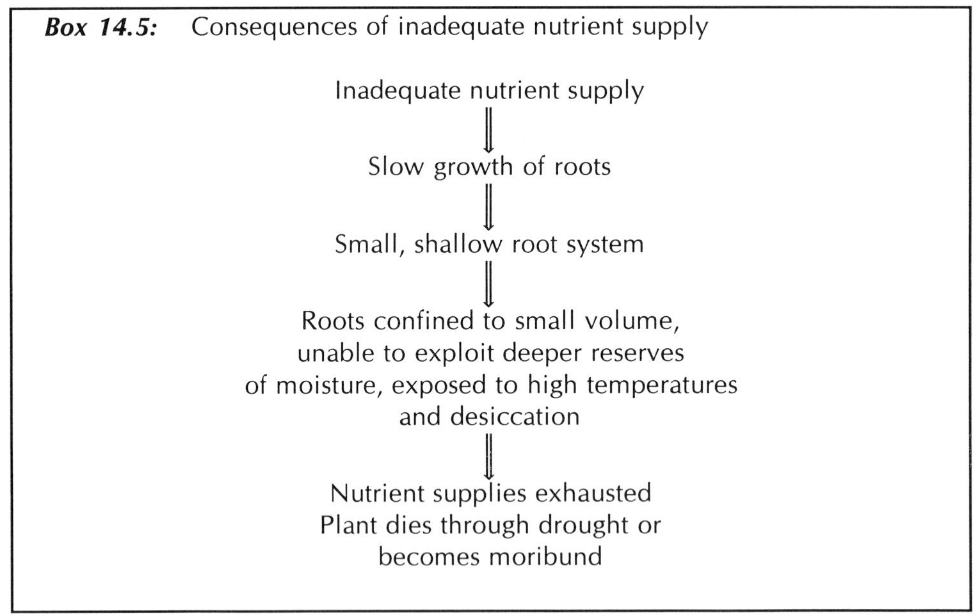

and decline (see Box 14.5). Legumes are particularly sensitive to a lack of available phosphate and the fixation of atmospheric nitrogen in the absence of adequate phosphate supplies will be poor.[185] Figure 14.3 shows the marked response of agricultural grass swards to treatment with fertilisers, and indicates the need to ensure that all major nutrients are in adequate supply. Although many non-agricultural species of herbaceous and woody plants will grow satisfactorily at much lower spoil nutrient concentrations the consequences of inadequate nutrient supply are the same - poor growth and susceptibility to drought.

14.3.4 Low organic matter content

Most of the nitrogen reserve in soils is in the form of organic matter containing typically 5% nitrogen which is mineralised at about 2% per year.[206] If organic matter is lacking it follows that the reserve of nitrogen is also poor. Organic matter contributes to the structuring of soils, particularly those with a high clay content, by stabilising aggregates of these fine particles as described in Section 5.3.3. Poorly structured soils, such as colliery spoil, will consolidate as they weather and the clay content increases. Slags, however, will weather very slowly and remain coarse-grained with large pore spaces. Consolidation of colliery spoil leads to impeded drainage, poor available water capacity and the restricted root extension of plants. As a result, extremes of summer drought and winter waterlogging are common on colliery spoils.[109, 205] Organic matter also absorbs moisture and so directly improves the available water capacity of the spoil. Colliery spoils and the wastes from iron or steel production are usually devoid of organic matter whereas a typical arable soil contains 0.5-2.5% organic matter. Soils under permanent pasture may contain much more organic matter than arable soils.

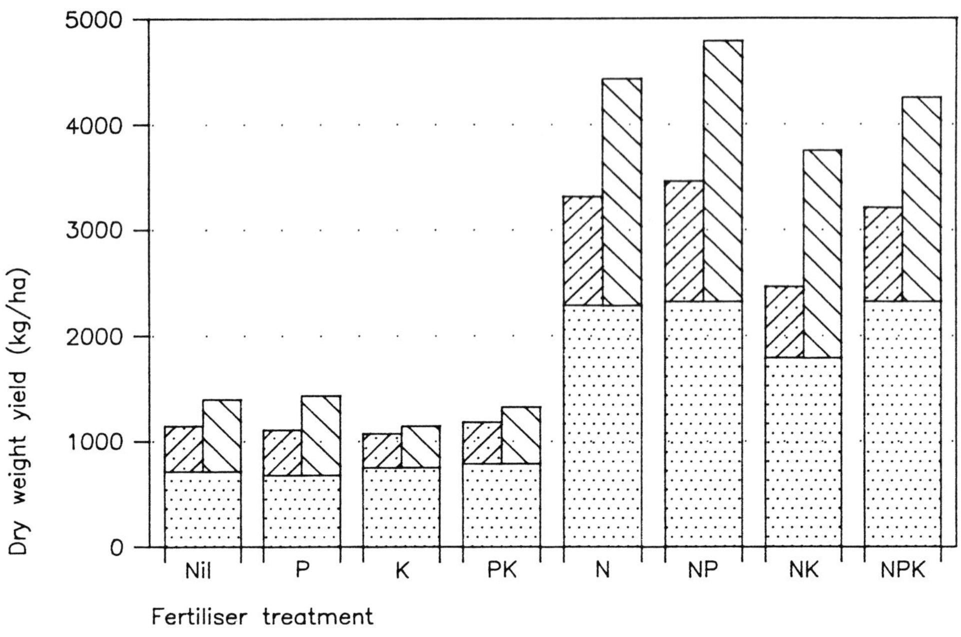

Second year's growth;
re–treated
untreated

Figure 14.3: The response of an established grass sward to fertiliser treatment (from Bradshaw and Chadwick, 1980[40])

14.3.5 Coarse-grained materials and compaction

Substrate particles larger than 2mm retain very little water and where these are abundant they can significantly dilute the finer-grained soil. Figure 14.4 shows how the available water capacity of soil is related to its texture. Very coarse-grained spoils and slags, and severely compacted or consolidated spoils, have few soil pores of the size needed to hold water against gravity drainage. Coarse-grained materials will therefore be excessively free draining and plants with roots confined to the upper layers will be subject to extreme water stress. However, once deep-rooting plants, particularly trees, become established, and their roots extend below the zone of greatest soil-drying, even very coarse-grained materials can support a vegetation cover.

Spoil compaction is a common result of the use of mining or engineering machinery to place or regrade spoils (see Section 5.3). Compaction causes poor drainage and waterlogging during wet weather, and creates a lack of soil pores able to store water for use by plants in dry conditions (see Photograph 14.3). Plants growing in compacted spoils produce very shallow root systems due to:

- the physical impenetrability of compacted spoil;
- the seasonal death of roots in the saturated zone above the compacted layer.

As a result, the root system:

- provides poor physical anchorage for the plant;
- is confined to the zone of greatest temperature fluctuation;
- is confined to the zone of greatest moisture stress;
- is unable to exploit nutrients beyond the surface layer;
- is unable to exploit nutrients in the surface layers during drought or waterlogged periods.

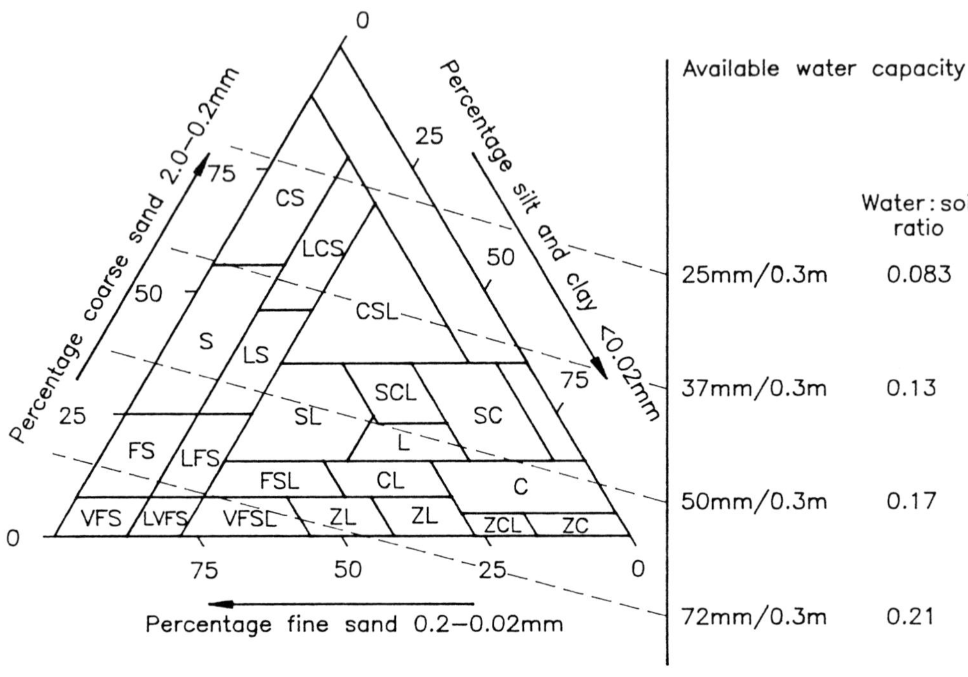

Prefixes: C – coarse
F – fine
VF – very fine
L – loamy

Symbols: S – sand
Z – silt
C – clay
L – loam
CL – clay–loam
SL – sandy–loam
ZL – silty–loam

Figure 14.4: The soil texture triangle showing available water capacity (after Winter, 1974[264])

Photograph 14.3: Waterlogging of the root zone, caused by excessive compaction, has killed the roots of this tree (source: Richards, Moorehead and Laing Ltd)

14.3.6 Invasive and persistent weeds

Many derelict sites contain stands of weeds which, if spread during reclamation works, will become extensive, detrimental to the desired land use, and very expensive to control. Particular problems have been experienced with:

- Japanese knotweed *(Fallopia japonica)* - spread predominantly by rhizomes;
- Himalayan balsam *(Impatiens glandulifera)* - spread by seed;
- Bracken *(Pteridium aquilinum)* - spread by rhizomes and spores;
- Horsetail *(Equisetum* spp.) - spread by rhizomes and spores.

The assessment of potential reclamation sites in order to identify invasive weeds before any ground disturbance commences will enable a weed control strategy to be developed (see Box 2.2). Guidance on Japanese Knotweed identification and control has been developed for the Welsh Development Agency[201, 203] (see Photograph 14.4).

14.4 The treatment of substrate characteristics

14.4.1 Introduction

The characteristics of the substrates found on abandoned coal mining and iron and steel making sites vary greatly from site to site and within individual sites. It is therefore essential that the different substrates are assessed carefully so that:

- the materials most suitable for plant growth are identified;
- the characteristics of each material are known;
- appropriate substrate treatments can be implemented;
- plant selection takes account of the substrate characteristics.

Two approaches may be considered once the substrate characteristics are known:

- selecting plants which will succeed in the substrates present. This approach is expanded in Section 14.5;
- treating the substrates to make them suitable for the plants which are desired.

Figure 14.5 summarises the techniques which are applicable to the treatment of common problems associated with colliery spoils and iron and steel-making slags. Many are standard practices in agriculture, forestry or civil engineering. Those techniques which have particular application to reclamation are described in the following sections.

Photograph 14.4: Japanese knotweed will spread rapidly if introduced to a site (source: Richards Moorehead and Laing Ltd)

14.4.2 Deep cultivation or ripping

Compaction and consolidation can be overcome by cultivation, but only if the spoil is loosened to create drainage pathways. The cultivation must therefore extend through the layer of impeded drainage to reach unconsolidated material beneath, or must be combined with the installation of a drainage system. Deep cultivation over an impermeable substrate will create a saturated marshy layer. This can be avoided if a minimum gradient of 1 in 10 is created during regrading or landforming operations,[32] with open ditches or other outlets for the drainage water (see Figure 14.6 and Photograph 14.5).

Deep cultivation within the restricted space available at developed sites can be difficult to achieve effectively. Main planting areas must be identified in the masterplan before reclamation, and these areas protected

Derelict land categories / Problems	Instability/erosion	Water shortage	Water–excess	Texture–coarse	Texture–fine	Compaction	Nutrient deficiency	Low ion exchange capacity	Alkalinity	Acidity–actual	Acidity–potential	Toxic–metals	Toxic–other	Salinity
Colliery spoil: opencast	●	⊠	○	○		⊠	●	○		○	○			○
deep mine	●	●	○	○		○	●	○		○	○			○
washeries	●	⊠	○	○	●		●	○						
Iron and steel: disused plant			●	○		⊠						○		
slag and spoil	●	●		●	○	○	●		●			○	●	
Coal carbonisation sites	○					○			○	○		○	⊠	
Railway land				●	●	○	●					○	○	
Demolition rubble			○	●		●	●					○	○	
Treatment														
Weathering				◆		◆	◆*		◆*	◆*		◆*	◆*	◆
Grading	◆	◆*	◆*											
Ripping	◆/◈*	◆	◈*		◆	◆								
Compacting	◆	◈*	◆*	◆	◈									
Liming							◆*		◈*	◆	◆*	◆*	◆*	
Fertilising							◆					◆*	◆*	◈*
Fine inorganic material		◆	◈	◆	◈			◆*						
Coarse inorganic material	◆	◈	◆	◈	◆									
Bulky organic material	◆*	◆		◆	◆	◆	◆*	◆*	◆			◆	◆*	◆
Drainage	◆		◆											
Irrigation		◆					◆/◈*					◆*	◆*	◆
Establish vegetation	◆		◆*			◆	◆	◆*						

○ : Present in some cases ⊠ : Serious problem in most cases ◆: Beneficial

● : Present in most cases * : In some cases ◈: Harmful

Figure 14.5: The treatment of substrate characteristics for revegetation (after Robinson Jones Partnership, 1987[209])

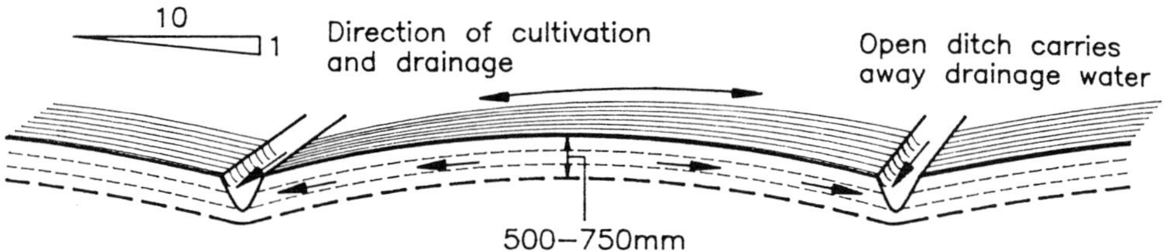

Compacted or impervious spoil

Deep cultivation of dry spoil coupled with an adequate gradient (*e.g.* 1:10) can provide a deep zone of well drained substrate in which roots can exploit the available water. Winged tines improve the shattering of compacted spoil between the tines. A 300hp (224kW) tracked tractor is needed to cultivate compacted spoil.

Figure 14.6: Deep cultivation to create substrate drainage

Photograph 14.5: Winged-tine ripper suitable for heavily compacted colliery spoil (source: the Forest Authority)

from compaction. Areas affected by building construction will require deep cultivation if ripping is impractical in confined areas.

Grass and planted areas should only be designed as soakaways for surface water runoff if adequate drainage capacity can be ensured, otherwise ponding will occur, resulting in the death of vegetation.

14.4.3 Use of topsoil and subsoil

Topsoil and subsoil are extremely valuable materials within reclamation sites, and should be conserved wherever practicable. The conservation and replacement of soils is a standard and critical practice in the restoration of most opencast mining sites, but very few abandoned deep-mine sites contain appreciable areas of undisturbed topsoil. The importation of soil to cover less suitable substrates is an attractive option which can overcome many obstacles to revegetation, but it is rarely adopted on a large scale since good quality topsoil is usually in short supply, is expensive to transport and can leave the source area in a derelict state. When soil is available locally from, for example, road or building construction sites, its use should be considered carefully:[34]

- topsoil is susceptible to a loss of structure, particularly if moved when wet (see Box 13.6);
- topsoil may contain a large number of weed seeds and fragments of pernicious weeds which will compete with young trees or shrubs in the placed material (see Section 14.3.6);
- topsoil may have a poor balance of nutrients, or may be too fertile for the establishment of low-maintenance swards or wild flowers;
- subsoil may be suitable for use if nutrients and organic matter are added;
- plant roots may not extend from the soil layer into the underlying substrate, leading to poor anchorage and the risk of drought.

14.4.4 Use of colliery spoil as a cover material

The fused slags from iron and steel-making are particularly inhospitable for plant growth. In some areas of Europe, coal mining and iron or steel making took place in close proximity and so the wastes from these activities occur together. Since colliery spoil is generally less difficult to make suitable for plant growth than slags, one option is to place a deep layer, 1m or more, of colliery spoil over the slag during reclamation, and to prepare this as the growing medium. This approach was successfully used in the preparation of the site for Garden Festival Wales 1992 at Ebbw Vale, South Wales. The site previously contained disused coal mines and a steelworks. Colliery spoil has also been used as a cover to reclaim other wastes (see Box 14.6).

Box 14.6: The use of colliery spoil as a cover material

Colliery spoil may be used to improve or cover other wastes during a reclamation scheme. This has been achieved at sites where colliery spoil heaps have been in close proximity to other wastes. Where the colliery spoil layer is used to cover a material such as unweathered slag which will not be exploited by plant roots, it is essential that the colliery spoil is sufficiently deep, and sufficiently improved and maintained to provide all the resources (water, nutrients, anchorage) needed by the developing vegetation. The substrate water regime may be predicted with the aid of spoil texture (see Figure 14.4) and climatic data.

At one site in Wales, a burning colliery spoil heap was removed and cooled and the material spread over metalliferous mine waste sealed with a polyethylene membrane. A 500mm deep layer of colliery spoil was used and vegetation was established directly onto the colliery spoil. Long-term trials had been established to test the method. In these trials, slag had also been used as a treatment, but vegetation failed because of the poor water holding capacity of the slag.

At a former steelworks site, colliery spoil from nearby tips was used to cover slag in order to establish vegetation. The slag was regraded to form benches to retain the colliery spoil. A depth of 1m was placed for tree planting; 450mm depth in areas prepared for grass (see Photograph 14.6).

Photograph 14.6: Placing colliery spoil over slag as a growing medium for trees and grass (source: Welsh Development Agency)

The physical and chemical characteristics of spoil heap materials can vary significantly with depth (see Section 5.3). The surface layer of a spoil heap may be much more suitable for plant growth than the bulk of the material, due to:

- weathering which improves texture;
- leaching which reduces acidity/alkalinity/salinity;
- microbial activity and plant colonisation which add organic matter and nitrogen.

Where it is proposed to use spoil materials as a substrate for plant growth the surface layer should be assessed separately and, if it is suitable, stripped, stored and replaced as a surface layer. If the spoil materials are mixed or used indiscriminately the benefits of the weathered surface layer will be lost.

Many other natural materials can be made into suitable substrates for plant growth, if the essential characteristics of available water capacity, nutrient supply, pH and soil aeration are provided. Soil-forming materials which have been used in land reclamation or restoration projects include shales, overburden, freshwater and marine silts, tailings from mineral processing, crushed brick and concrete, and a variety of organic residues.

14.4.5 Lime to correct acidity

The processes leading to spoil acidity are described in Section 5.3. Box 14.7 shows a method for calculating how much lime is required to bring the pH to the desired value and maintain it. The quantity of lime can be over 100t/ha if the spoil is highly pyritic, and so an accurate assessment of the requirement can also be economically important. Photograph 14.7 shows how pyritic acidity affects grassland.

The material most commonly used to raise the pH of colliery spoils is lime, in the form of ground limestone (calcium carbonate, $CaCO_3$). Magnesian limestone ($MgCO_3$) is avoided on sites where spoil salinity is high as the magnesium content contributes to the salinity.

14.4.6 Sources of nutrients and organic matter

The types and sources of chemical fertilisers are well known, but many cheap or waste materials can be used in their place as sources of nutrients and organic matter. The characteristics of some soil amendments are given in Table 14.1. Other materials which could be used may be available on a local basis. Small-scale trials of novel materials would help to identify of short-term toxicity or nutrient imbalance.

Box 14.7: The calculation of lime requirement in colliery spoil

Acidity in colliery spoil can be conveniently divided into:

- active acidity: the free hydrogen ion content which determines substrate pH;
- reserve acidity: the exchangeable hydrogen ion content which determines resistance to change in pH by buffering action;
- potential acidity: the acid generating power of the spoil.

Active and reserve acidity may be determined by standard soil science methods and a lime requirement calculated.

Should the spoil contain pyrite then the potential to produce acidity in the future could be considerable. An estimate of this potential and hence a lime requirement can be determined by analysis.[66, 96] The steps are:

(i) determination of pyritic sulphur;

(ii) calculation of the amount of lime the pyritic sulphur in the spoil would consume if it was all to be oxidised. 1% sulphur = 34.45 tonnes $CaCO_3$ equivalent per 1000 tonnes of material;

(iii) determination of the acid neutralising capacity of the spoil, in tonnes $CaCO_3$ per 1000 tonnes material;

(iv) subtraction of (iii) from (ii) to produce a lime requirement in tonnes $CaCO_3$/1000 tonnes spoil.

Figure 5.2 shows the levels of pyrite and acid neutralising capacity found in some colliery spoils and Figure 14.7 shows the effect of repeated additions of lime to a very pyritic colliery spoil when the lime added did not neutralise all the potential acidity.

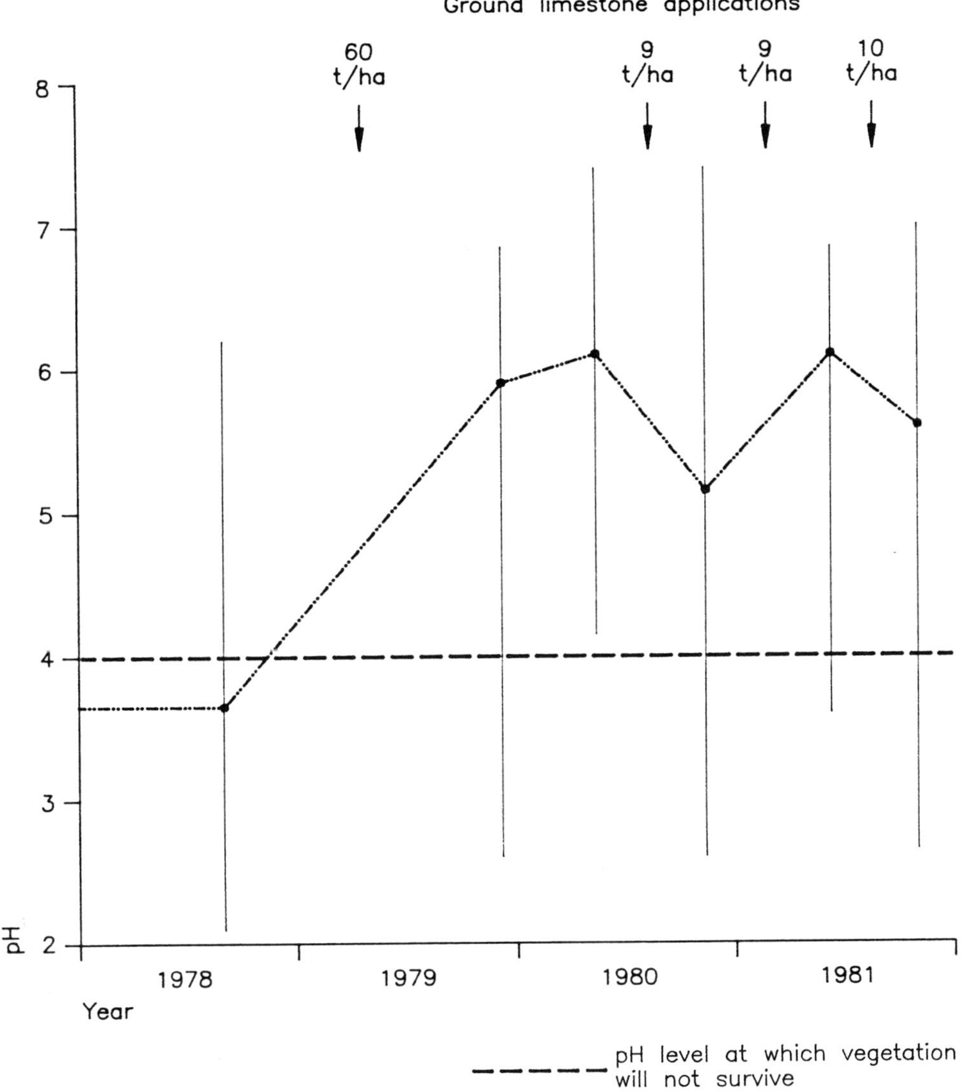

Figure 14.7: The effect of liming on mean pH and range on a pyritic colliery spoil heap. Even with repeated lime applications, the pH of parts of the site is too low to support plant growth

Photograph 14.7: Acidity from pyrite oxidation has completely killed the grass sward allowing erosion to begin (source: Welsh Development Agency)

14.4.7 The role of legumes

The use of legumes such as clover (*Trifolium* spp.) to provide a long-term supply of nitrogen is a traditional agricultural practice which is the subject of renewed interest due to the rising monetary and energy cost of fertiliser manufacture. These legumes, if provided with sufficient phosphate, moisture and temperatures for optimum growth, can fix 100kg or more of nitrogen per hectare per year.[67, 185] Much of this nitrogen is contained within organic matter in the soil where it contributes to the 'nitrogen capital' and is slowly mineralised for uptake by other plants. Figure 14.8 shows that although colliery spoil contains much nitrogen, very little of this is available to plants. If nitrogen fertiliser is applied to swards on colliery spoil the accumulation of nitrogen in organic matter is increased. However, the use of a legume in place of nitrogen

Table 14.1: Organic soil amendments (after Coppin and Bradshaw, 1982[56])

Material	Usual composition (% dry solids)				Usual application rates (dry t/ha)	Special problems or advantages
	N	P	K	Organic matter		
Farmyard manure	0.6-2.5	0.1	0.5	24-50	5-40	Variable
Pig slurry	0.2-4.0	0.1	0.2	3	5-20	High water content, possibly high Cu
Poultry manure, broiler	1.5	0.9-2.5	1.6-2.5	60-80	2-10	High levels of ammonia, odours
Poultry manure, battery	2.0-4.0	0.5	0.6	35	2-10	
Sewage sludge, digested	2.0-4.0	0.3-1.5	0.2	45	5-50	Possibly toxic metals and pathogens, odours
Sewage sludge, raw	2.4	1.3	0.2	50	5-50	
Mushroom compost	2.8	0.2	0.9	95	5-20	High lime content
Domestic refuse, composted	0.5	0.2	0.3	65	20-70	Contains miscellaneous objects
Brewery sludge, digested	1.5	0.9	0.3		5-20	Uncommon
Peat	0.1	0.005	0.002	50	5-10	Variable, high carbon to nitrogen ratio. Production may cause destruction of wetland habitats
Straw	0.5	0.1	0.8	95	5-20	Decomposition uses soil nitrogen
Sawdust	0.2	0.02	0.15	90	10-30	High carbon to nitrogen ratio, requires pulverising, maturing or composting
Woodchips	0.2	0.02	0.1	90	10-30	
Bark	0.3	0.09	0.7	90	10-30	
Lignite, ground	1	0	0	0		High cation exchange capacity

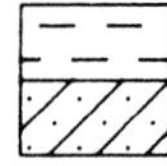

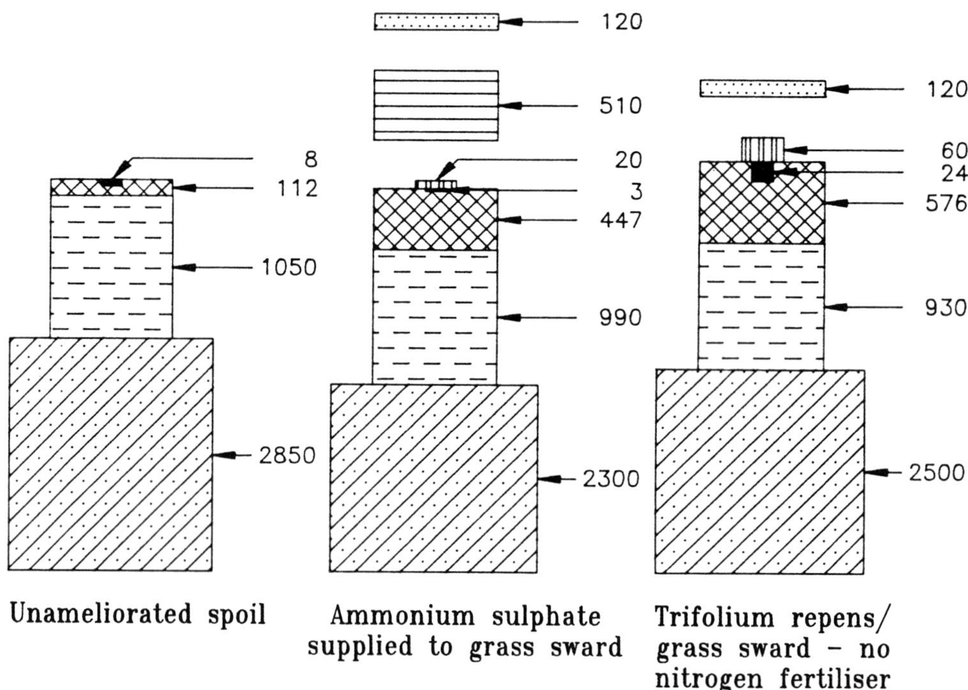

Figure 14.8: The accumulation of nitrogen in colliery spoil seven years after revegetation (after Palmer, 1984[184])

additions, over the same period, results in greater accumulation of organic nitrogen and more mineralisable nitrogen. Figure 14.9 shows that colliery spoils which are vegetated with legumes are able to supply nitrogen to the sward continuously throughout the season whereas colliery spoils which receive nitrogen only from fertiliser inputs quickly lose their power to supply nitrogen to the sward.

The legumes used in colliery spoil reclamation include agricultural herbaceous species; perennial shrubs such as lupin (*Lupinus arboreus*), gorse (*Ulex europaeus, U. gallii*) and broom (*Cytisus scoparius*); and trees such as black locust (*Robinia pseudoacacia*). alder (*Alnus glutinosa, A. incana, A. cordata*) although not a legume, also fixes nitrogen and is perhaps the most important tree used in land reclamation in Britain. Legumes will only fix nitrogen if inoculated with the correct strain of *Rhizobium* bacteria. As *Rhizobium* is likely to be absent from the substrates of derelict land, it is essential to inoculate the seed before sowing (see Section 14.6). *Alnus* and *Robinia* should become inoculated in the tree nursery, but if this has not occurred the soil from beneath established stands of these trees can be applied to the root zone at planting, to encourage inoculation.

14.4.8 Mycorrhizae

Mycorrhizae are naturally-occurring associations between the plant roots and symbiotic fungi, which have the effect of increasing the intensity and surface area of the root system within the rooting zone and therefore increasing the extraction of nutrients from a volume of nutrient-poor soil (see Box 5.2). Mycorrhizae are of particular benefit to species which typically grow on coarse-grained, impoverished and base-poor soils such as heathland and mountain soils. Pines and birches are frequently reported to benefit from mycorrhizae in such situations.[190] As many colliery spoils share the characteristics of these substrates there is considerable potential for the deliberate introduction of the appropriate fungi to trees planted on reclamation sites.

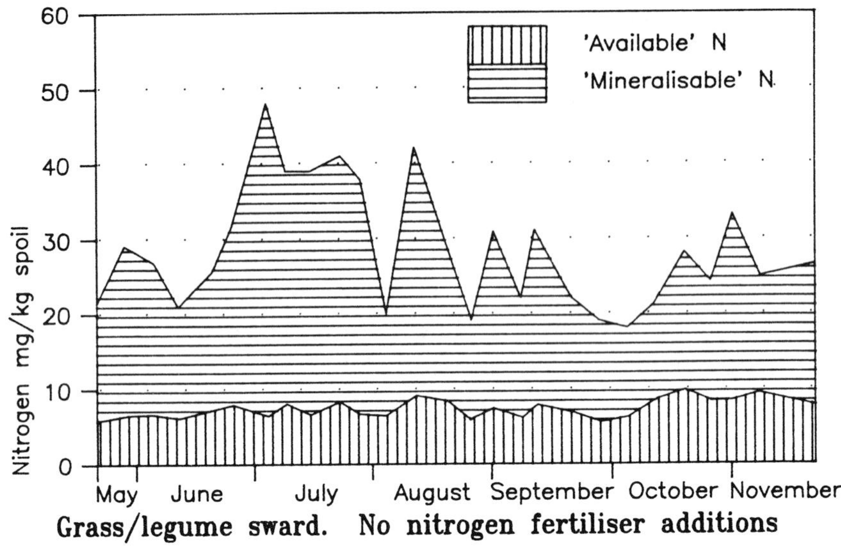

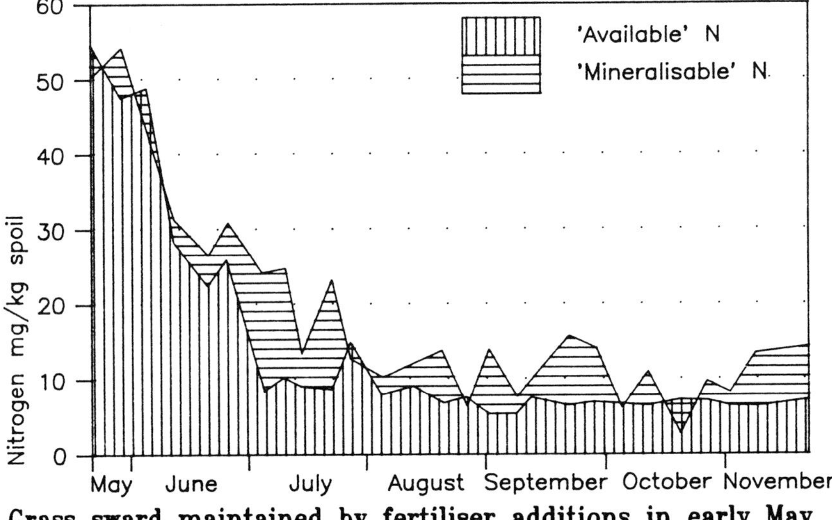

'Available' nitrogen is that immediately available in a mineral form (NH_4^+, NO_3 or NO_2^-). 'Mineralisable' nitrogen is that capable of being mineralised from spoil organic matter and is an indication of spoil fertility.

Figure 14.9: The short term effect of fertiliser nitrogen applied to vegetation on colliery spoil (from Palmer, 1984[183])

14.4.9 Water-storing polymers

The available water capacity of substrates can be increased by the use of polymers such as polyacrylamide gels. The use of these products has been reviewed.[128] The products were developed principally for use in arid zones, but offer a means to improve the water supply in individual tree pits or within the rooting zone. The gels absorb typically 100-400 grammes of water per gramme of dry granule when excess water is present. This water is then released as the substrate dries. The gels have a life of between 1 and 4 years in most substrates under temperate conditions, and can therefore buffer vegetation against drought periods.

14.5 Species selection

14.5.1 Site improvement with pioneer crops

The most commonly used procedure for the revegetation of derelict land has been to select the vegetation appropriate to the proposed land use and to establish it immediately after earthworks have been completed. This has entailed considerable efforts to modify the substrates to suit the vegetation. By contrast, those concerned with the restoration of opencast mined land, with afforestation and with agriculture frequently establish a pioneer crop of a short duration as part of a programme of soil improvement, before the final long-term crop or vegetation is established. Pioneer crops can provide:

- organic matter;
- improved soil structure;
- nitrogen (if legumes are used);
- shelter for exposure-intolerant plants (nurse-crop).

Nurse-crops should be selected with care to avoid excessive competition with the intended final vegetation. Annual grasses such as 'Westerwolds' rye grass have been used to protect trees and shrubs sown directly into colliery spoil. Short lived shrubs, particularly the legumes lupin, gorse and broom, have been used to protect seedling trees and to provide additional nitrogen.[151, 196] Vigorous, exposure tolerant pioneer trees such as alder and birch can produce a young woodland environment suitable for the growth of ecologically or commercially attractive canopy species such as oak (*Quercus robur, Q. petraea*) ash (*Fraxinus excelsior*) and beech (*Fagus sylvatica*).

It has become common practice to plant mixtures of woodland species regarded as pioneer, canopy and understorey species in order to create naturalistic woodlands. The practice is based on the expectation that the shade-tolerant canopy species will eventually emerge from the protection provided by the fast-growing, light-demanding pioneer species to form the eventual woodland canopy. Considerable success has been achieved by this means, when ground improvement and climate have been satisfactory.[105] This practice compresses the traditional succession pattern, by planting all three groups of plants at the same time. Significant management effort can be required if the pioneer species become too dense or if the canopy and understorey species are unable to survive exposed conditions until sufficient protection is provided by the pioneers. Where the work of vegetation establishment can be implemented in phases over say, 5-20 years, a system of site improvement by pioneer and nurse crops into which less robust but more desirable species are added when conditions are suitable, offers the potential for lower costs and more successful plant establishment. The creation of a diverse range of habitats is a common objective of many reclamation schemes (see Box 14.8). Diversity of species and structure can also be achieved by creative management over the medium term, as discussed in Chapter 15.

14.5.2 The performance of vegetation

Once the objectives of revegetation have been defined, the process of species selection and refinement is one of seeking to bridge the gaps between the required performance of the vegetation and the ability of each

Box 14.8: Habitat creation

Reclamation schemes offer an excellent opportunity for the creation of wildlife habitats to replace those which are being lost through Man's activities. Developing techniques for the establishment of diverse flowering grasslands, wetlands and waterbodies, woodlands and scrub continue to widen the range of possibilities.

In the majority of cases there will be no overriding reason for a single habitat type to be created over a whole site. More commonly, habitat creation will seek to produce a diversity of:

- vegetation types e.g. woodland, scrub, grassland, wetland;
- shapes and sizes of habitat;
- edges and transitions (ecotones);
- species within each habitat.

Such diversity is greatly assisted by appropriate landform design (Box 13.5) but can be achieved by :

- designing irregular shapes and mosaics to maximise the edges and transitions;
- designing species mixtures to create variety of light and shade;
- modification of the substrate pH and nutrient content.

candidate species to grow successfully on the site. Figure 14.10 summarises this process. As a general rule plants which can tolerate poor site conditions do not produce rapid growth or high crop yields. Where such performance is the objective and site conditions cannot practically be improved then a compromise must be found or the objective reviewed. A summary of the many criteria which contribute to the selection of species for a range of land uses is given in Table 14.2. Within each species the performance and characteristics of ecotypes, varieties or cultivated varieties (cultivars) may vary greatly. This is particularly true of plants used in agriculture, and of plants with a wide geographical range. The tolerance of climate and substrate, and the growth habits exhibited by different ecotypes, varieties and cultivars within one species

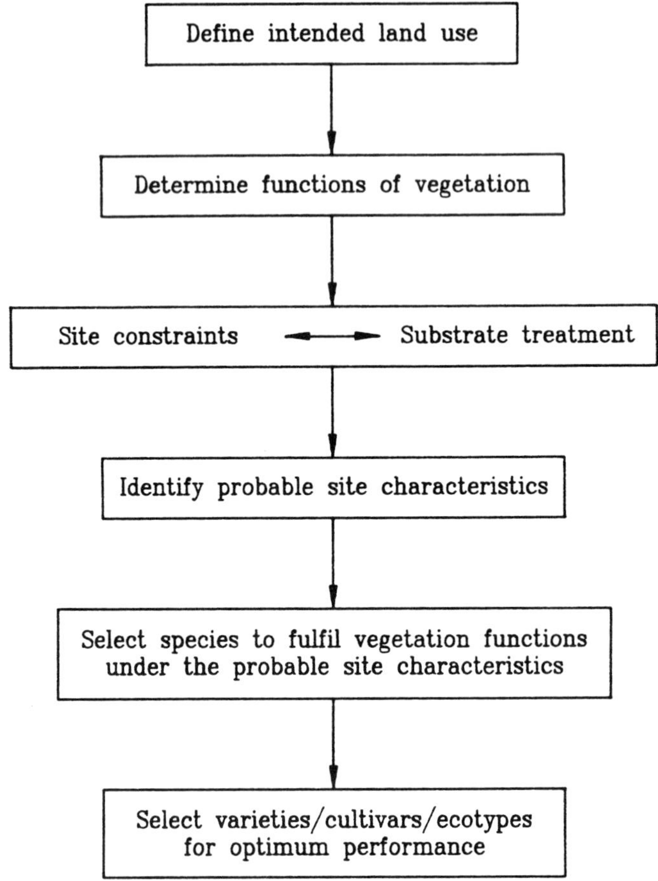

Figure 14.10: Species selection flowchart

Table 14.2: Criteria for the selection of plant species

Criterion	Agriculture	Ground cover				Trees and shrubs			
		Sports & Amenity	Wild	Under trees	Erosion control	Forestry	Amenity	Woodland (wild)	Screening
Climate:									
Drought resistance		X	X	X		X		X	
Frost hardiness		X	X	X		X	X	X	X
Exposure (especially coastal)						X		X	
Cold (short growing season)	X	X	X	X		X		X	
Land use:									
Wildlife value			X	X		X		X	
Nativeness			X					X	
Palatability	X								
Productivity, high or low	X	X	X	X		X			
Screening value							X	X	X
Timber quality						X		(X)	
Soil:									
pH	X	X	X	X	X	X	X	X	X
Fertility, high or low	X	X	X	X	(X)	X	X	X	X
Texture	X	X	X	X		X	X	X	X
Soil depth						X	X	X	X
Moisture availability	X	X	X	X	X	X	X	X	X
Flooding tolerance							(X)	(X)	
Pollution tolerance							X		X
Role:									
Pioneer/nurse				X	X	X		X	X
Climax			X					X	
Soil builder	X	X	X	X	X	X	X	X	
Quick establishment		X		X					X
Ecotypes:									
Cultivar	X	X	X	X					
Provenance						X	X	X	
Locally collected ecotypes	X		X					X	X
Plant habit:									
Height			X	X			X		X
Growth rate	X	X	X	X	X	X			X
Rhizomes/stolons		X	X	X	X				X
Suckering habit		X	X	X			X	X	X
Rooting depth					X		X		
Competitiveness			X					X	
Disease resistance	X	X				X	X		

X - important factor, (X) - in some cases.
From Coppin & Bradshaw, 1982.[56]

can be sufficient to govern the success or failure of revegetation. When the objective of revegetation is to create naturalistic habitats or to conserve wildlife, it can be particularly important that only local ecotypes are used. Where natural colonisation of spoil materials has already occurred the plants may be better adapted to the site conditions than commercially available plant material and so the propagation of colonising plants (by seed or vegetative means) should be considered.

14.5.3 Tolerance

In Section 14.4 it was shown that the selection of species or varieties which will tolerate the conditions of the site offers a complementary or alternative approach to treating the anticipated problems which the site presents. The following problems may be overcome partially or completely by the selection of tolerant species or varieties:

- exposure and extremes of temperature;
- substrate wetness;
- substrate dryness;
- substrate acidity or alkalinity;
- low nutrient status;
- contamination by heavy metals *e.g.* from flue dusts;
- salinity.

The use of tolerant species is likely to restrict the range of vegetation functions available since species which possess tolerance of extreme conditions are generally specialised in their adaptation. This approach may therefore require a review of the scheme objectives. The approach is considered to be particularly suited to schemes intended to improve the landscape, provide wildlife habitats and low-key informal recreation facilities, rather than for schemes intended to provide highly productive agriculture or forestry uses of sites.[34] Guidance on the use of metal-tolerant species is given in Box 14.9.

Box 14.9: The use of metal-tolerant species in land reclamation

Species of grasses growing on metalliferous mine sites have been selected because of their tolerance to high concentrations of heavy metals and some of these species are commercially available. These grasses can be used for the direct revegetation of soils and spoils with high concentrations of metals provided that the grasses are tolerant of the specific metals in the soil and of the other substrate and climate conditions. The advantages of using metal tolerant species are:

- they are relatively cheap to establish;
- it is not necessary to import covering material;
- it is not necessary to remove contaminated material prior to vegetation establishment;
- site disturbance is minimal, reducing the risk of metal dispersal;
- the vegetation cover will provide some erosion control.

Disadvantages are:

- if grazing is intended this will need to be carefully controlled so that grazing animals do not ingest large quantities of metals;
- run-off will be contaminated with metals and so vegetation establishment alone may not be a suitable treatment near watercourses;
- use of the site will need to be restricted so that sensitive members of the population (e.g. children) do not have prolonged contact with spoil materials;
- erosion may occur on steep slopes if the spoil material vegetated is fine-grained and erosion control measures (such as the use of geotextiles) are not used.

Local experience and knowledge of plant performance under the conditions expected to be present is particularly valuable since great climatic and substrate variation exists between the sites on which individual species have been used. Where little experience of colliery spoil revegetation exists in a region, much can be learned by characterising the spoils and examining natural vegetation development on other substrates with similar characteristics, under similar climatic conditions.

14.5.4 The implications of management

The selection of species and varieties to fulfil the desired purpose of the vegetation may commit future managers of the site to a particular method and regime of maintenance work. The management requirements of the vegetation *e.g.* cutting, fertilising, liming, thinning are therefore a significant factor in the choice of plant species or variety. Plants which are invasive or excessively vigorous have proved costly to control or eradicate.

14.6 Sowing and planting methods

14.6.1 Introduction

The sowing and planting methods used in reclamation works have been adapted and developed from agriculture and forestry practice, to deal with the coarser-grained substrates and other physical difficulties found on colliery spoil and steel-making slags.

14.6.2 Sowing methods

In all methods of sowing the objective is to place the seed where it receives adequate moisture and protection from desiccation but is able to emerge before the reserves of the seed are exhausted. In situations where a coarse-grained substrate, lack of rainfall and desiccating sun or wind are prevalent it is common practice to protect the seed with mulches.

The most widely used sowing method is broadcast sowing, in which simple agricultural equipment, or hand methods, are used to distribute the seed on to the prepared surface. The seed is protected from desiccation by:

- allowing rainfall to wash the seed into a rough substrate surface;
- covering the seed by very shallow cultivation (harrowing);
- covering the seed with an application of mulch.

In areas of very intense rainfall, many reclamation schemes have involved mulching the seed with poultry house litter which adheres to the substrate, protecting the seed and surface against erosion by water run-off. Litter has been applied at 10t/ha and acted as the principal nutrient source for grass (see Table 14.1).

14.6.3 Specialised techniques

Hydraulic seeding (hydroseeding or hydraulic mulch seeding) is used particularly to apply seed, fertilisers and mulch to inaccessible areas such as steep slopes where ground preparation and mechanical sowing are impossible (see Box 14.10). Hydraulic seeding has a number of drawbacks which restrict its wider use:

- it is more expensive than traditional methods;
- it can aid germination and establishment but will not overcome fundamental substrate problems;
- separate fertiliser applications are required after seeding.

Techniques for the establishment of trees and shrubs by sowing directly into derelict land substrates continue to be developed (Box 14.11). Straw and similar long-fibred mulches are used to protect the seed of trees and

Box 14.10: The use of hydraulic seeding

Hydraulic seeding ('hydroseeding') is a technique for the rapid application of seeds and fertilisers onto an area where, for reasons of access, speed of application or ground conditions, conventional seed drilling or broadcasting techniques cannot be used. It is not an alternative to the proper preparation of the growing medium although the use of mulches can improve the initial establishment of seedlings in situations where full seedbed preparation is not practicable. Specialised hydraulic seeding machinery consists of a tank fitted with a slurry pump, agitator paddles, hoses and a demountable jet. This machinery is usually mounted on a lorry or trailer. The application typically consists of the following:

- seeds;
- legume inoculum - *Rhizobium* bacteria;
- soluble and/or slow-release fertiliser;
- mulch - a bulky material to act as a carrier, protect the seeds, reduce soil moisture loss and provide initial erosion protection;
- stabiliser/binder - to protect the soil surface, seeds and mulch from erosion.

These are mixed with water to form a slurry of 10-15% solids, applied at 2-4 litres/m^2. Where substantial mulch protection is required *e.g.* 2-4 t/ha this is applied as a second layer.

Hydraulic seeding cannot overcome unsuitable substrate characteristics such as excessive droughtiness or infertility. The use of soluble fertilisers can be directly harmful to the seed in the slurry or to the germinating seedling, since evaporation can produce high concentrations of salts. Legumes, which are often essential to the long-term provision of nitrogen, are particularly sensitive to fertiliser damage.

If hydraulic seeding is to be used to vegetate inaccessible slopes, the vegetation type and its long-term maintenance should be considered with care.

Hydraulic seeding is a valuable technique for special situations; it is not a panacea for all unfavourable sites.

Box 14.11: Direct seeding of trees and shrubs

The natural colonisation of colliery spoil heaps by trees and shrubs, and their development into woodland or scrub communities has prompted researchers to develop techniques which mimic this process but take a much shorter time. Success was achieved in landscape schemes on natural soils in the late 1970s. In the early 1980s attempts were made to establish trees and shrubs from seed on colliery spoil and oil-shale heaps ('bings') in Scotland. These were also successful, but the techniques have not yet been applied on a large scale to reclamation sites in Europe.

Successful establishment requires that the substrate and any applied mulch protect the seed from seasonal desiccation and freezing over a prolonged period until dormancy breaks and the germinated seed establishes a root system. Consequently most derelict land substrates require preparation works, and the use of mulches such as chopped straw. At present the techniques are widely considered to be insufficiently reliable for large scale adoption, particularly on the more exposed sites, although demonstration plots have been set up.

The seeding strategy is to produce a rapid annual herbaceous nurse cover, which is succeeded by a shrub layer and eventually by the pioneer and climax trees. Nitrogen-fixing herbs, shrubs and trees are usually included to produce a steady improvement in the substrate.

The strategy requires close monitoring of the vegetation over the course of its development so that imbalances in the density of each layer can be corrected. Excessive herbaceous growth will cause competition for water, nutrients and light. Insufficient growth will leave the tree seedlings exposed to frost and desiccating winds. Management inputs such as selective herbicides, fertilisers and thinning are likely to be required.

shrubs against desiccation. Special machinery has been developed to chop straw and apply it pneumatically to seeded areas. Bitumen emulsion is applied to stick the straw in place. An alternative method, used where very steep slopes are to be grassed, involves the application of a sheet or mat of material over the prepared slope. The seed mixture may be sown before the mat is laid, or incorporated within the mat. Erosion control can be achieved with a woven mesh of coconut fibre, wood fibre or synthetic material. Where seeds require protection from desiccation, composite mats of straw, coconut fibre, paper and synthetic mesh are used.

14.6.4 Tree planting

It is generally accepted that young trees are better able to withstand the stress of transplanting than are older, larger trees, and that young trees are more adaptable to the harsh substrate or climate conditions of newly reclaimed sites.[34] For this reason it is sensible to transplant 1-3 year old trees, approximately 300-900mm tall, when establishing dense or extensive woodland areas. Larger trees of height greater than 2m are used only where good ground conditions and significant visual benefits justify the higher cost of the tree stock and planting.

14.6.5 Tree shelters

Tree shelters are used to provide protection for tree species that are intolerant of exposure, where circumstances dictate that they should be planted on exposed sites. Shelters consist of translucent polypropylene tubes of 80-100mm diameter and up to 1.8m height. They raise the temperature and humidity of the micro-climate around each tree, promoting growth. Tree shelters have been in common use since the mid-1980s[90] but the ability of young trees to adapt to the transition from the sheltered microclimate to the exposure at the top of the shelter is not documented for the severely exposed sites on which most benefit might be expected. An alternative approach of phased planting was described in Section 14.5.1. Tree shelters also provide protection against damage by mammals, herbicides and mechanical disturbance.

14.6.6 Weed control

Weed control is crucial to the successful establishment of transplanted trees and shrubs in many situations, and yet it is often neglected. Newly planted trees suffer significant root damage and can extract water from only a small volume of substrate. Grass and associated herbaceous vegetation is particularly efficient at extracting water from the soil, and can quickly create drought conditions which will kill or restrict the growth of the desired trees. It has been shown experimentally that most soil moisture is lost by transpiration through herbaceous vegetation, and that relatively little is lost by evaporation. A grass sward, whether mown or not, created soil moisture tensions sufficient to restrict water and nutrient uptake by the trees for most of the growing season (see Figure 14.11). This was reflected in the reduced survival and growth of the trees (see Table 14.3).

Infertile colliery spoil can provide a weed-free substrate for tree planting, but in many schemes a grass sward has been sown over the whole site immediately after regrading, in order to protect the surface against erosion. The vigorous varieties of grass used for this purpose are particularly detrimental to young trees (see Box 14.12 and Figure 14.12). Where erosion cannot be avoided by slope design or the use of mulch materials, the approach now adopted is to sow less vigorous grasses and to kill the sward in the planting area with a contact herbicide to produce a dead grass mat amongst which the young trees are planted. This approach requires sufficient time for sward establishment and thorough herbicidal control before planting. Dense planting which will rapidly produce a vegetation cover to protect the spoil surface is desirable. In some schemes the herbicide treatment has been confined to bands or spots so that much of the surface retains its grass cover. This practice reduces the risk of erosion but increases the need for subsequent weed control treatments around each tree. Weed control methods are described in Section 14.7.

14.6.7 Mulching

Mulching the spoil surface will aid the conservation of soil moisture, provided that grass and weeds have been removed. Mulches can also reduce the development of new weed growth but will have little effect on the growth of established or perennial weeds.

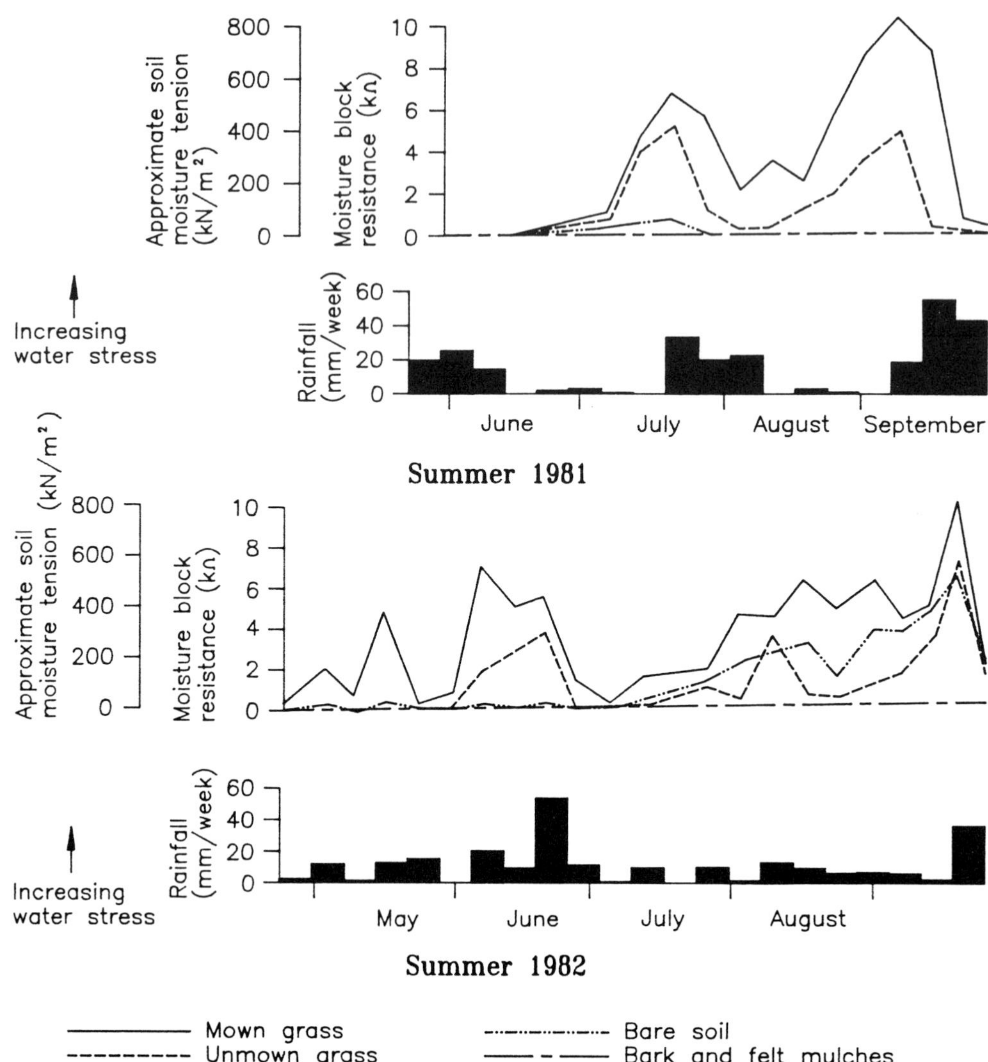

Figure 14.11: Moisture block resistances at 100mm depth under five ground covers in summer 1981 and 1982 (after Davies, 1987[68])

Table 14.3: The effect of grass competition on young trees

First year's growth and foliar analysis results for cherry (*Prunus avium*) at Alice Holt, UK, 1982 (after Davies, 1987[68]).

		Mown grass	Unmown grass	Bare soil
Height growth (cm)		9	31	80
Diameter growth (mm)		3	7	14
Leaf size index[(1)]		29	53	100
Foliar nutrient concentrations (% of dry weight)	N	2.6	2.5	3.4
	P	0.19	0.17	0.20
	K	1.1	1.1	1.6

[(1)] Leaf size index obtained by multiplying leaf lengths and widths.

Two types of mulch are commonly used in reclamation schemes:

- impermeable sheets *e.g.* polyethylene mats and strips;
- coarse granular materials *e.g.* bark, wood chips.

Granular mulches create a zone of still or slow-moving air at the soil surface. The humidity of this air rises so that the rate of evaporation from the soil is reduced. The mulch also intercepts solar radiation, thereby insulating the soil. Rainfall percolates rapidly to the soil surface where it is protected from rapid evaporation. Granular mulches only function if they are free-draining *i.e.* they are coarse-grained (free from particles of less than 5mm size) and relatively non-absorbent.

Box 14.12: Trees on colliery spoil (after Derelict Land Reclamation Research Unit, UK 1979[73])

The survival and growth of trees on colliery spoil are determined by tree species and site conditions. Figure 14.12 gives data on 12 tree species in a trial on colliery spoil established in 1971 in West Yorkshire. Some species (for example *Acer platanoides*) had high survival rates but did not grow whilst others exhibited both good survival and growth (*Betula pendula*, *Alnus* spp., *Salix caprea*, *Populus alba* and *Robinia pseudoacacia*). Survival and growth were particularly affected by competition from grass which itself was affected by fertiliser treatment as the following data shows.

Treatment	Average annual productivity (t/ha/year)		
	Betula pendula	*Alnus glutinosa*	*Salix caprea*
Grass removed, lime added	1	0.6	1.3
Grass removed, basic slag* added	0.7	0.3	0.8
Grass retained, lime added	0.4	0.9	0.5
Grass retained, basic slag* added	0.5	0	0.5

*Basic slag, a product of the steel industry, has a liming effect and is high in phosphate (Box 9.4).

The phosphate in the basic slag encouraged white clover (*Trifolium repens*) growth and the nitrogen fixed by the white clover caused vigorous grass growth. Where basic slag was applied grass growth was vigorous, even on those plots where grass had initially been removed. The result was that tree productivity was poorest where grass was sown and where basic slag was applied.

The results of this trial illustrate the need to match site conditions and treatment to ensure satisfactory growth of trees on colliery spoil.

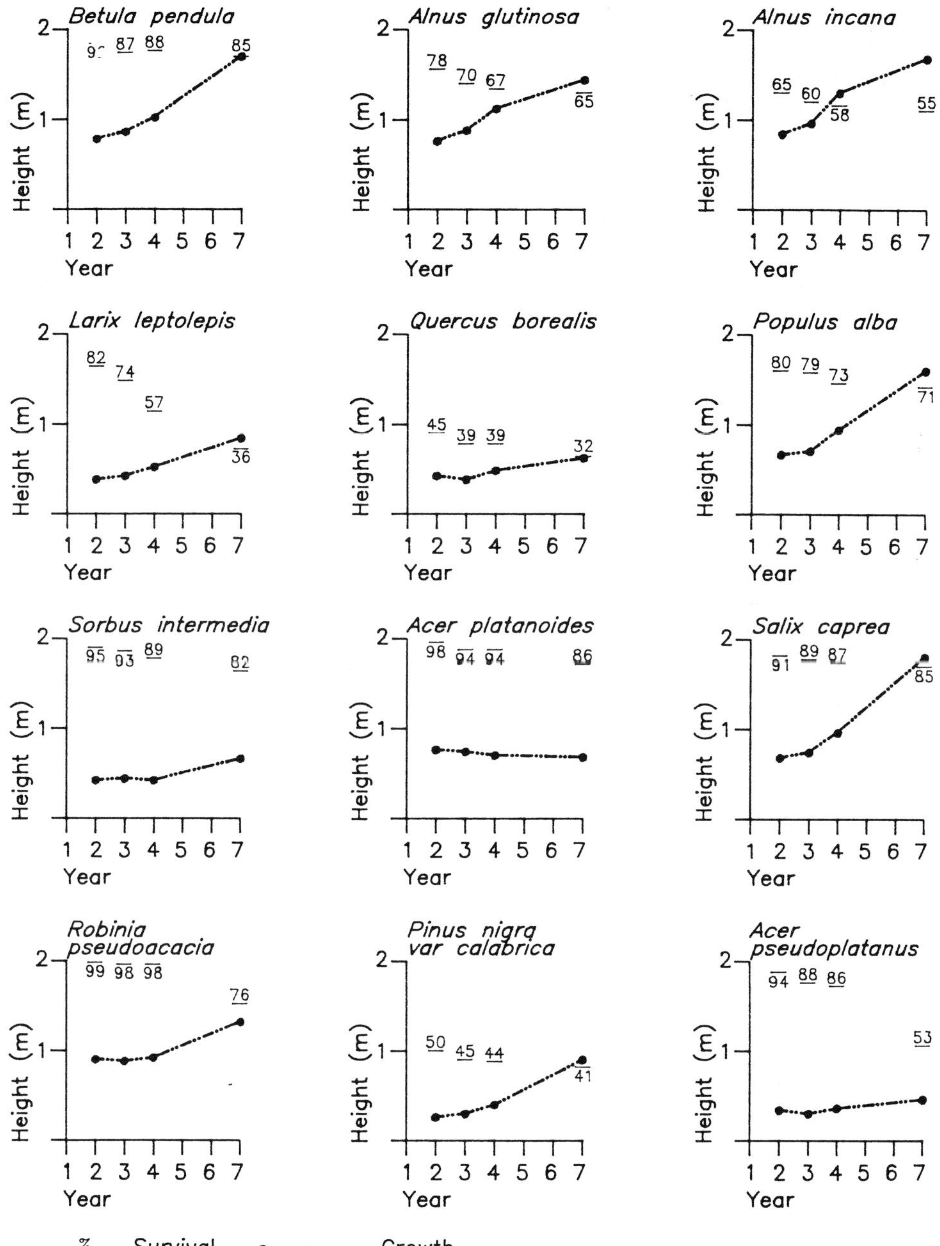

Figure 14.12: Growth and survival of trees on restored colliery spoil (after DLRRU 1979[73])

Tree bark is commonly used as a mulch. It should be prepared by chopping and screening to remove particles smaller than 5mm. Finely pulverised bark is not suitable as a water-conserving mulch since:

- it absorbs rainfall which then evaporates;
- in wet conditions it becomes saturated and anaerobic.

Many coarse-grained granular materials are produced as waste products from agriculture, horticulture or forestry. If they are non-absorbent and slow to degrade they can provide inexpensive mulch materials.

14.6.8 Irrigation

Irrigation water is commonly applied to transplanted trees during the first and sometimes the second, growing season. The primary purpose is to ensure survival during dry periods but effective irrigation will also increase the rate of root growth and therefore reduce the period during which the tree is unable to exploit reserves of water deeper in the substrate. Irrigation is generally used when water stress becomes apparent or the scheme manager believes that tree damage may occur, rather than in response to set rainfall or soil moisture deficit criteria.

Irrigation is routinely practised in agriculture and fruit growing. The equipment and techniques used locally in these industries can generally be applied to land reclamation schemes. Whatever techniques are used, irrigation achieves the best results when sufficient water is applied to wet the substrate thoroughly to the full rooting depth. Water at the substrate surface will encourage shallow rooting but is rapidly lost by evaporation. It is therefore largely ineffective.

14.7 The care of newly established vegetation

14.7.1 Establishment

Vegetation is usually described as established when it no longer requires special care to ensure that all parts of the plant are functioning normally and natural development will proceed. The requirement for works to care for vegetation is greater in the early stages of establishment and growth, and if sufficient inputs are not provided the establishment period will become extended (Figure 14.13). The principles and operations required to ensure the establishment and long-term development of vegetation on reclaimed land, including land which has been revegetated with a minimum of disturbance, are similar to the principles and operations applied to such vegetation on natural sites. The key difference is that vegetation on reclamation sites without a fully developed soil cover is particularly fragile and sensitive to any deterioration in substrate characteristics since nutrient supply, soil moisture and many other factors are poorly buffered by the substrate.

14.7.2 Weed control and trees

The critical importance of weed control, described in Section 14.6, extends through much of the establishment period, until the tree root systems develop sufficiently in zones of constant soil moisture. Weeds continue to reduce the uptake of nutrients by trees, by depleting soil moisture in the upper substrate where nutrient concentrations are highest. The early care of trees is usually directed towards achieving a closed canopy (*i.e.* touching branches) which will suppress weeds by shading (see Photograph 14.8). Contact herbicides are the most commonly used method of weed control in reclamation plantings. Residual herbicides such as simazine are not favoured since they bind poorly to the surface layer of spoil and can damage the tree. They are also susceptible to leaching and can contaminate watercourses. The use of herbicides is now closely controlled by legislation within the European Community. Mechanical vegetation control by mowing or strimming is effective

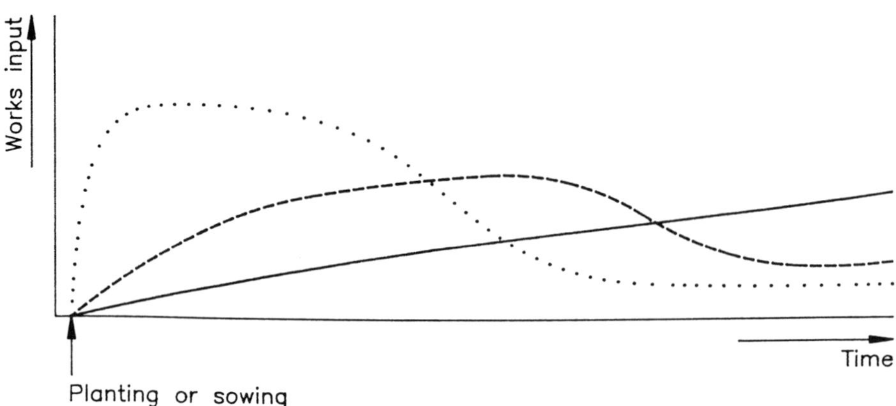

· · · · · · · · High works input initially. Minimum period to reach a steady state.
------- Medium works input initially. Extended period of input requirements.
——— Inadequate works input. Vegetation does not reach a steady state.

Figure 14.13: The care of new vegetation

against annual weeds but of little value in reducing competition from perennial weeds, particularly grass. Strimmers and mowing machinery have been found to cause considerable damage to young trees unless used with care.

14.7.3 Lime requirements

The pH of the substrate may fall progressively as shown in Chapter 5. Unless the substrate pH is monitored periodically and maintained within the range required for the vegetation, by the application of lime, the growth and development of the vegetation will be impaired. This process, termed regression, has led to established swards on reclaimed colliery spoil deteriorating to a moribund, patchy cover.

Photograph 14.8:
Rapid tree growth produces a dense, weed-suppressing canopy (source: Richards, Moorehead and Laing Ltd)

14.7.4 Nutrient applications

Colliery spoil and steelworks substrates lack reserves of nutrients, particularly nitrogen (Sections 5.4, 9.2.3). Applications of inorganic nitrogen fertilisers have only a temporary effect on the substrate nitrogen concentration and contribute very little to the nitrogen capital of the substrate. As a result vegetation, and in particular the vigorous, fast-growing species which demand high nutrient supplies, is particularly sensitive to inadequate nutrient additions. The application of organic nutrient sources can ensure a more consistent supply of available nitrogen than inorganic fertilisers. A programme of routine substrate sampling and analysis is valuable, particularly where nutrients are removed through agricultural cropping. In West Yorkshire all reclaimed colliery spoil sites used for agriculture are monitored annually.[215] Sewage sludge offers a cheap or free source of nutrients for maintenance purposes, but its use is restricted to sites:

- where odours are not a problem;
- where run-off can be avoided;
- where the heavy metal content of the sludge is within guideline concentrations;
- where there is no health risk.

14.7.5 Grazing

Perennial grasses sown on colliery spoil or slag may develop seed heads during the first growing season, as a response to moisture stress. This impedes their vegetative development towards a dense sward. Where rhizomatous or stoloniferous grasses have been sown to ensure the stabilisation of the substrate a dense, vegetative cover is particularly important. Mowing or grazing of the young grass is frequently practised. This encourages a vegetative response provided that the young plants are not damaged excessively. Grazing is a valuable tool for grassland management since it removes excessive growth and can recycle nutrients (see Section 15.5) but uncontrolled or excessive grazing has been a major cause of sward regression (see Photograph 14.9). Overgrazing causes regression through:

- damage to the growing point of grass plants;
- damage to the growing points of legumes, inhibiting nitrogen fixation;
- removal of nutrients.

Grazing also severely damages newly emerging grass plants which can be uprooted.

A key function of grassland management is the maintenance of a balance between nutrient inputs and removals, and a balance between the amount of grazing and the growth of the sward. This balance is also a vital part of the maintenance of nitrogen fixation by legumes (see Box 14.13).

Photograph 14.9: Severe sward regression caused by overgrazing (source: Richards, Moorehead and Laing Ltd)

Box 14.13: Key points in the maintenance of nitrogen fixation by legumes

- select legumes appropriate to the substrate and site use;
- inoculate with the correct *Rhizobium* strain;
- maintain near-neutral pH or select legumes adapted to the pH;
- maintain available phosphate concentrations;
- ensure adequate available water capacity in the substrate;
- avoid excessive salinity or soluble fertiliser applications;
- balance the growth of legumes and grasses by carefully monitored nitrogen applications;
- graze or cut to control grass growth;
- avoid overgrazing of legume swards.

15 MANAGEMENT OF RECLAIMED LAND

Chapter contents

continued...

15 MANAGEMENT OF RECLAIMED LAND

15.1 Introduction

15.1.1 Definition

Land which has been reclaimed from dereliction requires continuing management if it is to fulfil its required purpose. Land management encompasses all the activities which are required to ensure that reclaimed land can be used as intended in the long-term. These activities include legal, managerial, administrative, financial and practical matters covering the prevention and repair of deterioration, that is, maintenance activities, the progressive development of features such as vegetation and the use of the site for its intended purpose. Management at some level of intensity will be required for all reclaimed land in perpetuity.

15.1.2 Management and design

Thc considcration of sitc managcmcnt should bcgin at thc carlicst stagc of the process by which land use objectives are set and the reclamation scheme is designed. The land uses that are selected and the way in which site characteristics are modified in the reclamation scheme will determine the nature and intensity of management inputs which will be required. Conversely, the management skills and resources which will be available should be taken into account as reclamation proposals are formulated. There is little point in spending capital on land reclamation if the site is not managed effectively to achieve and maximise the benefits which were the objectives of reclamation (see Photograph 15.1). Management of revenue-generating land uses is essential if these revenues are to be maximised, either to repay part of the reclamation costs or to provide the funds for maintenance works. Many reclamation schemes have failed to realise their full potential as a result of the neglect of management activities. In the worst cases the land has reverted to a state of further

Photograph 15.1:
This neglected public open space provides little benefit to those it was intended to serve (source: Richards, Moorehead and Laing Ltd)

dereliction. The long-term management costs of reclaimed land can be allowed for by:

- selecting land uses which match the skills and resources available;
- including revenue-generating activities within the use of the site;
- establishing an endowment fund at the outset which will provide a continuing source of revenue towards management costs;
- designing details which minimise the requirement for maintenance and minimise the technical difficulty of such operations.

15.1.3 The value of management plans

The long-term management of reclaimed land is frequently the responsibility of individuals or organisations who/which were not those responsible for the objective-setting or scheme design (see Section 15.2). In order that the objectives of reclamation are carried through to the subsequent development and management of the site a management plan should be produced, as an integral part of the reclamation process (see Section 15.3). This plan may set out important constraints on the long-term development or use of the site (Section 15.4), and will guide the management of vegetation (see Section 15.5) and other facilities (see Section 15.6).

15.2 Management arrangements

15.2.1 Management planning

Suitable provision must be made for the implementation of a management plan. Whilst public authorities, such as the local or regional council or a government agency, most commonly carry out land reclamation work, such a body may not wish to retain ownership and control of the site in the long-term. The possible arrangements for long-term management vary according to the after-use of the site.

The following arrangements are typically established for the management of a site. On any one site, a combination of these may apply:

- management by the public authority;
- management by the private sector;
- management by voluntary organisation or charities;
- management by the private sector, voluntary organisations or charities through lease, licence or agreement, but retained in the ownership of the public authority.

15.2.2 Management by the public authority

Under this option, the public authority can retain full control over the use of the site and its maintenance, ensuring that the management objectives and operations remain consistent with the project aims so that the long-term development of the site is not compromised by short-term financial pressures. Management sensitive to the constraints imposed by the site and its intended use will ensure proper site development and functioning only if the authority has the necessary managerial and technical skills and resources. Sensitive management can be especially important in the early years after reclamation.

15.2.3 Management by the private sector

Many sites which have been reclaimed for financially viable uses such as industrial, commercial or residential development have been sold to the private sector (see Photograph 15.2). Uses involving active recreation, such as golf, dry-ski slopes, freshwater fisheries, theme parks and industrial heritage tourism have also been developed by, or sold to, the private sector for long-term management. In such cases it is common for a legal agreement to cover special restrictions on the use of the site to prevent damage to sensitive features such as shaft caps and contaminated areas. The private sector may also contribute to the management of part or parts of a site under lease, licence or contract agreements. Sites reclaimed for hard after-uses such as industrial development or housing, also fall within this category. Those involved in the initial development may sell off their interest, in which case the industrialist or householder will become responsible for continued management, with the public authority usually managing public open space, roads and utility services.

15.2.4 Management by voluntary and non-profit organisations

This arrangement for management can be especially appropriate for smaller sites, or parts of larger sites, and those with amenity after-uses such as nature reserves. Suitable organisations to take over site

Photograph 15.2: This reclaimed site was sold for the development of private housing (source: Welsh Development Agency)

management may include charities, wildlife organisations, sports clubs or historical societies. Providing grant-aid to enable a voluntary body to carry out management may prove far more cost-effective than management using a public authority's own staff and equipment, unless the site can be taken into an existing land management programme.[209]

A site may be retained in public authority ownership but be managed by a voluntary organisation, perhaps under the guidance of a management committee.

15.2.5 Management through lease, licence or agreement

Management through lease, licence or agreement allows a public authority to retain ownership and control of a site, but to engage additional expertise in the management of a site on a contract, franchise or mutual basis. On sites where a commercially viable use is desired, the landowner may share in the returns by leasing the site or by developing the facilities and engaging professional management by contract. This arrangement now applies to many publicly owned recreation facilities (see Photograph 15.3).

For sites where there is a formal sports after-use, for example sailing or motor sports, a lease to a recognised club may be appropriate. For sites of wildlife conservation interest, a long-term management lease can provide the managing organisation with an assurance that their long-term efforts will be worthwhile.

The most common arrangement for agricultural grazing on reclaimed land in the United Kingdom is the short-term let or grazing licence. Under United Kingdom law the landowner retains full control of the land with these arrangements (which can be renewed annually) whereas longer agreements can create a protected agricultural tenancy. Advantages and disadvantages of short-term and long-term arrangements for grazing of reclaimed land are given in Box 15.1.

Photograph 15.3: A committee of public authorities and user groups manage this country park, created by reclaiming old coal workings (source: Welsh Development Agency)

Box 15.1: Arrangements for grazing on reclaimed land

	Short-term lease	Long-term lease
Advantages	• Landowner retains close control • Maintenance assured • Freedom of choice over who uses site and when • No long-term commitment by user	• Improvement of soil and sward • Maintenance controlled by user • Income generated regularly through rent • Security for user of site
Disadvantages	• Little incentive for user to promote soil and sward development • Lack of long-term security for user • Rent may be high to reflect short-term lease	• Over-grazing and bad management may occur • Landowner may lose close control of site • Weaker legal position for owner over change in site use or user

15.3 Management plans

15.3.1 Definition

A management plan is a document which provides long-term guidance for the future managers of a land-holding. It contains:

- a reasoned, clear statement of the aims and objectives;
- information for reference;
- a programme of work for implementation;
- a mechanism for periodical review.

The preparation and implementation of a management plan is the means by which the integrated approach to the planning and management of land is applied to a specific site. By providing a framework within which all future management is carried out, the plan enables any person involved to understand how and why decisions are taken, in relation to the reasoning behind the policies and proposals for action.

Box 15.2 gives a generalised description of a management plan which can be applied to most types of post-reclamation land use.

Box 15.2: The content and operation of a management plan

RECLAMATION AIMS
A statement of the purpose of reclamation, and the broad policies which will underlie the management of the various land uses and interests.

REFERENCE
A comprehensive record of the land before and after reclamation, forming a basis for the analysis and objectives. External influences and constraints are included.

⇓

ANALYSIS
An examination of the options for management of the land and the relationship between potential land uses. The identification of potential problems and conflicts, and weighting of various interests. Objectives are formulated from the decisions reached.

⇓

MANAGEMENT OBJECTIVES
Specific statement of the land uses and interest to be achieved and promoted, their priorities, and targets for physical/biological/financial performance.

⇓

MANAGEMENT PRESCRIPTION
An outline of the work required and the resources needed to achieve the management objectives. A long-term outline of the programme for implementation.

⇓

IMPLEMENTATION
Details of the operations to be carried out within the review period in order to achieve the management objectives. Statements of resource requirements related to yearly programmes for the review period.

⇓

MONITORING AND REVIEW
An assessment and record of management achievements, with arrangements for a periodic review of the plan and renewal or revision of the rolling implementation programme.

15.3.2 Plan preparation: new projects

The failure to produce a management plan for reclaimed land invites undirected and wasteful management which may unintentionally conflict with the original aims of reclamation. For this reason the preparation of a management plan should begin when the aims of reclamation are set, so that the designer's long-term intentions are clearly recorded. The reference material can then be selected from documents prepared during the design and construction process. The analysis should be commenced in parallel with the scheme design, since the long-term management of a site is an important design consideration, but further analysis will be required once site works are completed.

15.3.3 Plan preparation: completed projects

Many reclamation schemes have been completed without the preparation of a management plan. In these cases preliminary research will be necessary to determine the reclamation aims and to obtain the reference material. In some cases these aims may be absent or fragmented and original survey work will be required to facilitate the analysis and formulation of the management objectives.

15.3.4 Monitoring and review

The management of land requires the flexibility to respond to changes in circumstances within and outside the site. External pressures may, for example, increase the importance of revenue-generation or reduce the resources available for land management. A review of the management prescription may confirm that the management objectives can still be met, or may indicate that a more fundamental review is required. Monitoring of the site to identify, for example, the development of vegetation or the deterioration of structures, will influence the management prescription and provide a basis for the implementation programme.

15.4 Site records and reference material

15.4.1 Purpose

The characteristics of a derelict site and the way in which those characteristics were modified during reclamation, can be essential information for the subsequent owners or users of a site. Large sums of money have been spent on the investigation of sites abandoned by mining or industrial activity. The conservation of records can avoid the need for similar investigations of sites in the decades after reclamation. Accurately recorded information is valuable for:

- the routine maintenance of structures, utility services and vegetation;
- the location and identification of buried hazards and utility services;
- planning and controlling the development of the site;
- the design of new structures;
- the monitoring of ground conditions such as contamination;
- the assessment of the site for new uses.

Site records are particularly useful if new land uses are proposed. Ground contamination, for example, may have been reduced to meet the criteria for the initial after-use, but not to meet the more stringent criteria for uses which are subsequently proposed.

15.4.2 Material for retention

Material that is likely to serve the purposes identified in 15.4.1 should be retained; for example:

- feasibility studies and design drawings;
- site investigation data including physical and chemical analyses;

- photographic and documentary evidence of the site before reclamation, and in particular the previous uses of land;
- the specifications for the works as carried out;
- accurate 'as-built' drawings and records of the landform, surface features and sub-surface details such as utility services;
- details of the location, and characteristics of all buried contaminated material, demolition debris, unbroken foundations, abandoned underground services and drainage systems;
- details of the location construction and contents of all waste containment cells;
- details of the location and treatment of shafts, shallow mine workings and other buried voids;
- the methods and results of all verification testing carried out during and after construction.

The majority of this material will be prepared or collected as a routine part of the reclamation process. Provided that a management plan is commenced at an early stage, it is possible to collate reference material with a minimum of additional effort.

15.4.3 The maintenance of records

The safe keeping of records and information over long periods is essential if future land managers and site users are to benefit. Systems which automatically warn of hazards or other vital information are more reliable than systems which rely on positive action or enquiry by land managers. For this reason it is good practice for features such as shaft caps, containment cells for contaminated material, and utility services such as electricity to be permanently indicated on site by exposed marker posts, and by plaques or plastic warning tapes buried in the ground above the feature to be protected (see Photograph 15.4). Such on-site warnings can refer attention to documentary details stored elsewhere.

Photograph 15.4: This raised brick cylinder will indicate the location and size of a capped and buried shaft once the surrounding land is regraded (source: Richards, Moorehead and Laing Ltd)

Documentary material should be stored where it is safe but accessible. Systems of cross-reference which automatically alert new users/managers to the location of documents, are particularly valuable. Such systems include annotations on land ownership deeds, references in management plans, and registers of reclaimed land held by public authorities. Multiple cross-referencing provides a safeguard against the loss of documents when the responsible organisations are reorganised or relocated.

15.5 Vegetation management

15.5.1 Requirement and purpose

Vegetation will develop and change according to the site characteristics and to the balance of external influences such as grazing, public use and management activities. The purpose of vegetation management is to ensure that the vegetation fulfils the needs of the intended site use, through manipulation of the various external influences. This purpose will be one of the objectives set out in the management plan for the site. Management operations may seek to develop certain features, such as species diversity or soil fertility, or to maintain the vegetation as use of the site continues.

15.5.2 Existing vegetation

Significant areas of vegetation are sometimes retained through the reclamation process, to add maturity or conserve wildlife habitats. This mature vegetation may require different management activities from those applied to newly established vegetation on the site. The management plan should recognise the particular characteristics of the mature vegetation, and any statutory protection or other special status, in prescribing management operations and site use.

15.5.3 Naturalistic vegetation

A natural approach to vegetation design is described in Box 14.4. One benefit of the establishment of a naturalistic vegetation which is suitable for the site and its use is that the requirements for vegetation management are usually considerably lower than those for more formal or productive vegetation. Figure 15.1 illustrates this relationship.

Naturalistic vegetation requires particular management skills and methods which may not be part of the landowner's existing resources. For example, management to enhance species and habitat diversity (see Box 15.3), requires a good understanding of the ecology of the site.

15.5.4 Grassland management

The grasslands which have been established on reclaimed colliery spoil range from productive grazing swards to species-rich swards for amenity and wildlife conservation purposes. Productive swards, when

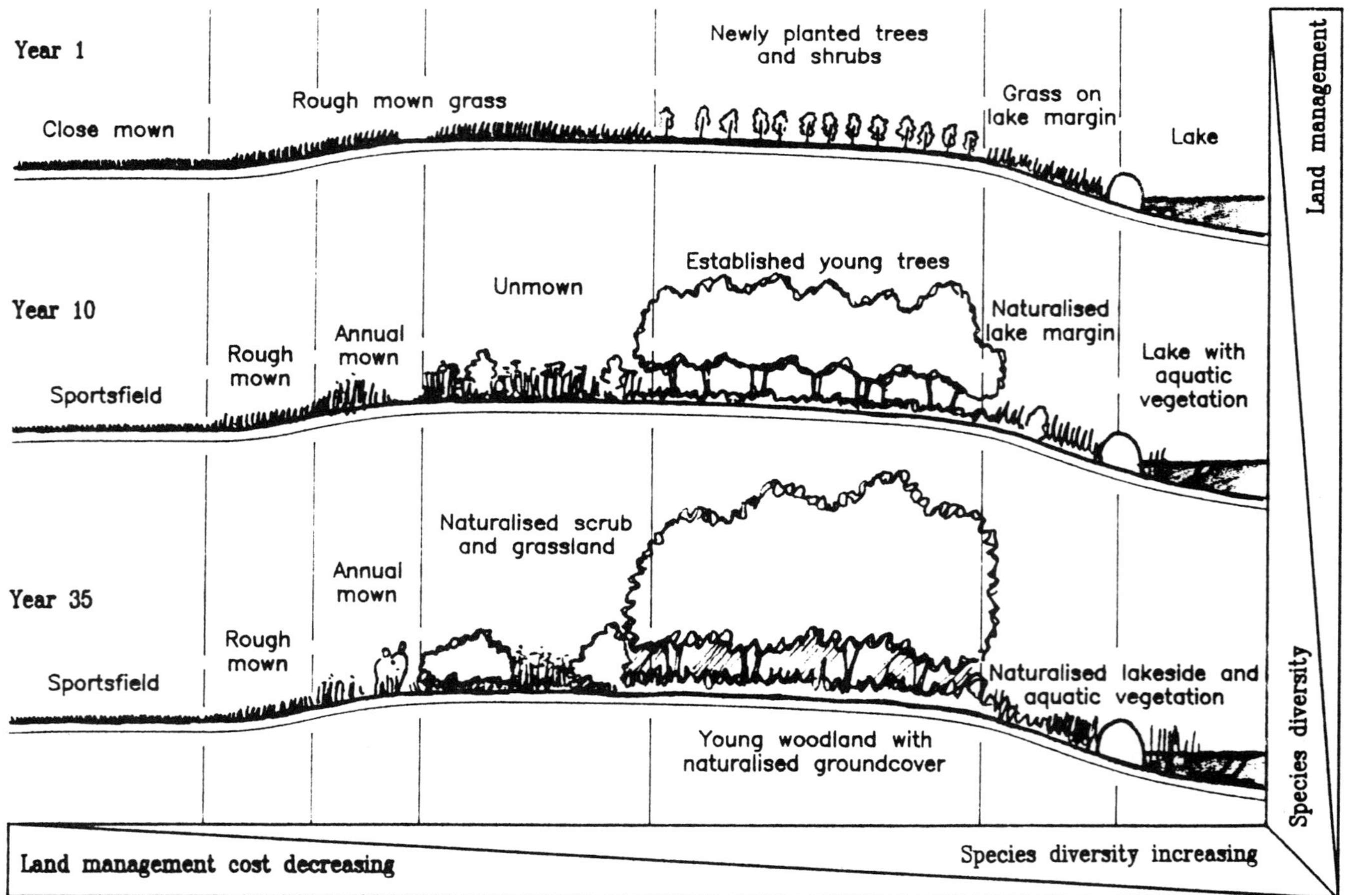

Figure 15.1: Changes in species diversity and land management with time

Box 15.3: Management for species and habitat diversity

Natural colonisation of regraded colliery spoil or iron and steel-making slags can be used to generate a diverse vegetation, if this does not conflict with other objectives of reclamation. If the physical conditions resulting from landform and earthworks (Box 13.5) are right then natural colonisation, guided by an overall strategy for land use and judicious management intervention only where necessary, will produce the most ecologically effective results.

Whether natural colonisation or deliberate planting and sowing are the means of revegetation, the role of the ecologist/land manager is to ensure that the desired habitats and species develop and are sustained. This may involve the deliberate introduction of desirable, appropriate species if monitoring indicates that they are unlikely to colonise naturally, and in many cases will require intervention to prevent or limit natural succession towards a more mature vegetation type. For example, most grasslands are unstable and will be invaded by scrub unless managed by grazing or cutting. Conversely, herbaceous species and trees will only colonise grassland if patches of bare soil are maintained or created by grazing, livestock trampling or cutting. Coppicing or repeated thinning of woodlands are required to maintain a light-demanding ground flora. A management plan is essential if the land manager is to know the original intentions of the reclamation scheme designer and to maintain consistent objectives.

successfully managed to provide worthwhile grazing, hay or silage cropping, can generate significant income towards the cost of management. Swards which are managed primarily for amenity purposes can be cropped at low intensity for hay or silage as a means of off-setting management costs and encouraging biodiversity.

Grasslands established on reclaimed blast furnace and steel-making slags have been restricted to very low productivity swards, managed for aesthetic, wildlife conservation and low-intensity recreational uses.

15.5.5 Agricultural management of grassland

The management of grasslands in grazing use should follow good practice for agriculture, but must take particular account of the poorly-developed soils of reclamation sites (see Section 14.3). The grazing period may be restricted by excessive ground wetness in the winter and by lack of grass growth in dry summers. Swards often lose productivity despite good management, and can be renewed by recultivation and reseeding on a regular basis (see Box 15.4). The selection of grazing animals also has a significant effect on the sward (see Box 15.5).

The presence of heavy metals in the substrate, *e.g.* from flue dusts, should be clearly stated in the management plan since there is a risk that livestock will ingest heavy metals via plant material. Box 15.6 summarises the measures which will ensure that unacceptable metal uptake by animals is avoided.

15.5.6 Management of grassland for amenity purposes

Amenity grassland established on colliery spoil or the slags from iron and steel making can be managed generally in the manner of any other amenity grassland, provided that the objectives of management are guided by the aims of reclamation and take note of any particular circumstances identified in the management plan. These may include:

- the need to ensure a vegetation cover for surface stabilisation;
- the need to restrict nutrient levels to encourage species diversity;
- the need to time mowing to allow seed production by particular species;
- the need to maintain an unmown buffer to restrict access to sensitive habitats.

Box 15.4: Recultivation systems for grassland productivity

Colliery spoil sites which have been successfully reclaimed to grassland frequently become less productive after 5-6 years despite regular ameliorative treatment.[216] The deterioration follows weathering of the spoil, a process described in Chapter 5. Weathering leads to:

- increased acidity;
- consolidation of the upper 150-200mm profile;
- impeded microbial activity;
- nutrient deficiency;
- impeded drainage;
- thatch development;
- impeded root penetration;
- ineffective lime and fertiliser applications;
- extremes of soil moisture availability.

These factors combine to produce an unproductive sward. This has been overcome by aeration and cultivation methods. Aeration and subsoiling temporarily improve the physical condition of the substrate until further weathering leads to a recurrence of the problems listed. Full recultivation, consisting of rotavation to 75mm followed by deep subsoiling, seedbed preparation and reseeding, has a much longer lasting effect through:

- aeration and improved drainage;
- deep incorporation of lime and nutrients;
- effective pH control;
- incorporation of organic matter from the sward;
- increased microbial activity producing nutrient mineralisation and humus;
- formation of a soil crumb structure;
- improved moisture retention;
- deeper root penetration.

Recultivation can be used in conjunction with short-term pioneer crops, such as green manures and legumes, grown specifically to develop soils from spoil and as part of the long-term management of land on a 5-7 year grass ley system. Recultivation will be impeded unless care is taken to eliminate buried obstructions and debris from the upper layer of the substrate during the initial earthworks.

Box 15.5: The management of grazing

The nature, intensity and duration of grazing affect the ability of the sward to recover, and its species composition.

Grazing should be managed to crop the sward evenly and closely, without leaving clumps of weeds or unpalatable grasses and without damaging the growing points of the grass. Clover is less tolerant of close grazing than is grass.

Sheep graze closely and evenly. They bite through the grass rather than pulling at it, and so do less damage than horses provided that grazing is stopped when the sward has been cropped. Sheep grazing favours a dense sward of low-growing grasses, and causes little damage to the soil surface. In urban fringe areas dogs cause problems by worrying sheep.

Cattle tear at the grass rather than biting through it. This is damaging to poorly rooted or poorly established grass. Cattle do not graze the sward as closely as sheep, and if stocking rates are low they will avoid the less palatable grass species which can spread as a result. Cattle can cause considerable surface damage on wet or soft land. Calves, being lighter, cause less damage to the ground.

Horses graze very selectively, avoiding unpalatable grasses and weeds. They also tear at the sward, uprooting and damaging the grass plants. The sward rapidly becomes untidy and unproductive and open to weed invasion. Horse hooves do a great deal of damage, particularly on sloping, wet ground. Horses are a potential danger to the public on open land.

Practical considerations can also limit or guide management operations, *e.g.*:

- physical obstructions to close mowing;
- the need to mow regularly to control invasive weeds;
- the need to remove thatch build-up, to control pH and to promote nutrient cycling (see Box 15.4 and Figure 15.2);
- the need for fertiliser to ensure growth to repair sward damage caused by intensive use;

Box 15.6: Limiting the intake of heavy metals by grazing animals

High concentrations of inorganic contaminants (except cadmium and zinc) in soils have little influence on the concentrations in herbage provided that the soil pH is maintained at 6.5 or above. In these circumstances the risk to livestock depends almost entirely on the amount of soil ingested and toxic element concentrations in that soil. The soil contamination of the herbage can be minimised by:

- placing uncontaminated material as the uppermost layer of substrate;
- excluding livestock until a close-knit sward has developed;
- mowing the new grass to encourage a close-knit sward;
- maintaining a close-knit sward;
- maintaining a mat of vegetation debris at the soil surface;
- harvesting silage with a pick-up harvester and leaving a long stubble where contamination of the soil is highest;
- managing grazing to avoid overgrazing and poaching of the surface.

The potential for metal uptake by livestock should be monitored and controlled, by analysis of the herbage and careful rotation of the grazing periods allowed for each group of animals.

Detailed guidance is given in ICRCL Guidance Note 70/90.[122]

- the availability of a free source of nutrients *e.g.* sewage sludge, where quality, run-off and public health considerations permit its use.

The costs of grass mowing are significantly influenced by the design of the site, the complexity of shapes and edges, the size and type of machine which can be used, the finish and use of the sward, the species and cultivar mixture established and many other factors. The cost of management should not dictate the design and objectives but should be carefully considered at the design stage so that objectives can be achieved and maintained within the available resources. Many interesting landscapes have been simplified and made bland by measures taken to reduce the costs of maintenance.

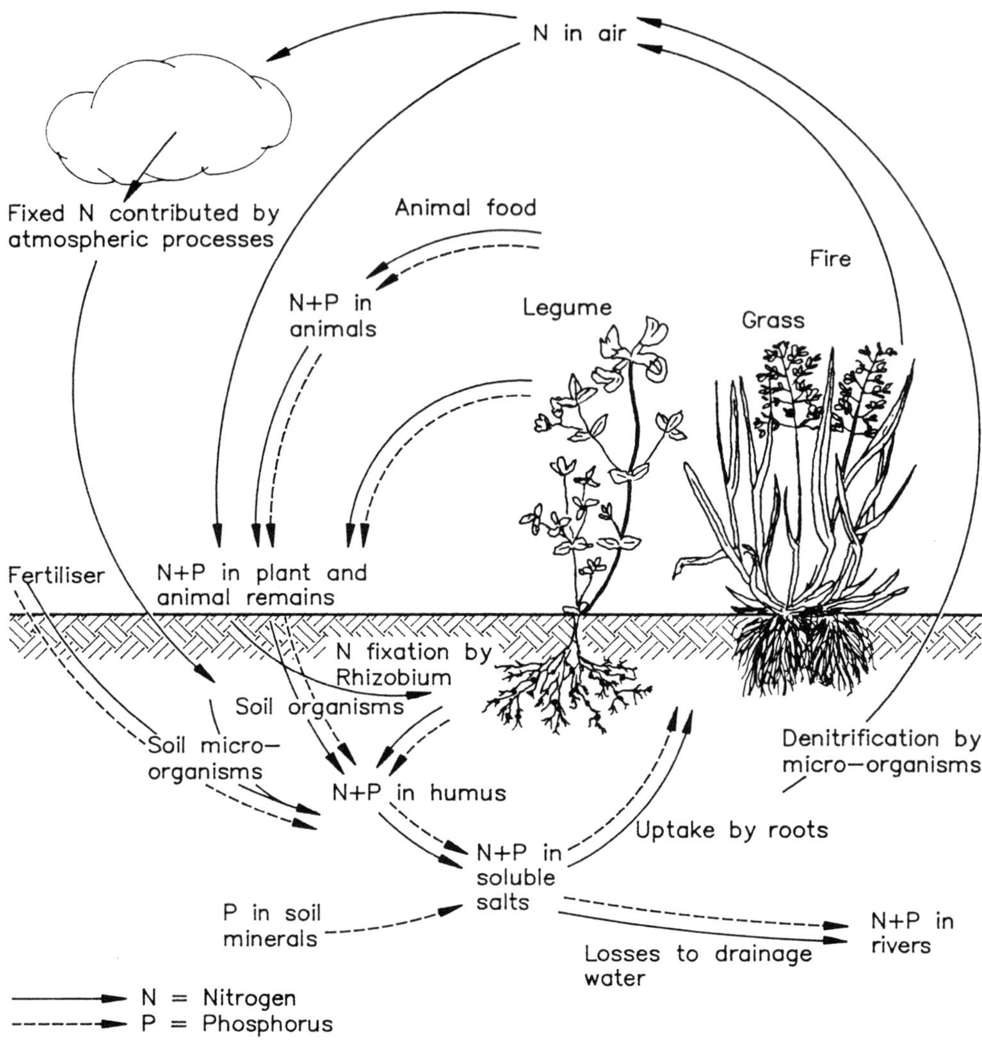

The greatest stores tend to be in the organic matter, released by the activities of micro-organisms; there is input of nitrogen by fixation and phosphate by release from the soil minerals, and little is lost.

Figure 15.2: Nutrients circulating continuously in the soil/plant ecosystem (after Bradshaw and Chadwick, 1980[40])

15.5.7 Management of woodlands

Woodland and forestry plantations established on colliery spoil can be managed in the same way as those on natural ground, provided that the objectives of management recognise the particular requirements and circumstances of the site. These may include:

- a greater need for nutrient additions after the establishment phase. Such additions are uncommon in commercial forestry;
- the need to retain nitrogen-fixing nurse species. *e.g.* alder and lupin, as a component of the maturing woodland on nutrient-poor substrates;
- the need to monitor the growth of vigorous nurse species and to thin selectively to ensure the development of the desired ultimate species mixture;
- the greater risk of fire damage at urban-fringe and recreational sites. This can be reduced by rigorous control of grass and scrub around the woodland;
- the greater risk of fence damage at public sites, allowing livestock access to young woodlands;
- the greater risk of windthrow where tree rooting is restricted to shallow layers of improved substrate. The selection and management of short trees or scrub will reduce this problem.

Sewage sludge is a source of nutrients and organic matter which can be applied to developing young woodland using irrigation or soil injection equipment if practical difficulties such as access, odour and temporary exclusion of the public can be overcome.

The costs of woodland management can be offset by the harvesting of timber products and the conversion of wastes such as thinnings or brash into useful wood chips, mulch or composts.

'Urban forestry' *i.e.* the management of urban and amenity planting in a semi-commercial way, has been used successfully.

Voluntary groups, such as those concerned with wildlife and woodland conservation, assist with labour-intensive tasks of management in some woodlands where conservation and public amenity is the primary management objective.

15.5.8 Management of wetlands

Where wetland systems have been established to treat acid mine drainage water, as described in Chapter 12, the primary objective of their management is their efficient functioning. The management plan should clearly show that any secondary objectives such as wildlife conservation should not conflict with the water treatment objective. These wetland systems, if properly established, require minimal maintenance but routine monitoring is required to identify any decline in the effectiveness of water treatment. Remedial action may then be required in order to renew plant growth, correct substrate conditions or to restore correct water flow patterns (see Photograph 15.5).

15.6 Management of engineering structures and utility services

15.6.1 Introduction

Reclaimed sites typically contain many structures which protect the integrity and use of the land. If these structures fail to function as designed *e.g.* through deterioration or damage, considerable disruption of the land surface or its use may result.

15.6.2 Drainage

Reclamation landforms are highly susceptible to erosion by flowing water, particularly in the stage before a vegetation cover is fully established.

Land drainage systems which concentrate surface water into ditches and channels increase the damage which can be caused should the system fail through inlet blockage, ditch overflow or other causes. The management plan should contain a programme for the routine inspection of drainage systems, based on an analysis of the likely consequences of failure.

Drainage systems are at risk from:

- blockage of inlet pipes and screens by litter, plastic sheets, debris from subsequent development and deliberate obstruction;
- blockage of silt traps and catch pits by eroded material;
- blockage of french drains, filter-fabric and perforated pipe drains by silt particles;
- erosion damage in exceptionally high flow conditions.

Photograph 15.5: This drainage system is completely blocked by silt. Regular inspection and clearance is essential (source: Welsh Development Agency)

15.6.3 Contaminated areas

If contaminated materials have been disposed of in on-site containment areas, placed in a manner where dispersal is prevented, or otherwise retained on site, the integrity of the disposal system should be monitored by periodic sampling of groundwater, surface water, vegetation or other indicators. A monitoring programme is often a formal requirement of statutory consents for the disposal of contaminated materials (see Section 11.9).

15.6.4 Structures

The management of new structures such as buildings and bridges which are erected as part of the development of reclaimed sites will typically form part of the routine management programmes of the responsible authorities or landowners. Reclamation schemes may also involve the retention of earlier structures which continue to serve a purpose *e.g.* dams and retaining walls, or have historic interest *e.g.* disused buildings, bridges and industrial relics (see Photograph 15.6). Structures of historical relevance will require periodic maintenance and repair which, because of their age, may be frequent and/or costly. The responsibility for the care of these structures should be clearly identified at an early stage so that appropriate resources can be made available. Without adequate maintenance these structures can become a significant hazard to the public.

15.6.5 Subsidence and settlement

Shallow mine workings, underground voids and significant depths of fill all have the potential for subsidence or settlement which can affect structures or the use of land. The effectiveness and long-term integrity of the engineering treatment of mine workings, voids and filling should be confirmed and monitored (see Chapter 3).

Photograph 15.6: This engine house has been carefully restored. A long-term programme of care and maintenance will ensure that it is used and enjoyed by future generations (source: Merthyr Tydfil Heritage Trust)

15.7 The management of site use

15.7.1 Litter, rubbish tipping and vandalism

Litter, tipping of rubbish and vandalism frequently occur on derelict land which may be regarded by some sections of the public as waste land, uncared for and therefore open to misuse. Once such use patterns have built up they commonly continue after the reclamation of a site. Designs which leave areas of unused land or apparently unmanaged land are more likely to prolong the abuse of land than designs which show the land to be in use. If litter and vandal damage are left untreated a pattern of misuse will quickly develop. Land uses which encourage the presence of people throughout the day tend to discourage misuse of the site.

15.7.2 Public safety

Reclaimed sites contain hazards which are common to many other types of land, for example;

- lagoons, balancing ponds and other waterbodies;
- abrupt changes in level;
- electricity sub-stations and similar service installations.

Protective measures such as fences and barriers, lifebelts and safety equipment, and warning notices need to be inspected and maintained on a routine basis. Such inspection may be a legal requirement.

15.7.3 Public access

The newly established vegetation of reclaimed sites is generally less tolerant of disturbance and wear than similar vegetation on more fertile substrates. Areas of wildlife interest may also be intolerant of disturbance. For these reasons the control of access can be an important part of the management of site use. Access can be encouraged or discouraged by design details such as paths, signs, interpretive information, physical barriers, landform and dense vegetation. The general pattern of access will be decided during site design, but the detail will commonly be refined in response to developing patterns of site use (see Photograph 15.7).

Areas of more intensive use, such as sports pitches and recreation facilities, require control so that the intensity of wear does not exceed the capacity of the facility (particularly the vegetation) to regrow and repair the damage caused.

Photograph 15.7:
Paths, signs and information boards help visitors enjoy the busy Brynbach park, but create an informal atmosphere. Low-key signs, using sympathetic materials, are more appropriate for informal and rural areas.

15.7.4 Wardening and site staff

Staffing by a warden or ranger service, possibly on a voluntary or part-time basis, may be necessary for the satisfactory management of a site. Full-time staffing provides a means for collecting revenue from users of the site, by charging for the use of facilities.

16 A FRAMEWORK FOR SITE REGENERATION

Chapter contents

16 A FRAMEWORK FOR SITE REGENERATION

16.1 The concept of framework

This book is concerned with producing guidance on the methods that are available to the practitioner for the successful reclamation of derelict and despoiled land. Whilst the book focuses on the coal and steel industries in Europe, many of these techniques are applicable to a variety of former industrial sites. The methods highlight the increasing requirement for a multidisciplinary approach to reclamation, in order to satisfy local expectations and future users of a given site. However, none of these methods can be practised without the enabling facilities of:

- aim: a well defined requirement level for the work to be undertaken;
- investment: funds with which to undertake reclamation;
- planning: detailed plans for proposed schemes;
- approval: agreement, at local level, on a chosen scheme.

These four key elements form the route towards implementation, and should be seen as the essential elements, or foundation, of a reclamation framework. After satisfying these elements, a reclamation scheme may then proceed through the processes of:

- implementation;
- control;
- validation;
- aftercare.

These essential features of a reclamation scheme are incorporated into the flow chart shown in Figure 16.1, which illustrates the framework for reclamation, around which the precise methods used will fit.

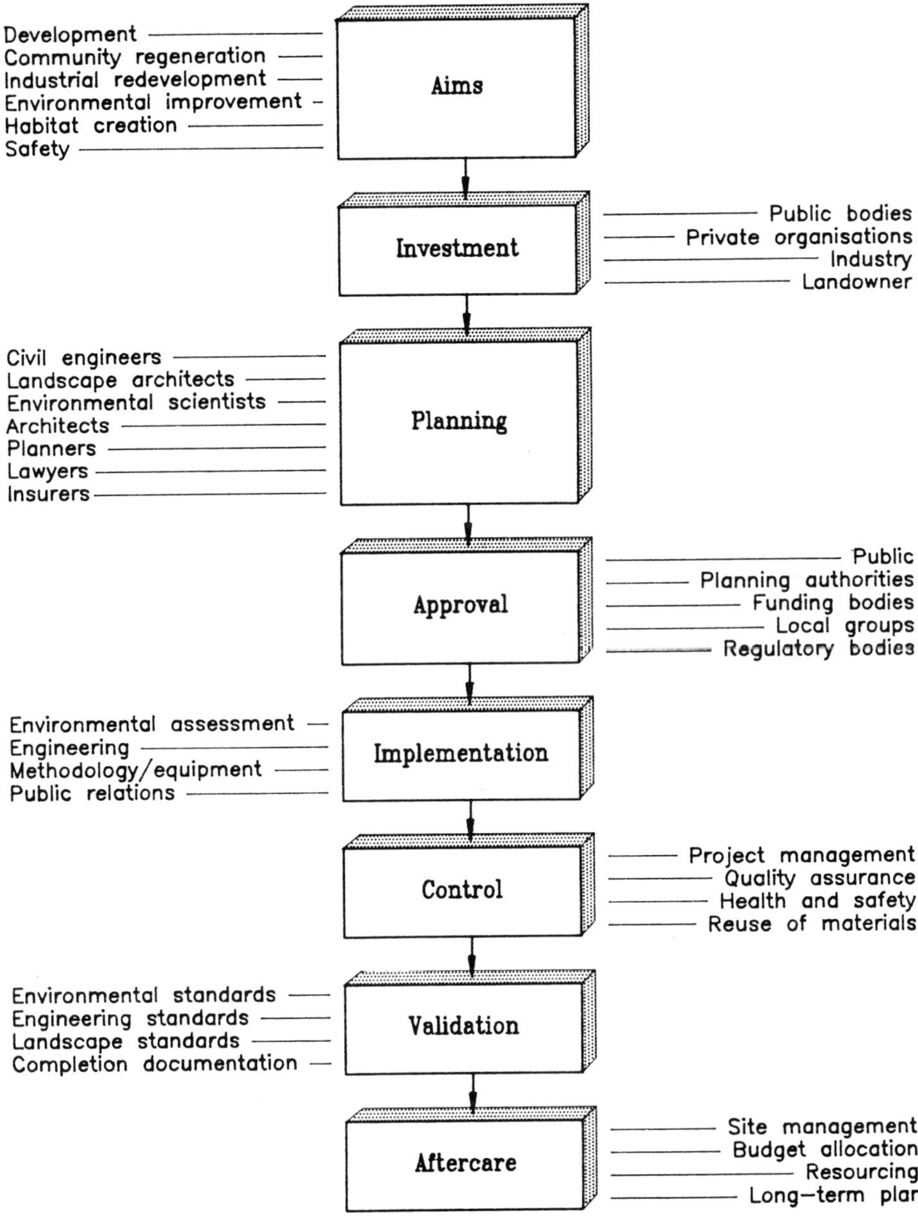

Figure 16.1: Critical path flow chart and principal considerations for a typical reclamation scheme

16.2 Planning for regeneration

16.2.1 Introduction

In general terms there are four situations in which dereliction and despoilation may be met by a reclamation team. These are:

- pre-closure;
- post-closure, but prior to demolition;
- post-closure and demolition;
- post-closure and reclamation, but where reclamation has been inadequate for current requirements, or is incomplete.

The stage at which the need for reclamation is identified is crucial to the planning of restoration, and will dictate the approach which is taken in the first instance. Experience gained by industry with the implementation of Integrated Pollution Prevention Control within the European Community and similar national systems, and by the use of environmental auditing as a management tool, has demonstrated the latent benefits of good environmental practice. These benefits are felt by both operating industrial concerns, and by potential purchasers of land.

Often however the luxury of detailed information is not available at a given site, whether operational or closed, and initial steps towards regeneration involve detailed archival searches and site investigations (see Chapter 2). The careful planning and execution of these investigations at an early stage in the redevelopment of industrial and potentially contaminated land, can clarify the requirements for site treatment prior to and during clearance and redevelopment. The expense involved in these investigations, whether or not they are undertaken by a third party (often a consultant), may be viewed sceptically by developers, but their importance, and the potential for cost savings during on-site works, should not be underestimated.

16.2.2 Pre-closure scenarios

It is likely that the developments made by industry towards greater environmental accountability during their operation of processes, will have a positive effect upon the amount of relevant site information available after the cessation of operations. For facilities in the throes of closure there is now every opportunity to assess:

- the environmental effects of closure;
- the need for precautions during site decommissioning and site clearance;
- the liabilities presented by the site, and therefore the true value of the site to a buyer or developer.

In the future it is quite likely that those sites with more detailed information available, accrued through diligent record keeping and environmental management during operation, will tend to attract buyers and developers more readily than those sites where little information exists.

16.2.3 Post-closure scenarios

Much reclamation work has been undertaken on former steel and coal sites that have been vacated for some time. The avoidance of risk from ground conditions remaining after vacation by industry is increasingly important in terms of legislative compliance in Europe. The legal principle described in Europe as the 'polluter pays' (Council Directive 75/442/EEC on Waste), emphasises the increasing liability vested with industry in terms of chemical disturbance of the environment (soil, water and air). Where there is potential for contamination to be present as a result of previous industrial activity, but where the polluter is either no longer liable or no longer exists, the principle of *caveat emptor* (or 'buyer beware') applies to the purchaser. In this case, lack of information concerning a site prior to acquisition can lead to the purchase of significant environmental liabilities, which may sometimes outweigh the asset or development value of a site.

Thorough site investigation is therefore important for the following reasons:

- for assessment of environmental, health and safety liabilities;
- as a basis for the assessment of legal liabilities;
- for the production of physical and chemical information on which to guide subsequent reclamation;
- to aid definition of the quantities of material that may require special treatment during site reclamation;
- to provide information on the opportunities the site presents for regeneration.

Where sites have been reclaimed in the past, records of the reclamation activities may also be lacking. Should these sites be required for development it would be prudent to undertake investigations to ascertain the degree of success and longevity of the reclamation works which have been carried out.

16.2.4 Investigation and reclamation

Whilst information on ground conditions will affect the future reclamation and development of coal and steel sites, the focus for reclamation is often largely shaped by financial considerations, local socio-economic pressures, and the views of the design team. Investigations may therefore be carried out on site primarily in order to fulfil a legal obligation or to accord to a particular standard rather than properly address the constraints a site may impose in the context of its intended end-use. The result is that the investigation may not provide the information the designers of a reclamation scheme need, and the reclamation scheme will not fully address the constraints imposed by the site.

In some instances site investigation has been undertaken during or after the development of a site. This wholly unsatisfactory scenario highlights a lack of understanding and/or flagrant disregard for the intended purpose of site investigation, and may lead to severe legislative and financial penalties if problems are subsequently encountered.

Innovative approaches to reclamation, involving the reuse, retention and/or treatment of materials on site offer schemes new possibilities, and the opportunity for designers and contractors to create valuable selling points for completed sites. Increasingly the market for land and buildings will be manipulated by environmental factors such as the embodied energy of building materials and the environmental approach taken towards development. A more environmentally aware approach towards development may include not only considerations of the success of the finished scheme itself, but also the measures taken towards treating or controlling contamination on site, and the minimisation of local environmental disturbances during the reclamation.

16.3 Site clean-up

16.3.1 Pressures for decontamination

Chapters 2, 11 and 12 in particular have highlighted the importance of ensuring the chemical stability and quality of the ground during reclamation and after development. The need to set higher environmental standards has been largely been brought about by legislation following health, safety and environmental issues, both at a national and a European level. In turn this pressure has led to investment in the development of new technologies for dealing with contaminated land and waste materials, and this trend is likely to continue, with country by country demonstration of new technologies and the gradual acceptance of a wider range of choice in this field. This choice in itself is making the decision processes behind reclamation design more complex, and a broader understanding of decontamination technology more essential.

16.3.2 Publicly-funded reclamation

Whilst there is a justifiable reluctance for financial assistance from national governments and the European Commission for the decontamination of contaminated sites in Europe on an *ad hoc* basis, there are circumstances under which public funding for site clean-up will continue to occur. These circumstances are likely to be confined to those sites which:

- were publicly-operated industrial concerns;
- were operated by industrial parties who are now bankrupt, extinct or otherwise untraceable;
- are sites of national importance, where the polluter is not in a position to pay for remedial action;
- are sites that have been purchased by public bodies, often regional councils, for future development.

Often public funds for the reclamation of derelict and contaminated land are facilitated through governmental departments or agencies. In the United Kingdom, for example, funds may be made available through regional councils via the Department of the Environment in England and Northern Ireland, the Welsh Development Agency, or Scottish Enterprise. Similarly, in the eastern part of Germany the federal government has operated a scheme whereby certain organisations are exempt from the costs of clean-up of sites which they have occupied. These costs are then covered by the Treuhandanstalt and the region involved. France too administers public funding for reclamation through its regional Etablissements Publics, and in Denmark sites contaminated prior to 1972 are eligible for remedial action funded by the government's Environmental Protection Agency.[117]

Increasingly, publicly-funded schemes in Europe reflect interest in the use of new technologies to overcome difficult ground conditions, and many states have funding schemes that specifically recognise and fund these new technologies.

Publicly-funded schemes often demand a proportion of the funding to be supplied from local government budgets in the areas concerned, which thus provides impetus to projects through local incentives.

16.3.3 Insurance

The issue of insurance has not been dealt with in specific chapters of the book, as it permeates almost every aspect of reclamation, from client-contractor liability during on-site works to long-term liabilities after reclamation has been completed.

Increasingly, insurance advice is likely to be taken by landowners and industrial operators in order to safeguard environmental liabilities on reclaimed land, and it is pressure such as this that is likely to require permanent, failsafe options for reclaiming despoiled land in the future.

16.4 The product of reclamation

Any reclamation is geared towards the creation of an improved environment, usually linked to a specific end-use of the site. Such end-uses range from low impact, vegetated open space to high impact building developments, such as for housing or industry. These end-uses are in themselves the ultimate test of the success or failure of a reclamation scheme, and so success becomes a function of site use.

Poorly designed schemes have failed. These failures teach hard lessons regarding the consequences of low standards of reclamation. Conversely, successful schemes are often overlooked because they have achieved their aim of creating a new environment that bears little resemblance to its former condition. The results of past reclamation practices serve as monuments to successes and failures in this respect, and this should not be forgotten.

In order to illustrate some of the approaches to reclamation on land used by the coal and steel industries, Chapter 17 of this book illustrates some specific case studies from member states of the European Community.

17 CASE STUDIES

Chapter contents

continued...

17 CASE STUDIES

The process of reclamation within the coal and steel producing regions of Europe is ongoing. The earliest schemes have now stood the test of time, and the lessons learned are gradually being disseminated. The following collection of case studies within the European Community aims to illustrate practical experiences of reclamation in the coal and steel regions, and approaches to some of the challenges these former industrial sites impose on the communities and reclamation teams concerned.

The case studies have been selected to illustrate the variety of scenarios that exist, and examples have been chosen from the UK, France, Belgium, Italy, Germany and Spain.

17.1 Scotland

17.1.1 Introduction

The central belt of Scotland, stretching from the coast west of Glasgow to Edinburgh in the east, has a long industrial history, the mainstay of which was the coal and steel industries. There was one deep coal mine still operating in Scotland in 1993 and the remaining steel production, at Ravenscraig near Glasgow, was being shut down.

A great deal of reclamation has been carried out to remove the scars left by former industry with the aim of bringing economic rejuvenation to the region. Public funding for reclamation was administered by the Scottish Development Agency (SDA), whose Land Engineering Division was set up in 1975 with the aim of reclaiming key areas of derelict land for new uses. In 1991 the SDA merged with the Training Agency to form Scottish Enterprise, which operates through 13 Local Enterprise companies which are associated with local government areas. The aim of each enterprise company is to develop the economy, enhance skills and improve the environment in Scotland.

17.1.2 Easton Bing

Background

Easton Bing (spoil heap) formerly dominated the town of Bathgate, situated almost halfway between Edinburgh and Glasgow in West Lothian District. Bathgate experienced a serious decline in employment in the early 1980s due to a decline in local industries. It suffered from dereliction caused by extractive industries such as coal, oil shale and fire clay which had prospered in the nineteenth century but have since ceased operating. The town is in a prime location in relation to transport infrastructure and urban centres, but development was hindered by the poor quality of the general environment and the physical constraints associated with former industrial sites.

Environmental quality

Easton Bing, the site of the former Easton colliery, covers an area of 30ha. In the early 1980s it consisted of heaps of unvegetated colliery spoil and lagoons containing fine coal waste and water. A series of fires within the spoil caused air pollution.

Site restoration

Restoration of Easton Bing was considered in conjunction with that of another site 2km away, near Bathgate town centre. This second site, Little Boghead, consisted of 44ha of low-lying, boggy ground which, despite its poor ground conditions, was the only area within Bathgate designated by the local plan for private housing. An economic assessment of various reclamation options for these two sites was carried out by a firm of chartered accountants. This assessment took into account the costs of various alternatives and the likely financial return from the sale of developable land. All schemes were found to have a net cost, but this was substantially reduced for the option which used material from Easton Bing to create developable land at Little Boghead, compared with options which reclaimed the two sites independently of one another. The Easton Bing scheme was approved by the Board of the Scottish Development Agency, who provided funding in 1984. The Land Engineering Division of the SDA acted as the client/employer for the scheme and design and supervision was carried out by the Landscape Development Unit of Lothian Regional Council.

To ensure that the reclaimed land at Little Boghead would be acceptable to housing developers the National House Builders Council were consulted prior to design of the scheme. They had several stringent requirements intended to ensure that materials, such as burnt colliery spoil with a high sulphate content, were not adjacent to the concrete of house foundations and that a house fire could not cause combustion of the underlying spoil. The requirements included independent supervision of earthworks operations and provision of a 600mm deep layer of inert, structurally adequate material above the spoil. Fortunately, a source of suitable clay material was found nearby.

The fires at Easton Bing resulted in the bing being composed of a mixture of burnt, unburnt and partially-burnt colliery spoil. British Coal advised that the partially-burnt spoil was likely to be unstable, liable to swelling and therefore only burnt or unburnt spoil would be used in the fill at Little Boghead. These two kinds of materials were distinguished by their appearance, well-burnt spoil being a bright red colour and unburnt spoil, black. Selection of materials at Easton Bing required constant site supervision. The coal content of unburnt spoil was measured; the risk of spontaneous combustion was considered to be low when coal content was less than 10%, and only spoil which satisfied this condition was placed in areas of the site designated for housing.

At Little Boghead, construction of trial embankments had shown that existing material on the site, consisting of top soil, peat and sandy clay, had to be removed to avoid unacceptable settlement. After their removal to stockpiles, infilling with materials from Easton Bing was carried out. Materials were placed in layers 200mm thick. A minimum of 400mm of burnt spoil and 600mm of clay was placed over unburnt spoil in order that new concrete foundations did not come into contact with the sulphate-rich burnt spoil. The structural fill was protected by 600mm of the stockpiled material, and high standards of compaction were required for the structural fill (see Box 17.1). The latter was ensured by an extensive supervision and testing programme which required six full-time staff.

After completion of infilling at Little Boghead a surface combustion test was carried out in which a fire, the size of a typical house fire, was lit on top of the structural fill. Thermocouples inserted into the infill layers below showed a negligible rise in temperature, confirming that heat would not penetrate through the fill.

At Easton Bing far more spoil with a coal content of greater than 10% was encountered than had originally been estimated. Dealing with this spoil greatly increased the cost of the contract. Some of these problems could have been foreseen if the original borehole investigation of the bing had been more extensive. A more thorough investigation may also have given a more accurate estimate of the coal content of the spoil, which was found during the work to be far greater than had been estimated. As a result, a coal recovery scheme was carried out after the removal of material to Little Boghead. This scheme was successful in reclaiming

Box 17.1: Compaction requirements for colliery spoil at Easton Bing

The requirements for compaction of colliery spoil are:

- air voids ratio should be less than 10%;
- compaction for unburnt spoil 90%;
- compaction for burnt spoil 95%;
- coal content should be less than 10%;
- moisture Condition Value should be greater than 8.5 %;
- settlement: by Plate Bearing test at:
  100 kN/m^2 8 mm max after 15 minutes
  200 Kn/m^2 16 mm max after 15 minutes.

coal but unsuccessful financially because of the royalties charged by British Coal.

On completion of the coal recovery scheme the bing was regraded and treated with processed sewage sludge cake, followed by direct sowing and planting to form grazing land and woodland. The site now resembles a natural hill. The spoil is, however, generating some acid mine drainage water, which can be seen in a stream on the north side of the bing. Despite the compaction achieved at Little Boghead, some acid mine water has appeared in the culvert crossing the site.

The general air of dereliction at Bathgate has largely disappeared since the reclamation of Easton Bing and similar sites. The reopening of the local railway line to Edinburgh has also made the area a more attractive place to live. Part of the justification for reopening the line was the anticipated increase in population arising from new house building at Little Boghead. As a result of these improvements, prices for development land in Bathgate have risen substantially since the economic appraisal of the reclamation scheme. The revenue that had been anticipated from sale of the whole site has now been received from sale of just one fifth of the area to a private housing developer.

17.1.3 Hallside

Background

The Hallside steelworks, occupying an area of 32ha, was situated on the outskirts of Glasgow on the border between town and country. The works closed in 1978, buildings and structures were demolished and the site was acquired by the Scottish Development Agency (now Scottish Enterprise). An aerial view of the site is shown in Photograph 17.1.

The Hallside site is adjacent to the Strathclyde greenbelt. Several after-uses were considered, of which residential development is no longer seen as either viable or practicable.

The site will be leased or owned by the Strathclyde Greenbelt Company, a charitable company set up in 1991 to sustain and enhance the landscape and ecological value of the Strathclyde greenbelt. The company was set up jointly by Strathclyde Regional Council, Scottish National Heritage and Scottish Enterprise National, and aims to be a partnership between people who live and work in the greenbelt, public authorities with responsibilities for land in the greenbelt and voluntary organisations. Over one tenth of the area covered by the Strathclyde greenbelt consists of derelict, despoiled and unmanaged land. One of the objectives of Strathclyde Greenbelt Company is to restore such land in order to improve environmental quality.

Environmental quality

In 1993 the Hallside site had a derelict appearance, with extensive concrete foundations and remains of slag deposits clearly evident on the

Photograph 17.1: Aerial view of the Hallside site (source: Strathclyde Greenbelt Company)

surface. Much of the slag that was present was removed for use in road construction.

Analysis of water draining from the site indicated little leaching of contaminants.

The area surrounding the site appeared generally degraded in terms of environmental quality. Underdeveloped areas forming a rural fringe were frequently unmanaged and the urban environment was subject to vandalism and decay.

Site restoration

The original proposals for Hallside, drawn up upon purchase of the site by the SDA, involved development for housing or industry. Investigations revealed these proposals not to be financially feasible. In 1990, within the context of the Strathclyde Structure Plan, negotiations took place between the authorities and a developer during which it was agreed to allow residential development on adjacent land in return for the restoration of Hallside to agriculture, recreation or forestry. However, these proposals floundered when it became apparent that conventional restoration of Hallside for such uses would involve importation of massive quantities of good quality subsoil and topsoil at prohibitive cost. The Strathclyde Greenbelt Company designed a scheme along similar lines, where the housing developer will cover the site to a depth of 1-2m with colliery waste from two nearby colliery spoil heaps, freeing the land beneath the heaps for housing development. The greater part of Hallside will then be used for short rotation forestry to provide wood for energy production, with the remainder planted and managed as community woodland.

The short rotation forestry will consist of fast-growing tree species such as willow and poplar. These will be harvested by coppicing every four years. A staggered planting programme will ensure that some harvesting is carried out every year. A processing plant for the harvested wood was included in the plans for the site.

The community woodland will consist of mixed broad-leaved amenity trees which will provide a screen between the energy forestry and residential areas. Pedestrian and cycle routes through the site will be provided within the areas of community woodland.

An important aspect of the proposals was the use of sewage sludge to improve soil conditions for tree growth. Sewage sludge from Strathclyde Region is currently discharged into the sea west of Glasgow. This disposal route will no longer be available after 1998, when all sea dumping of sludge is to cease within the European Community. Strathclyde Greenbelt Company have produced a strategy to use some of this sewage sludge to grow short rotation energy forestry on derelict sites in the greenbelt. Hallside is seen as a testing ground for this strategy.

During spreading of colliery waste, 9,000t of dried digested sewage sludge cake will be incorporated into the surface layers. Liquid digested sewage sludge will then be applied to the short rotation coppice, as required. The community woodland will not receive liquid sewage sludge. Breakdown of sewage sludge components and build-up of a soil structure will, it is hoped, be encouraged by the introduction of earthworm cultures.

Current leaching of pollutants from the site is considered to be negligible and the colliery waste to be imported contains low levels of hazardous metals and is not acid-generating. The trees will also take metals out of the soil and accumulate them in the roots. However, as a precautionary measure all drainage from the site will pass through a wetland treatment area before discharge to a local stream.

17.2 England

17.2.1 Introduction

England is rich in coal and iron deposits, and has a long history of associated industries. The largest coal field extends from the East Midlands to the North-East coast. The coal and steel industries have been in decline for several decades, and the economic life of several cities and towns has depended upon development of the vast areas of derelict land formerly occupied by these industries. Reclamation has therefore played on important role in environmental and economic regeneration.

17.2.2 Orgreave

Background

This site lies to the south-east of Sheffield, South Yorkshire, one of the heartlands of the coal and steel industry in England. A varicty of activities related to the coal and steel industry have taken place at the site. All these activities have now ceased and the area is derelict, except for some small chemicals works which formerly utilised by-products from the coke works, but now use other sources of raw materials. British Coal plan to reclaim the area by extraction of coal from near-surface seams beneath the site by opencast mining, with restoration to enable industrial regeneration of the site. An area to the west, known as Waverley East, was formerly used for the disposal of wastes from the steel industry (slags and flue dusts). Opencast coal removal from this area is nearing completion. Restoration is planned to enable industrial regeneration on the site.

The principal areas of the Orgreave site are as follows:

- **Orgreave colliery:** this mine was sunk in the late nineteenth century and continued production until 1982. However, the coal processing plant of Orgreave colliery continued to treat run-of-mine from the nearby Treeton colliery until the latter ceased production in 1991.

- **Orgreave tip:** a large colliery spoil heap, covering 36ha, up to 43m deep and containing 12 million cubic metres of material. The tip received the discard from the coal processing plant. A large tailings lagoon is located near the summit of the tip.

- **Orgreave coking works:** located to the north-west of the colliery from which it received coal. The coking works was owned by a steel company, which had purchased Orgreave and Treeton collieries in the early twentieth century, when the coking works were under construction. The coking works produced coke for steel production and the gas was also piped to nearby steelworks. A wide variety of by-products, such as benzene, toluene, xylene, naphtha, ammonium sulphate and resins were produced from the crude tar and ammonia liquor. Production ceased in 1990. Photograph 17.2 shows a view of the coking works.

The area also includes a former domestic landfill site and Woodhouse Mill tip, a 17ha area to the east of Orgreave tip, adjacent to a lake. Woodhouse Mill tip was the site of a slag reduction works and also received a variety of waste from the chemical and metallurgical industries.

Environmental quality

Colliery spoils in the Sheffield area contain high concentrations of sulphate and other salts. Deposits of salts can be seen on the surface of the spoil and plant growth will not occur until most of the salts have been removed for example by leaching. Acid mine drainage is not a major problem. Ammoniacal liquor from the coke works was at one time disposed of on the tip, where it drained through the spoil into the adjacent river.

Orgreave tip is predominantly unvegetated, but liming and seeding with grass and clovers has achieved an effective cover on one side of the tip and a clump of birch has become established naturally at the north end.

The coal preparation plant was demolished to ground level, and the coal preparation plant area is contaminated by sulphates.

The coke works was demolished soon after closure. Care was taken to preserve records of the location of processes and to obtain information from former workers at the site. Underground pipes which carried organic liquids were then purged during demolition to avoid future loss of their contents into the ground through leaks or through disruption of pipes during later excavation works on the site. A detailed site

Photograph 17.2: Flooded gas holder base at Orgreave coke works, with Orgreave tip in the background (source: Richards, Moorehead and Laing Ltd)

investigation was carried out under the supervision of experts in the field to determine ground contamination. The Draft British Standard for Development Code of Practice for the Identification of Contaminated Land and its Investigation (DD175: 1985)[43] was followed. The works were found to be built upon made ground, contaminated with a wide range of hydrocarbons, including polyaromatic hydrocarbons and phenols.

The Woodhouse Mill tip consists of various types of wastes, including slags and flue dusts from the steel industry. Elevated arsenic concentrations were found in groundwater though the arsenic did not appear to be migrating into nearby surface water. The wide range of materials have varying pH values, resulting in a sparse but diverse flora which includes some locally rare species. Unvegetated areas are important nesting sites for birds.

The River Rother, which flows past Orgreave tip is heavily polluted before it reaches the site. However, the National Rivers Authority plan to substantially improve the quality of the river and have advocated treatment of the contaminated material at the coking works to remove any input of pollution to the river from this source. The Local Authority, Rotherham Metropolitan Borough Council, have aspirations to see the area cleaned up and returned to beneficial use.

Site restoration

British Coal Opencast has plans to commence the Orgreave Reclamation Project by continuing the existing opencast operation at Waverley East in the westerly direction, by removing coal from beneath the chemical works, the former Orgreave coking works, coal preparation plant and Orgreave tip.

Several treatment technologies have been considered for use on the contaminated materials present in the former coking works. Biological treatments, which remove the hydrocarbon contamination, were found, using laboratory tests, to be unable to deal with the wide range of compounds present, including complex hydrocarbons and heavy metals. Soil washing, followed by biological treatment of the waste water, was also investigated at laboratory scale. However, large volumes of contaminated water would have been produced, with little reduction in the contamination status of the soil material. These remediation technologies were therefore rejected in favour of disposal in an engineered landfill within the site boundary. The landfill is to be lined and located above the water table.

Other materials on the site will be mixed with the overburden and replaced after coal extraction. Very little natural soil exists on the site and the final surface cover will consist of the natural sands and gravels currently beneath the site, rather than soil.

Plans for the opencast working have included consideration of the after-use of the site. Architects have been employed to produce a masterplan for the site which envisages a business park in the north-west of the area, housing, and then open green space in the south, with shallow wetland areas for migrating birds adjacent to the existing lake.

A minimal reclamation treatment is planned for the Woodhouse Mill tip, to maintain and enhance the nature conservation interest of this area. This will involve prevention of fly-tipping, planting of native species in areas of poor vegetation cover and retaining bare areas for nesting birds.

Reclamation of derelict coal and steel sites by opencast coal mining has been practised at several other sites in Britain. It is considered by most local authorities to be more acceptable from an environmental point of view than opencast operations at greenfield sites and provides a means by which derelict land can be returned to beneficial use at minimal cost to public funds.

17.3 Wales

17.3.1 Introduction

The valleys of South Wales have a long history of coal mining and iron making. In addition to the natural deposits of hard coal and ironstone, the densely forested hill sides helped the development of these industries in the eighteenth century by providing valuable wood and timber. The use of charcoal pre-dates coal for the smelting of iron, and when coal mining started to provide a material to replace charcoal as a source of energy for iron-making, the timber which was available provided building materials and pit props for the collieries.

The topography of the South Wales valleys led to the area supporting a large number of small, deep collieries and iron works dotted along the valley floors. Small communities, subsequently evolving into towns and villages, marked the positions of these industries. Large waste heaps dominated the sides of the hills, because of the limited space available for tipping colliery spoil and iron slags on the valley floors.

Coal mining in South Wales has been reduced to a few deep mines and a number of opencast sites, as the remaining coal deposits have become more difficult to mine, and United Kingdom coal production has been reduced in the face of changing world markets.

Since the 1960s, and particularly since the huge colliery spoil tip slide at the village of Aberfan in 1966 (see Photograph 1.1), Wales has implemented an extensive policy of reclamation at colliery, coal processing and associated sites throughout the Principality. This case study concentrates on a large reclamation scheme undertaken in the Ebbw Fach valley, located in the north-eastern corner of the coalfield.

17.3.2 The Nantyglo and Blaina valley

Background

In the eighteenth and nineteenth centuries, the area between Nantyglo and Blaina was densely industrialised. The site contained an ironworks, a number of coke ovens, brickworks, lime kilns and collieries. The Ebbw Fach river passes through the valley, and was culverted underground over long stretches to make more room for industry.

Many of the ironworks closed at the beginning of the 1900s, and the last colliery closed in 1975, following an underground fire. It was this cessation of the local industry, and the vast area of dereliction that had resulted that gave rise to the Rising Sun reclamation scheme, named after the nearby Sun Pit coal mine and the former Rising Sun public house.

At the time of reclamation, the 74ha Rising Sun site, was one of the most dominating examples of industrial dereliction in the South Wales valleys.

Environmental quality

Intense coal washing operations had been undertaken during the early 1970s in the northern part of the site. These operations had left behind numerous large slurry ponds, which formed unsightly and dangerous features of the site. In addition, many large tips were present, consisting of colliery spoil, ironstone shale, iron slags and coal washing slurries. A variety of derelict buildings and structures remained. Only a small proportion of the original valley floor remained unaffected by industry.

Surveys and preliminary reclamation design work commenced in 1973, and a masterplan was presented to the local authorities in March 1974. Reclamation of the area started in 1977; two large tips on the hill side, known as Red Ash and Inkerman, were unstable at this time, and these and other tips were monitored for movement. The history of the site showed that a number of tip slides had occurred in the area previously, and that some of the unstable tips remaining lay directly above housing on the valley side. A detailed geomorphological survey indicated that, in addition, natural soil slippage was taking place and would require attention if land lying on the valley floor was to be made available for development.

The culverted stretches of the Ebbw Fach river were also found to be in poor structural condition in many places.

Site restoration

The Rising Sun scheme was formulated by collaboration of the local council (Blaenau Gwent), the regional council (Gwent County) and consulting engineers and environmental scientists, and was fully funded by the Welsh Development Agency (WDA). The WDA gave approval for the work to commence in June 1977, and the main contract reclamation works were completed in October 1983. Other work carried out included the construction of bypass and access roads for the area. The reclaimed site was officially opened to the public in April 1984.

The consultants were responsible for the geotechnical, civil and structural engineering and landscaping aspects of the design, as well as supervision of the works.

The aims of the reclamation were wide-ranging. Primarily the scheme was designed to bring the area back into beneficial use, and this involved the demolition of old structures, regrading and stabilisation of unstable waste tips and natural landslips, revegetation, and the provision of building land suitable for both housing and industry.

The final ground profile was achieved by balancing the excavation and filling operations wholly within the site boundary, and this involved the movement of approximately 1.5 million cubic metres of material. The Inkerman tip was completely removed and much of the colliery spoil from this was used to raise the level of the valley floor. A new concrete pipe river culvert replaced the old stone arch culvert. This new culvert was designed to withstand the substantial surcharge loads from the regraded waste tips.

Nineteen mine shafts were made safe by a combination of grouting and capping with reinforced concrete. Several mine adits along the valley floor were secured by plugging with concrete and/or brickwork, and provisions made for draining these adits. Surface drainage took the form of contour ditches with lined inverts, and some of these were subsequently filled with free-draining stone after the establishment of grass on the site. The newly regraded slopes along parts of the mountain side were drained using stone-filled buttress drains to stabilise the landslip, and open channels diverted run-off water down the slopes towards the Ebbw Fach river. The increased drainage across the site was accommodated in the enlarged storm water drains and river culvert, and also by the construction of two regulating/settlement ponds at the foot of the hill side.

Following the engineering works, a long-term programme of revegetation was undertaken. This consisted of broiler house litter application followed by grass seeding to produce grazing and amenity grass areas. Well-established grass on the slopes has helped to prevent surface erosion, and in places wild flowers, both seeded by hand and naturally colonised, can be seen in abundance. Extensive planting was carried out to provide shelter and enclosure in two recreational 'buffer zones' between the central industrial area and existing and new housing at the north and south ends of the site. Further planting adds interest and shelter belts to the hillside grazing land, and divides lower grazing by hedges. Species were chosen which tolerate the elevation and rainfall, grow on the infertile spoil of the site, and provide food and shelter for wildlife. From these species, mixtures were composed to produce woodland, understorey, edge, waterside and hedge characters.

Particular care was taken to rip the surface of the ungraded colliery spoil in order to counteract the compaction introduced by the civil engineering machinery. This loosening enhanced root growth and subsequent tree establishment.

Due to the complexity of the scheme and extremely competitive tendering by contractors, three firms and a receiver were forced into receivership during the course of the works. These events had the effect of complicating and extending the time taken to complete what was already a large and complex project.

Costs and after-use

The completed scheme provided 15ha of land for industrial development, 8.5ha for housing, and a further 11ha for amenity use and open space. In addition, 33ha of mountain side were made available as agricultural grazing land. The remainder of the site area was accounted for by roads and further recreational land, including a school nature study area.

The final cost of the reclamation scheme, in the early 1980s, was £7.25 million, which included all the undertakings mentioned here, including the acquisition of land and the ground preparation for building. Today the area comprising the Rising Sun reclamation scheme is a well balanced community of housing, light industry and public open space, where investment has provided new hope for an area once severely affected by industrial dereliction and decay.

17.3.3 Brynbach Park

Background

Brynbach lies within the Borough of Blaenau Gwent, in the north-western corner of the County of Gwent. Blaenau Gwent occupies the north-eastern corner of the South Wales coalfield. The site forms part of an elevated open landscape at the heads of deeply incised valleys which descend to the south.

Shallow coal measures and ironstone deposits were once worked extensively at the head of these valleys. A large number of small reservoirs were constructed to supply water to the iron industry which was established in the eighteenth and nineteenth centuries.

Environmental quality

Brynbach Pond was one of the water supply reservoirs which were situated on land underlain by coal measures and surrounded by waste from many small mines which were once active in the area. In the late 1960s and early 1970s the National Coal Board were mining the coal outcrop by opencast methods, and proposed that one mine should be extended by 80ha including the area occupied by Brynbach Pond.

Site restoration

In 1972 the idea of reinstating Brynbach Pond as an amenity lake was proposed (see Photograph 17.3). Further, an opencast mining operation was proposed which could create the landform for the new lake as part of the restoration process, at no additional cost to the mining operation or the community. This principle was accepted by the National Coal Board. Planning permission was granted by Gwent County Council for the mine extension, incorporating a preliminary plan of the final landform and lake, which had been designed by consulting engineers employed by the Borough Council. The engineers and National Coal Board then agreed a mine backfilling programme in which the basin to accommodate a 20ha lake would be formed by 1978. The lake was to have a maximum depth of 6m and an average depth of 2m, which was to be controlled by an overflow and draw-down system. Islands would be constructed to dissipate wave action and reduce erosion. The lake would be filled with water draining from the surrounding hillsides. As these hillsides were also on land reclaimed after deep mining during the previous hundred years, extensive silt traps were designed to intercept the sediment from this drainage water.

Photograph 17.3: Visitor centre and sports clubhouse at Brynbach Pond and Country Park (source: Richards, Moorehead and Laing Ltd)

The parties involved in the planning phase of this project were:

- The Welsh Development Agency;
- The National Coal Board and their contractors;
- Gwent County Council;
- The Forestry Commission.

The landform was prepared by the National Coal Board, who also established grass and trees and maintained the restored mining site for five years. The Welsh Development Agency funded the elements which make up the lake:

- the 1500 gauge polyethylene sheet liner;
- the drainage system;
- the overflow system;
- the lake edge formation, erosion protection and margins;
- aquatic and marginal planting.

The polyethylene membrane liner was required since the mine backfill was not considered sufficiently impermeable to prevent leakage of water. The lake was prepared, lined and filled between 1978 and 1980.

Site management

A masterplan was prepared in 1978 for a wider area, centred on the new lake but extending to surrounding reclaimed land across the local government boundary. This plan proposed that 600ha of land, reclaimed to grassland and young woodland, would become a public amenity and recreation area, in order to:

- improve the landscape in an area of derelict and degraded landscapes;
- increase opportunities for recreation for the people of the densely urbanised valleys;
- provide alternative recreation facilities to those already in the area, particularly new water-based opportunities;
- assist in the development of tourism;
- increase employment locally.

A management scheme was also prepared in 1978 to establish guidelines and responsibilities for the management of the park over a 7 year period. Eight governmental or non-governmental organisations were involved as well as two neighbouring County and two Borough Councils. Brynbach was by this time recognised as potentially important sports and amenity facility.

After-use and success

The Brynbach reclamation scheme is seen as one of the most successful in South Wales. By the early 1990s the landscape had become well developed despite the elevation (360-430m above sea level) and rainfall (1470mm per year) largely due to good management under the direction of the County Council's forestry officer. In the early years of its existence the park received few visitors, and these were mainly local people, but in 1989 and 1990 well over 200,000 visitors were recorded. The lake facilities are used by water skiing, sailboarding, angling, canoeing and subaqua clubs. A development study indicates that Brynbach is an important asset to the region with significant potential for growth.

The 1978 management scheme was reviewed and updated by management plans in 1985 and 1990. The important factors illustrated by Brynbach are:

- an early identification of the main aim *i.e.* to create a new lake;
- the stimulation of the plan for a 600ha park;
- the cooperation between many diverse organisations;
- funds were contributed by many organisations;
- perseverance by particular members of the design team who were determined to overcome technical problems;
- the acceptance of a long-term management plan;
- the implementation of a long-term plan with modest funds;
- successes gradually increased the recognition of the park as a real asset.

The 1990 management plan indicates that the objectives set out in the earlier reports have been achieved and goes on to outline a further 10 years of work in management and development on the site.

17.4 France

17.4.1 Introduction

France played an important part in the instigation of the ECSC in 1952. Iron ore was the main mineral resource of the country, helping to rank France third among world steel producers in 1993. The iron ore deposits are mainly concentrated in the north of Lorraine near Metz which was a major steel producing region in France. The other major steel-producing area of France is in the north, particularly in the Dunkerque area.

Coal deposits in France lie in 3 main regions: lignite is found in the south of France at Gardanne, Provence, while hard coal is found in the Nord-Pas de Calais and Lorraine, with some deposits scattered around parts of Centre-Midi. The coal industry in France, like many other areas of Europe, is in decline. Mines are closing and as a result, coal is now produced only in Lorraine.

The case studies in France will cover the following regions:

- Lorraine;
- Nord-Pas de Calais.

17.4.2 Lorraine

Background

Lorraine is situated in the north-eastern part of France, bordering Luxembourg and Germany. It was a region rich in coal and iron ore deposits. Mining for the main mineral, iron ore, occurred across the region and continued into Luxembourg. All the iron mines in the region have closed leaving behind an underground labyrinth which has given rise to major subsidence problems.

The coal deposits in the Lorraine region are found in the Upper Carboniferous and Lower Permian series, and reach depths of 1700m. They are a high bituminous type with a low sulphur content.

Environmental quality

Derelict land first became a problem in Lorraine during the 1960s with the slump in the textile market and iron mining. The problem has since increased with the decline of the steel and coal industries. A total of over 3000ha of land and buildings have been left degraded.[87] By 1985 the problem of dereliction had become serious enough for the State and the Regional Authority to plan and define a strategy aimed at reclaiming the abandoned land in order to change the image of the area and restore its economic and social status.

Site reclamation

The Etablissement Public de la Métropole de Lorraine (EPML), a publicly funded organisation, began implementing a reclamation programme in 1986. The role of EPML is to purchase derelict land, reclaim it and prepare it for new use. If, following feasibility studies, the site is found to be contaminated, the principle of 'polluter pays' is adopted wherever possible.

No national standards for reclaiming derelict land or treating contaminated land exist in France; those of the Netherlands are generally followed. These standards are, however, often not suitable to the land type in Lorraine and are usually modified. The EPML are currently working with DRIR (an administrative body), USINOR (a steel company) and Charbonnage de France on the geological context of sites and are developing guidance on safe concentrations of contaminants in relation to background levels, which will be used as reference values. A pragmatic approach.

One coal site, Falquemont, and two steel sites, Micheville and Homécourt, are described in the following paragraphs.

17.4.3 Falquemont

Background

Coal was deep mined at Falquemont and provided the major fuel source for a power station 15km away. The mine closed in 1974. However, coal dust reclaimed from the settling lagoons continues to provide fuel for the power station.

As the mining company has progressively withdrawn, the local council has become involved in promoting new uses of the site for industrial, educational and recreational purposes. A large part of the site is now used as a training college for operators of heavy earth-moving equipment.

The site, which occupies an area of 160ha, is divided into three parts:

- former mine buildings;
- spoil heaps;
- settling lagoons.

Falquemont provides a good example of a long-term project to reclaim derelict land for a variety of new uses.

The former mine buildings

The main reception building serves a double purpose for teaching and student accommodation. Some of the remaining buildings are occupied by small industrial users. However, some of the buildings are in a poor state of repair and only the ground floors are suitable for occupation.

The spoil heaps

The spoil heaps are used as the main driver training ground, and have been found to be ideal for earthmoving operations in all weather conditions. The spoil heap material is continually moved around by the students practising their driving and operational skills.

The settling lagoons

HBL (Houillière de Basin Lorraine), the original mining company, is exploiting the settling lagoons for recovery of coal dust and the empty lagoons are being filled with spent ash from the power station. HBL are working in conjunction with the EPML to reclaim the lagoon area, and further studies are to be carried out as to how the alkaline nature of the ash will affect the reclamation proposals:

- upper lagoon - this is to be completely filled with ash and will serve as an additional training area for machine operators (see Photograph 17.4);
- middle lagoon - this lagoon is to be partially-filled and a fishing pond/wildlife area is to be created;
- lower lagoon - plans to recover coal dust from the lagoon have been halted due to the low coal content. It is likely that the lagoon will be drained to improve the stability and then planted to create woodland.

Landscape work

Some initial landscape work has been carried out along the old railway line at the periphery of the site to improve the outlook from the nearby residential area. This will also help to reduce the noise from the ongoing earthworks operations on the site.

Shafts

There are two shafts on the site, both of which have been capped. It is planned to utilise water from the flooded workings for local industrial purposes.

Photograph 17.4: Falquemont. Upper lagoon being filled with ash from the power station (source: Richards, Moorehead and Laing Ltd)

17.4.4 Micheville

Background

Micheville is situated in the north of the region, 1.5km from Luxembourg. Five communities, with help from the EPML, are combining their forces in an attempt to restore this derelict iron mine and steelworks, in order to promote a more attractive image for the area and to discover new potential for the site (see Photograph 17.5).

The Micheville site first opened in 1871. It covers a surface area of 370ha; 180ha are the former iron workings in the west of the site and 190ha the steelworks area in the east. Installations on the site included a crushing mill, blast furnaces, rolling mills, a coal carbonisation plant, and a power station. At the height of its operations in 1964, Micheville

Photograph 17.5: Micheville. Aerial view of the steelworks site following reclamation and initial landscape works (source: EPML)

Photograph 17.6: Micheville landscape works in the area of former iron workings and slag heaps (source: EPML)

employed 4100 workers. The site finally closed down in 1986 as a result of the decline in the steel industry.

The EPML acquired the site in two parts in 1986 and December 1989 at a total cost of 6,300,000FF. Demolition of the site began in 1987. Landscape work was carried out between 1989 and 1990 and the restoration of some of the buildings in 1991 (see Photograph 17.6). The cost of the feasibility studies undertaken totalled 1,010,000FF, whilst the reclamation and landscape work cost 31,700,000FF.

Micheville is a large site. A symptom of the obvious deprivation its closure caused is the reduction in the population of the town as people moved elsewhere for work. The reclamation of this site is an ambitious project intended to attract new industry and reintegrate the site into the local community.

Former iron workings and slag heaps

Much of the large area of the site left derelict by the opencast iron workings has been landscaped and reclaimed for recreational use with the introduction of tree planting, circular walks and activities areas. In some parts of the site original mining features such as opencast workings and slag heaps have been retained in order to emphasise the mining history of the site. At the foot of a cliff on the mine plateau, a natural amphitheatre provides a striking setting for summer outdoor theatrical events.

Blast furnaces and coal carbonisation plant

Demolition works on the site were completed and redevelopment work commenced. Initially ground contamination was not considered to be a problem. However the steelworks site is in a limestone area, and it was subsequently discovered that there was a risk of contamination from the coal carbonisation area leaching into the groundwater and adversely affecting the water supply for the town. This was overcome by excavating and removing the contaminated material off site to a special waste tip. In order to avoid contamination of the groundwater during earthworks operations the area of contaminated ground was first encapsulated by the use of a bentonite cut-off trench (see Figure 17.1). The uncontaminated overburden was removed and stored for re-use and a membrane was introduced to prevent access by surface water during the progressive removal of the contaminated material.

Rolling mill

The rolling mill site underwent major changes; all of the buildings and structures have been demolished. The waste materials from the iron

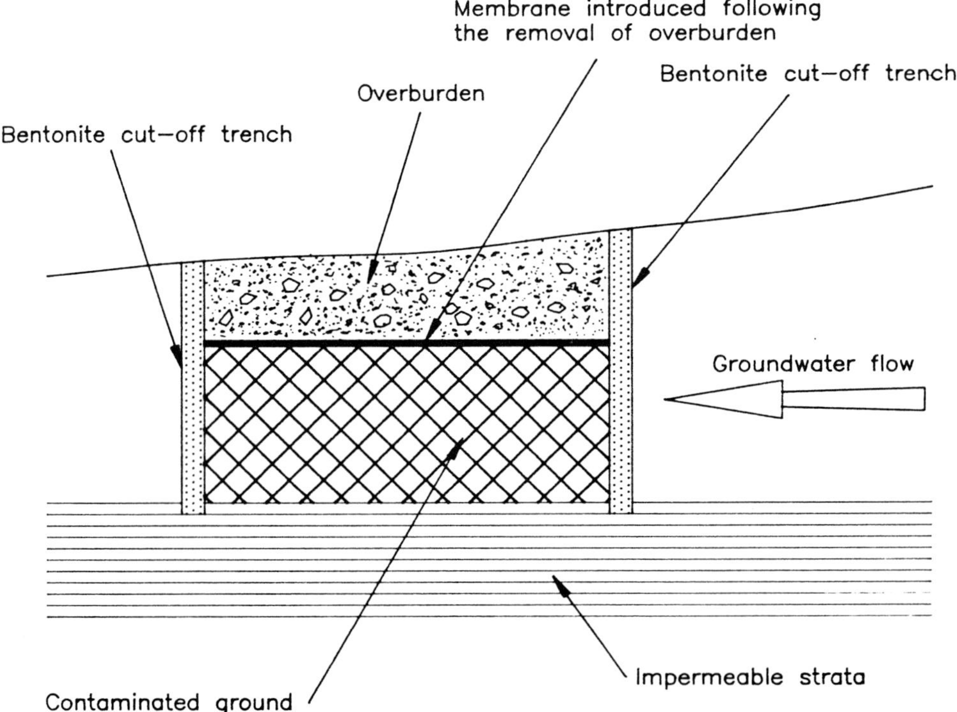

Figure 17.1: Procedure for encapsulation and removal of contaminated material

workings were used to raise the site levels by 2m, and the area is intended for a light industrial after-use. Topsoiling and grass seeding was carried out in some areas, and planting pits were used where trees have been introduced.

Buildings

Many of the buildings on the site were demolished, although a number have been retained and have been converted for industrial use. The original administration and management buildings near the site entrance are of architectural interest and it is intended to convert them for use as apartments.

17.4.5 Homécourt

Background

Homécourt, situated at the bottom of the Orne valley, was once a fishing village. With the construction of the steelworks in the nineteenth century the village expanded and was transformed.

The Homécourt steelworks site, which included the villages of Homécourt and Joeuf, opened in 1881 and covered a surface area of 89ha. The initial activities of the site were iron smelting, coal carbonisation and iron processing. By 1913 production of steel had reached a level of 346,000t/year. Steel products from the site included, railway lines, sleepers, steel plate and steel billets. The site contained iron workings, blast furnaces, finishing works, a coal carbonisation plant, a rolling mill and a power station. The highest number of workers employed at any one time was 5,000, at the end of 1969. Sacilor were the last operators of the site before it closed down in 1983. The EPML acquired the site in December 1987 at a cost of 6,000,000FF.

The reclamation work on the site has been carried out in a series of stages. All the buildings with the exception of one were demolished; the

Photograph 17.7: Homécourt. View of the steelworks site after reclamation works and initial landscape works (source: Richards, Moorehead and Laing Ltd)

retained building is of architectural interest and has been refurbished for industrial use. Feasibility studies cost 600,000FF, reclamation and landscape works 13,800,000FF, and work on the buildings 1,300,000FF.

A new road has been built across the former steelworks site to provide direct communication between the two ends of the town.

At the site of the coal carbonisation plant, which lies at an altitude overlooking the town, an access road has been constructed to allow the site to be redeveloped. The material from the slag heaps has been used for the road construction.

Pollution from hydrocarbons was discovered to be a problem near the coal carbonisation plant. Treatment methods under consideration included the use of a membrane or a clay capping layer.

Some of the slag heaps on the site are still being exploited for use as aggregates and fill material. Where appropriate, the iron-rich slag is subject to metal extraction for reprocessing. Other slag heaps have already been reclaimed and landscaped, mainly with pine trees. Whilst planting pits were used for large trees, a thin layer of topsoil was found to be adequate for small trees and shrubs.

17.4.6 Wingles

Background

The region of Nord-Pas de Calais in the northern, industrialised corner of France includes a large coal field with many abandoned colliery sites. Reclamation has been carried out in the region for at least 20 years but to date the proportion reclaimed remains small. As a result, prominent spoil heaps dotted around a flat landscape remain typical of the region.

This case study concerns a colliery site which has been transformed into a very successful public park. It is located on the edge of Wingles, a small mining town situated centrally in the coal basin, 20km south-west of Lille, the regional capital

Site description

The site comprises around 200ha of land and water, and is open to the public as a recreation park. Numerous activities are catered for, including sailing, fishing, motor-cross, cycling and jogging (see Photograph 17.8). There is also a campsite, an information centre and various play areas. There is a substantial amount of woodland, with large areas reserved for wildlife.

The park has been developed progressively since reclamation. The 12ha watersports lake, excavated for the purpose in an area of subsidence, was completed in 1986 (see Photograph 17.9).

Photograph 17.8: A network of paths provide for walking, jogging and cycling. The information centre can be seen in the background (source: Richards, Moorehead and Laing Ltd)

Photograph 17.9: A large watersports lake has been created by excavation in an area of mining subsidence. The colliery spoil heap on the right is untreated, but partially covered in self-sown vegetation (source: Richards, Moorehead and Laing Ltd)

Photograph 17.10: These fishing ponds, formed by subsidence, pre-date the reclamation works (source: Richards, Moorehead and Laing Ltd)

The 20ha of fishing ponds, formed by mining subsidence, were already in existence when reclamation commenced (see Photograph 17.10).

Site history

Three Communes, the lowest tier of local government in France, combined their resources to reclaim (initially) 66ha of derelict colliery and swampy land. The Communes of Wingles and Billy-Berclau collaborated from 1973 and were joined by Douvrin in 1975. The financing of the scheme was divided between the Communes on a population basis. The total population is approximately 17,000.

The Département, the second tier of local government, agreed to finance by 75% the purchase of more land, to give a total of around 200ha.

By 1988, 19,800,000FF had been spent on the site, with a further 14,500,000FF predicted to complete the work.

Site reclamation

The site consisted of many spoil heaps and areas of subsidence. Some of the tips were flattened or reduced, filling low areas, and those which could be utilised were exploited. This practice still continues.

No colliery features, such as pithead gear, were retained.

Studies were carried out regarding ground contamination and none was found. A stream which runs through the site is visibly polluted, but this is said to originate from local agriculture and industry. Acid mine drainage is not an issue as the geology of the site is alkaline.

Other studies considered the soil materials on site for their potential to support vegetation. The site is well-vegetated, although there are some bare and eroding areas where colliery spoil is visible. In many areas, naturally colonised vegetation was retained.

A tip adjacent to the watersports lake continues to burn. Nothing has been done to deal with this and it is not considered to be a problem.

The Département was in charge of the reclamation design and the work was carried out in eight phases.

Management and after-use

At an annual cost of around 400,000FF, the Communes maintain and clean the site and also provide the staff who organise events in the summer. Maintenance of vegetation is carried out by private contractor.

By collaboration of local authorities, a large area of derelict land was reclaimed and transformed, over a period of many years, into a very successful and attractive public facility. Existing site features, such as ponds and self-sown vegetation, were incorporated into the scheme.

The local population is not very large, and it is hoped that people will travel from further afield to visit the site and so encourage new public interest in the area.

17.5 Belgium

17.5.1 Introduction

The province of Liège is the major coal and steel producing area of Belgium. Since 1961 economic development in the region has been promoted by the Société Provinciale d'Industrialisation (SPI), an association of local authorities and other economic and social institutions in the province. The members of SPI now include 68 of the communes of the province, representing 91.5% of the population.

To help bring the increasing amount of derelict land back into beneficial use, SPI set up an affiliated organisation, SORASI (Société de Rénovation et d'Assainissement des Sites Industriels) in 1988. The objective of SORASI is to reclaim derelict land and then sell it for a suitable after-use. Site investigations are carried out prior to purchase of sites by SORASI. If significant problems of contamination are discovered SORASI will not buy the site. Treatment of contaminated sites is the speciality of another public company.

The general approach of SORASI to treatment of sites is to demolish unsafe buildings and structures and remove foundations to 0.5m below ground level. Prospective purchasers are given details of any significant obstructions below this depth.

Two sites, one former coal mine and the other formerly part of a steel works, were visited in the town of Seraing near Liège. The two sites are known as Colard and Aciérie L.D. respectively.

17.5.2 Colard

Background

This is the site of a former coal mine which ceased production in 1979.

Environmental quality

There are two shafts on the site. Prior to reclamation a colliery spoil tip was present at one end of the site, with a few remaining buildings. The spoil is not acid-producing.

Site restoration

A reclamation scheme has been carried in which the spoil heap was levelled to provide land suitable for the construction of light industrial units (see Photograph 17.11). Access roads have also been provided. No building is allowed within 25m of the shafts and this restricted area has

Photograph 17.11: New industrial units at Colard (source: Richards, Moorehead and Laing Ltd)

been fenced. Scrub vegetation is developing within the fenced area. A small trolley from the coal mine has been used as the centrepiece of a roundabout adjacent to the site, as a monument to the coal industry.

17.5.3 Aciérie L.D.

Background

This 14ha site was originally a coal mine, then part of the steelmaking facilities of the local steel company, which still has active blast furnaces adjacent to the site. After closure in 1985 the steelmaking plant was dismantled and transported to China.

Environmental quality

The site is covered with several metres of made ground, consisting of wastes from the coal and steel industries. A site investigation was carried out to determine the ground conditions and contamination status of the waste. It was concluded that the majority of the wastes could remain *in situ,* but should be covered by clay to exclude water. One small area of more contaminated material has been removed. Two shafts are present on the site.

Site restoration

The two shafts have been capped and at the time of the site visit the site was in the process of being restored by infilling with imported materials (Photograph 17.12). This infilling is necessary to give a gentle slope from one end of the site to the other, enabling construction of an access road across the site. Currently the site is at a lower level than the surrounding ground.

The site is to be divided into four zones, for housing, commerce, light industry and recreation. The zones are to be separated by green areas and the recreation area, a sports field, will be a buffer between the existing active steelworks and the other zones.

Photograph 17.12: The Aciérie LD site part way through reclamation. The drainage system is being installed (source: Richards, Moorehead and Laing Ltd)

17.6 Italy

17.6.1 Introduction

Italy is not richly endowed with natural deposits of the raw materials for steel making. The country has, however, developed one of the most significant steel industries in Europe during the twentieth century, through a programme of infrastructural development and importation. This industry, as in most other parts of Europe, is undergoing severe decline, and Italy faces the consequences of industrial dereliction at steel facilities occupying vast areas of urban and urban fringe land. In the latter part of the 1990s action will be taken to move or reduce the size of the steel sites, which often tend to be centrally located in towns and cities. Amongst the most significant areas, many of which are coastal, are Genoa, Naples, Milan, Piombino, Terni and Taranto. In the early 1990s

Taranto was the second largest steelworks in the Community, and although many of the other steel plants are substantially smaller, collectively they are of considerable significance.

Coal mining activities are now largely confined to the south west corner of the island of Sardinia, around the Sulcis area of south-west Carbonia.

17.6.2 Steel sites - general

Background

Eleven priority sites have been identified by the Italian central government as requiring remediation and redevelopment. These include the huge Naples steelworks, which has been in decline since the 1960s.

Environmental quality

The larger steel manufacturers already have much in-house expertise on recycling materials and environmental issues within the industry, and many by-products are recovered. Effluent and flue gas treatment plant, and waste management systems are important support industries to the steel manufacturers. Nickel, zinc and chromium, for example, are recovered from demolition materials and dusts by plasma furnace treatment.

The steel sites, however, are burdened with many of the obstacles to reclamation and redevelopment that are typical of the industry, including contaminated land and the presence of large structures.

Site restoration

There appears to be little direct experience with the reclamation of steelworks sites in Italy, as the industry is relatively recent, and planned reclamation schemes are only just starting on a small scale. Planning permission for development in Italy can take years to finalise, unless a site is identified as an urgent case.

The introduction of high technology processes is creating smaller steel sites, and consequently more derelict land. Genoa is no exception, and now has a regional development plan which includes the centrally situated steelworks, where it is planned to build a technology park on part of the site. Huge steel plants also exist in Turin and Massa Carrara. The Massa Carrara redevelopment project involves the retention of some existing buildings of industrial and archaeological interest, and the building of others.

In Piombino the local authority seems less positive about redevelopment of the area, despite fears of high unemployment. In Piombino, 860ha is occupied by steel processes, including central urban land and coastal areas owned by the state and leased to the Ilva steel corporation. Here, it is planned to move steel production out of the town, to the east, where it will take a condensed form but maintain jobs.

17.6.3 Sesto San Giovanni, Milan

Background

The Falck steelworks in Milan is at Sesto San Giovanni, outside the city centre, in the industrial region. In 1993 it employs 2500-3000 people, and occupies 150ha. Like many other steelworks, it is undergoing changes that will condense its activities in terms of both site area and employment; in 1970 it employed some 70,000 workers.

Operations at the steel plant involve the production of coils/springs and sheet metal. Much of the work is commissioned, and uses specialised steels.

Environmental quality

The industry, like most, is based on scrap metal, and has one of the largest mills in Italy. Oil and metal contamination is a potential problem at such sites. The provision for recycling the metal parts of scrap cars at the plant indicates the likelihood of other wastes being generated. Since its foundation in 1906, annual production reached about 2 million tonnes in the 1990s, based around two gas operated furnaces. Each furnace handles about 140t/h.

Site restoration

Of the total site area, 80ha is to be retained for industry, and will incorporate part of a new autostrada to take traffic away from the centre of Milan. The other 70ha will be available for development. Many of the workers flats in the area are in a poor state of repair and will require demolition.

Falck is planning to build a 50MW power station at the site, and has experience in the power generation industry through its hydroelectricity operations in northern Italy.

A subsidiary of Falck has a waste immobilisation (solidification) facility which handles industrial waste from the steelworks and elsewhere, and this may be used in the treatment of any contaminated soil materials encountered during the redevelopment works. At this site, inorganic waste is mixed with concrete and calcium silicates, which is pumped, as a slurry, to a landfill, where it cures. The facility treats about 50t/h of waste, and 20,000 - 24,000t/year. Similarly waste water treatment is performed on site by sedimentation, flocculation and sand/charcoal filtration, and water is reused for industry.

17.6.4 Pietrafitta

Background

The Pietrafitta site lies in the Umbria region, approximately 30km from Perugia, near the Nestore river, and comprises a 400ha area of past lignite extraction. Extraction took place between the mid 1950s and the mid 1980s.

Site restoration

Plans have now been detailed for the area, and the University of Venice is working on landscape design. ENEL (national power company), the local authorities, and various landscape architects are working together to initiate the reclamation scheme.

The main features of the site are to be a central facility for power generation, a public park, a palaeontological museum, and a cultural museum dedicated to the steel and coal industries in Italy. The latter is thought to represent an important cultural focus for the area, because of its industrial past.

The parkland is planned to include areas of open water using depressions arising from the coal extraction, artificial oases, and botanical greenhouses, which will be heated directly from the adjacent power station. The power station itself is central to the socio-economic future of the area, and its location in the parkland aims to lower its visual impact on the environment.

Plans for the site have been published in a well-produced book[3], showing artists impressions of the various aspects of the park and the power station. This book is seen as an important part of the programme needed to advance the reclamation scheme.

17.6.5 Santa Barbara

Background

The Santa Barbara site occupies an area of some 2700ha in the Valderno area of Tuscany. It is approximately 50km from Arezzo, near the Tiber river. Opencast coal extraction in the region has largely ceased but some areas are still being worked, although this will end in the latter half of the 1990s.

Site restoration

Many of the tracts left after extraction have already been filled and are now used for agriculture, which was the original land use in many parts of the site. Some zones have been revegetated, and the reclamation plans for the entire area includes further agricultural land and reforestation.

The project has been initiated by ENEL, in conjunction with the University of Bologna, a petroleum chemicals manufacturer, the local authorities and the forestry institute of Arezzo.

17.7 Germany - Saxony (Eastern Länder)

17.7.1 Introduction

The Eastern Länder of Germany have a long and diverse history of coal mining and steel production. The coal industry is typified by opencast brown coal (lignite) fields and associated coal processing industries. The production of iron and steel relied on local deposits of raw materials, and although the industry is in decline, steel is still manufactured from scrap metal.

The climate is typical continental European, with warm summers and cold, wet winters. The topography of the area is generally fairly flat, becoming more mountainous towards the south, where a metal-rich mining belt occurs.

Industry in the region is in a state of flux due to the consolidation of operations since the reunification of East and West Germany in 1990. This is resulting in high unemployment in industrial areas, where often a single operation or industry has dominated the local social and economic environment. Consolidation is also resulting in further decline to add to the legacy of dereliction left by former industrial activities.

Current German legislation makes it easier to extend the boundaries of an existing mine rather than to open a new one, and thus the brown coal fields continue to expand. After the cessation of mining in an area, the planners take over from the mine legislators, and restoration can proceed.

Whole towns have grown up around the steel industry, relying on the industry for employment and income. As the steel industry declines, the situation is rapidly changing and economic planners look to regenerate the old steel areas for new industry.

Four case studies are presented in Sections 17.7.2 to 17.7.5: coal despoiled areas around Witznitz and the town of Espenhain and the steel towns of Gröditz and Riesa.

The area south of Leipzig is rich in near-surface deposits of brown coal. These have been exploited by opencast mining since the turn of the century. The Witznitz and Espenhain case studies are within this area. Leipzig is approximately 100km to the north-west of Dresden.

The steel town of Riesa lies about midway between Leipzig and Dresden; Gröditz is approximately 20km north of Riesa.

17.7.2 Witznitz

Background

The Witznitz field was first mined in 1922, and continues to be exploited by traditional methods of shallow opencasting. The company MIBRAG (Vereinigte Mitteldeutsche Braunkohlenwerke AG) operates the extraction and processing industry in the area, and gradually towards Leipzig where the coal deposits become progressively deeper. Coal production at the site is being reduced from a previous peak of 100Mt to a projected 12Mt by the mid 1990s.

There are two or three seams of lignite being worked by large bucket and conveyor machines. The seams are worked in strips, and the overburden used to backfill the previously worked strip. The ratio of overburden to lignite is approximately 3:1. In the 1950s the topsoil in the areas to be mined were moved to one side and stockpiled for reuse when the next strip had been mined. Unfortunately, this practice was stopped in the 1960s and 1970s because of economic and political pressures in the area. Therefore, overburden and topsoil were not separated prior to replacement, and the area suffered a serious decline in the quality of surface materials suitable for the establishment of vegetation. The older areas of the site are therefore relatively well colonised with plants, for example, silver birch, but there are now vast areas of extremely poorly vegetated land in the wake of the mining operations.

The opencast site was once surrounded by 28 lignite processing facilities, which are now in decline. A total of 27 villages and local communities have been abandoned and destroyed during the life of the mine, although there is a town on the edge of the current workings that has been saved, and appears as an 'island' on the perimeter of the field.

Environmental quality

The coal contains between 2 and 3% sulphur and about 50% water, and is used to produce briquettes, with the remainder used to feed local power stations. The calorific value of the coal is relatively high at approximately 2400kJ/kg.

The burning of coal in the area leads to the deposition of significant quantities of dust and sulphur compounds on the land. This has led to a decline in the quality of the soil in the locality.

Groundwater in the area is naturally near the surface, and is pumped to keep the water table depressed. This causes problems when mines are closed, and there are areas where large lakes have appeared after pumping has stopped.

Site restoration

There are plans to hold a national garden festival at the Witznitz site by the year 2000, and new industries are to be located in the area. There are many areas where soil improvement will be required prior to revegetation, and although there is potential for the incorporation of the artificially created bodies of surface water into a scheme, it seems likely that these will need treatment to improve water quality.

17.7.3 Espenhain

Background

The community of Espenhain lies to the north of the town of Borna, south-east of Leipzig. The area is worked for lignite, although production has declined; from 40 million t/year and 20,000 workers in 1988 to 14.5 million t/year and 6,700 workers in 1992. This decline continues.

A total of approximately 750Mt coal has been extracted from the Espenhain area. Two pits in the north of the area, as well as others to the south and east are being closed, and there is a corresponding migration of workers out of the area, mostly to the west. The borders of the coal activities in this area are only 30km from the city of Leipzig.

Environmental quality

Much of the reduction in the mining activities is due to the high sulphur content of the brown coal, and the corresponding difficulties in achieving air emission quality standards that the state demands. Therefore, by 1996, two large coal processing plants will close and the operations will terminate.

Over the whole mining and coal processing area there is believed to be many hectares of contaminated land arising from coal processing by-products, and from desulphurisation processes which result in the production of gypsum as a waste product. Coal by-products are said to have been used efficiently in the past and therefore the volumes of waste products were kept to a minimum. However, the fall-out of materials from emissions in the area is thought to have significantly raised the background levels of coal-derived chemicals in the region. Work has been carried out to define the areas of Espenhain that are contaminated, and this data will be used in the future to match the after-use of each area with the levels of contamination encountered. Different areas are likely to be cleaned to different standards according to the after-use of the given area.

Site restoration

Plans for the regeneration of the Espenhain coal field are not yet well defined. However, certain issues appear likely to be addressed.

Two new power stations are to open at nearby Lippendorf, each of 800MW capacity, and these will take coal from a different field. Between 2000 and 2500 jobs are planned to be recreated in this operation.

There are plans to reclaim the land in the area of Espenhain, and this will secure further employment in the area. Soil improvement works are likely to be needed in order to establish vegetation.

By the year 2000, it is likely that there will be three or four large lakes created as a result of the cessation of the mining activities, and the problems or opportunities posed by these do not appear to have been addressed.

Amongst the plans for this area it is hoped to construct a decontamination plant to serve the needs of the area, although the nature of this facility is not finalised. In addition it is planned to use some of the former opencast pits for the disposal of municipal waste.

17.7.4 Gröditz

Background

The iron and steel activities at Gröditz began in 1797, using surface deposits of iron nodules in the area. To this day the iron and steel works dominates the town of Gröditz, which has a population of about 10,000. The steel works employed around 5000 people prior to German reunification, but operations are being slimmed down so that 850 staff will be needed to operate the plant competitively.

Today, the plant produces only 30,000t/year of forged items, including rolled rings for trains, cranes and large machinery. Previous production was in the region of 40 million t/year. The company has been supported by the Treuhandanstalt, and 72% of its products are sold to customers in the west, including: USA, UK, Scandinavia and the Netherlands.

The plant contains one 6000t, one 2000t and two 1000t presses, which are fed by scrap iron and steel.

The site of Gröditz steelworks was approximately 150ha in area, but only about 50ha is now required. The rest of the site is now to be developed for new industry, and awaits Treuhandanstalt funding for support of the architectural bureau's plans for the site. Before this can happen however, the city must become the owner of the land, to allow the subsidy to be provided. The site has the advantage that infrastructural services such as railway and water supply are already in place in the area.

Environmental quality

Contaminated areas at Gröditz are thought to be small. Five sites of between 1500 and 3000m^3 have been identified as contaminated with oil, and these are to be excavated and treated by thermal and biological means. Environmental surveys have been undertaken in order to waive environmental liability in the Saxony region, and the estimate is now 30% of the area previously thought to be contaminated.

Site restoration

Plans aim to encourage new industries to the site, once buildings and space become available during the closure and redevelopment programme.

The buildings at the steelworks are to be reused where feasible, and building materials arising during demolition works *e.g.* stone cobbles, are being recycled where possible.

17.7.5 Riesa

Background

The ironworks in Riesa was founded in 1843, and began to cease operations in 1990. The steelworks operated under the name 'Flik' in the 1920s, employing 12,000 people and producing approximately 1.2Mt/year of raw steel. The site is bordered by the River Elbe to the east.

Operations were based around nine water-cooled Siemens-Martin and four electric furnaces on the plant. Lignite was used to fire the furnaces until the 1950s (as evident from sulphur deposits on building rubble at the plant), but since then natural gas has been the source of energy. Most of the machinery was installed in the 1940s and 1950s, and required three reheatings of the steel raw materials.

Environmental quality

Prior to the closure of the plant a huge tar pit was discovered beneath one of the main buildings. This tar was excavated and burnt in the Siemens-Martin furnaces prior to their demolition. Other contaminated material was largely excavated and removed from the site to a landfill site. A landfill in neighbouring Brandenberg (40km away), received some of this waste in the past, and is now causing pollution problems. There are plans to secure this landfill rather than to treat the wastes within it. New German legislation on waste disposal in the 1990s has now stipulated that different classes of waste have to be disposed of at particular classes of waste sites.

On an area of oil-contaminated ground at the Riesa works two biological aeration treatments have been tested by environmental remediation companies. These have both failed to adequately treat the soil, and a third attempt is planned for the near future. The worst oil contamination tends to be at around 7m (approximate depth of the fill). There is a natural clay layer at approximately 15m, and there is another layer of oil contamination just above this clay.

Site investigation and restoration

Development plans for the area include retention of many of the former steelworks buildings, many of which are visually attractive (see Photograph 17.13). Redevelopment started on 4 March, 1991, and it is planned to create 1500 jobs in the 'recycling' of the site. The idea is to make the area into an industrial park for Riesa, for small to medium-sized businesses. Already there are 38 new enterprises to be located on the site. These include a Swiss marble and stone preparation company (using one of the main steel mill buildings, retaining an existing crane, and an office block which has been restored), a small steelworks (Italian steel company), heating equipment manufacturer, and a tin plating works. A technology museum for the steel industry is also planned for the area.

Photograph 17.13: Former steel mill undergoing refurbishment (source: Richards, Moorehead and Laing Ltd)

All new businesses are required to provide environmental compatibility studies prior to development.

New investors in the area are keen to know if there is contaminated ground in the vicinity of their planned new developments. The German Federal authorities want the area to be free of liability for clean-up as far as possible. An archival survey of the site (going back 125 years) was undertaken initially, and this focused subsequent soil and groundwater sampling and analysis in areas considered to be of high risk. Contamination was found in these areas. Much of the contamination was due to mineral oils, which are viewed as of high risk to groundwater, and are particularly evident in the vicinity of the old railway lines. Up to 70,000mg/kg oils were found in soils.

PAHs and heavy metals were also found in soil and debris. Metals tended to be in areas of high pH, and were therefore considered to be less of a problem, although high concentrations of zinc and cadmium were discovered on brickwork from the chimney stacks. Five out of six of these chimneys have been demolished, and metals discovered up to 60mm into the bricks, which has consequences for the disposal of demolition waste. Chromium (VI) was used in one part of the area, although no such contamination of ground was discovered.

Boreholes of between 6 and 10m in depth have been drilled in the area for the purposes of sampling.

There are no set soil chemical standards for the area, and those involved use the Dutch (national), Brandenberg and Berlin (regional) soil chemical quality lists.

Due to the quick action of the local District Offices, Riesa secured early funding from the Treuhandanstalt for redevelopment of the steelworks. This situation is unusual, and is thought highly unlikely to happen elsewhere within such a short time-frame.

A video recording was prepared of the closure, demolition, restoration and future development of the Riesa steelworks site. This has been used as an aid to the promotion of the area.

17.8 Germany - The Ruhr (Western Länder)

17.8.1 Introduction

The area around the rivers Ruhr, Emscher and Lippe, where they flow westward into the Rhine, forms the largest industrial region in Europe. The area known as the Ruhr is administered by the Kommunalverband Ruhrgebiet (KVR), the Municipal Association of the Ruhr area. This was formerly known as the Siedlungsverband Ruhrkohlenbezirk (SVR), the Settlement Association of the Ruhr Coal Area, the first German association of local authorities formed in 1920 to control industrial and urban development in the already heavily industrialised region. The KVR encompasses 11 independent cities and four counties with their districts. In all there are 53 separate political areas. The population of the area has increased by twenty-fold in 160 years, with particularly rapid expansion in the last few decades of the nineteenth century. Despite the massive industrial output the Ruhr is not one vast urban conurbation but a series of towns and cities, intermingled with green spaces. Over 50% of the area is designated as 'communal green space'. The decline in the coal and steel industry has brought about high unemployment and the associated problems of an industrial area in decline.

17.8.2 IBA Emscher Park and Duisburg-Nord Country Park

The Internationale Bauausstellung (IBA), *i.e.* International Building Exhibition, of Emscher Park is an extension of the tradition of presenting innovations in building technology at building exhibitions. It covers an area of 802km² along the length of the Emscher river, an area with severe problems of industrial decline. The subtitle of the IBA is "Workshop on the future of old industrial areas" and the emphasis of the IBA is on technological, social and organisational innovations to achieve ecological, economic and social renewal of the Emscher area.

A memorandum setting out the scope of the IBA was issued by the Land of North-Rhine Westphalia in 1988 after consultation with local authorities and other interested parties in the Emscher area. This memorandum sets out seven spheres of activity which include:

- the creation of a landscape park along the length of the Emscher area by extending and linking together existing open spaces (approximately 320km^2);
- registration of important industrial buildings and structures to ensure their preservation by reconstruction and finding new uses;
- creation of working parks for industry and commerce by reclamation of former industrial sites;
- ecological and technical reconstruction of the Emscher system (over the next 20-25 years);
- modernisation and building of houses and dwellings.

The Emscher system presently consists of 350km of canalised watercourses which, historically, carried the waste water from the region (while mining was taking place beneath the area underground pipes could not be laid because of subsidence problems).

The emphasis is on innovation and quality. Each scheme has to demonstrate an ecological net gain and high standards of design. For industrial/commercial parks created on former industrial sites at least 40% of their area must be green to create a pleasant working environment. Parks are being marketed to particular industry groups, with the aim of diversifying the unbalanced economic base of the Emscher area. The registration of industrial buildings of high architectural quality or historic importance is in recognition of the contribution such buildings or structures make to the identity of the region and their role in providing inspiration for new architectural design.

These ideas are captured in the title for this part of the work of the IBA, which translates as "industrial monuments as bearers of culture". In many cases new uses are found for old buildings but where this is not possible buildings may be preserved simply as monuments.

The IBA Emscher Park is not a funding organisation. A planning company, which receives state funding, has been formed to develop ideas, promote exchange of experience and organise the transformation of draft proposals into plans which are capable of implementation. Individual projects are implemented by agents, who may be local government organisations, private companies or citizens groups. These agents are responsible for securing funding for projects from normal private and public budgets and financial sources.

Environmental quality

Duisburg-Nord Country Park is one of the pioneering projects of the IBA Emscher Park. It covers an area of 200ha between the suburbs of Meiderich and Hamborn in the north of Duisburg. This was the first of the integrated sites of August Thyssen, founder of the Thyssen group of companies. Colliery, coke works and iron and steel works were all built on the same site, in the 1890s, to ensure the security of supply of coke for iron smelting. The site has seen many innovations and developments in iron smelting technology, adding to its historical importance.

The coal mine closed some decades ago. The coke works stopped working in 1977, with demolition occuring in 1981, though contaminants have been left in the ground. Smelting of iron ceased in 1985 and much of the blast furnaces and associated plant are still standing. Photograph 17.14 shows the blast furnace area of the park.

The project to convert the site into a country park is the responsibility of the Landesentwicklungsgesellschaft (LEG), a development agency of the Land of North-Rhine Westphalia. Thyssen have provided information on the history of the site and the wastes and contamination present. They have also carried out work to dispose of hazardous wastes.

Photograph 17.14: The blast furnace area, Duisburg-Nord Country Park (source: Umgang mit der Hochofenanlage Bericht der Expertenkommission, IBA)

Site restoration

A competition was held to find a plan for development of the site into a Country Park. Five planning teams, including one from the United Kingdom and one from France, were commissioned to produce a model for development and design of the park. Each team had expertise in urban planning, architecture and landscape architecture. The winning concept was chosen by a jury of architects, landscape planners and representatives of the city of Duisburg. A group of local inhabitants, former workers and neighbours of the company and experts, Interessengemeinshaft Nordpark, was also involved. This group have a continuing interest in the site. The jury was advised by experts on specialist matters. The winning design was by Latz and Partners of Freising. This design builds on existing features of the site to a large extent. The blast furnaces area is largely preserved, as a museum of industrial history and the focus of the Country Park. Over the remainder of the site a series of parks and gardens are to be formed, in some cases using existing features such as the canalised Old Emscher river and former railway lines. Work is to be done in stages. Work has already been carried out in the area of the blast furnaces to make safe structures and preserve what could be retained. Hazardous wastes from the area, such as flue dusts and ashes have been disposed of in the iron ore bunker, to create a raised area which provides a viewpoint over the site. This project was used to provide training and work experience in the building trade for 60 long-term unemployed, and in landscape design and maintenance for 48 long-term unemployed people.

Some parts of the site are already open to the public. New uses of existing structures include a climbing wall on an old wall in the blast furnaces area. Several publications on various aspects of the site are available. These include descriptions of walks through the site which point out aspects of industrial or natural history. Much of the site, which has been undisturbed for ten years or more, has been colonised by a wide variety of plant species. Some of the plants and animals are on the North-Rhine Westphalia 'red list' of protected species. In some cases, such as in the area of the former coke works, the presence of unusual plants results from soil contamination. Guided tours on various aspects of the site are available.

17.8.3 Sachsen mine, Hamm

The former Sachsen coal mine and coke works covers an area of 50ha near Hamm, in the east of the Ruhr area (see Photograph 17.15). The majority of buildings and structures on the site were demolished in 1979 and the site has been abandoned. The Landesentwicklungsgesellschaft (LEG), a development agency of the state of North Rhine Westphalia, intends to redevelop the site for industrial and commercial use, but with an ecologically orientated design *i.e.* a high proportion of green areas linked to other open, green spaces in the locality to provide a pleasant working environment, minimisation of areas with an impervious covering to maximise infiltration of rainwater, and provision of areas of open

Photograph 17.15: Aerial view of Sachsen mine, Hamm. The large building is the machine hall, the only building on the site which has not been demolished (source: AHU)

water. AHU-Büro für Hydrogeologie und Umwelt GmbH in Aachen, together with various working groups consisting of geologists, planners, civil engineers, chemists and toxicologists from several organisations, have been commissioned to carry out a use-related redevelopment study. The study has included research into the industrial history of the site and investigations of soil and groundwater contamination.

Environmental quality

The site is covered with mine wastes and building rubble to a depth of up to 10m. The uncontrolled manner in which demolition was carried out has resulted in contaminated wastes being distributed over wide areas. Contaminants include polyaromatic hydrocarbons (PAHs), cyanides, ammonia, sulphates, chloride, sulphide, phenols and aromatic compounds.

Disturbance of the ground resulted in substantial odour problems in several areas.

Site restoration

The site has been divided into several areas according to the degree of contamination and proposed use. Contamination has been assessed using an adaptation of the Eikmann and Kloke system (see Section 2.6.5), with addition of values for total and free cyanide concentrations, as shown in Table 17.1. Where the BWIII levels are exceeded, near-surface soil will be excavated, treated to remove contaminants and replaced. In the area of the former gas purification plant-contaminated materials are to be removed and treated off site by thermal methods. In the former by-products plant area groundwater is contaminated with phenols, aromatics, ammonium, PAH, chloride and sulphate. Costs for various options involving different degrees of encapsulation of contamination, protection of buildings and treatment of groundwater have been considered. Costs were least for options which minimised the amount of contaminated groundwater extracted. Treatment methods being considered for extracted groundwater include adsorption on to activated carbon and biological degradation.

17.9 Spain

17.9.1 Introduction

Little reclamation of former coal and steel sites has taken place in Spain, despite the importance of these industries, which are now in decline. This case study will concentrate therefore on the nature of the industries and the problems that have been caused, or may be caused, on their abandonment. The following are described:

- Iron mining sites in the Basque Country;
- Steel producing sites in the Basque Country and Asturias.

Table 17.1. Use-related standards for Sachsen mine

End-use of Kloke and Eikmann (1991)[139]	End-use at Sachsen mine	Soil depth (m)	Soil value	Free cyanide mg/kg	Total cyanides mg/kg	As mg/kg	Pb mg/kg	Cd mg/kg	Hg mg/kg	Zn mg/kg	BaP† mg/kg
Multifunctional use possible			BWI	1	5	20	100	1	0.5	150	1
Children's play areas		0.35	BWII BWIII	1 10	5 50	20 50	200 1000	2 10	0.5 10	300 2000	1 5
Houses and small gardens	Area for purification of domestic waste water using plants, open water areas, demonstration of a farmer's garden	0.35	BWII BWIII	4 40	20 400	40 80	300 1000	2 5	2 20	300 600	2 5
Parks and open spaces	Semi-paved areas, grassed areas	0.10	BWII BWIII	5 50	25 250	40 80	500 2000	4 15	5 15	1000 3000	3 6
Industrial /commercial	Areas of impervious hardstanding	0.10	BWII BWIII	10 50	100 500	50 150	1000 2000	10 20	10 20	1000 3000	5 10

† Benzo-a-pyrene

17.9.2 Iron mining - Basque Country

Iron mining has been carried out in the Basque country near to Bilbao for centuries, first by open pit mining (see Photograph 17.16) and then by deep mining. Only one mine area still operates via a drift entered at the base of an open cut, but a substantial number of these mines have also closed in recent years.

Some of these mines are very close to Bilbao itself; for example, those on the Monte de Miribilla and the Mina del Morro. These mines are complexes of pits and spoil heaps, shafts and ruined buildings and have been subject to the fly-tipping of wastes, some of which are toxic. The less-toxic materials have become partly vegetated. The mined areas take up land which could be used for development. Bilbao is short of development land because of its location in a relatively narrow river valley.

The spoil produced from these mines is not particularly toxic and the mines and spoil heaps have become well-vegetated. The spoil and mines are not apparently responsible for pollution of the nearby river Nervion. The deposition of wastes from other industries on iron mining spoil heaps is, however, more of a problem since the deposited materials include wastes with high concentrations of heavy metals and other contaminants (see Photograph 17.17). During the 1980s, large areas of the former opencast workings were filled with waste materials from construction activities elsewhere in the region in order to provide level ground. The materials dumped included toxic industrial wastes.

The principal constraints on development of the area are therefore:

- the topography of sites;
- the possibility of unstable ground due to mining;
- unsuitable materials on which to construct buildings without prior treatment;
- the presence of potentially contaminating materials due to fly-tipping and the filling of mining voids.

Photograph 17.16:
Open pit iron ore mine near Bilbao. Mining is still carried out underground with access down the haul road (shown in the photograph) (source: Richards, Moorehead and Laing Ltd)

Photograph 17.17:
Eroding illegally dumped wastes at a former iron mine in Bilbao (source: Richards, Moorehead and Laing Ltd)

17.9.3 Steel producing sites - Basque Country and Asturias

The principal steel producing areas of northern Spain are in the Basque Country and Asturias. Steel production in these areas commenced in the nineteenth century. In Asturias the steel industry developed after 1930 on previously non-industrial sites. So whereas in the Basque Country the developments in steel production have been superimposed on one another at the same site, in Asturias much of the steel making is in single generation plants. Both areas, however, support very large steel industries and their economies are dependent on steel production. In the Basque Country there were nearby iron ore and limestone deposits, and in Asturias, coal and limestone.

There was considerable expansion of steel production in the third quarter of the twentieth century. In Asturias in particular considerable expansion took place after 1950 when the Empresa Nacional Siderúrgico S.A (Ensidesa) was formed. A major steelworks was commissioned at the small town of Aviles in 1956, and a further integrated plant at Gijon in 1971. Older steelworks in the area were closed by 1984 so that those in Aviles and Gijon are now the only ones in operation in Asturias. Characteristics of the steelworks are provided in Table 17.2.

In Asturias steel is produced only from iron ore in blast furnaces. In the Basque country there are both blast furnaces and electric arc furnaces which produce steel from scrap metal. In both regions there are coking plants to produce coke for the blast furnaces. There are also secondary industries, such as chemical industries utilising by-products of coke production, and aggregate industries producing road building materials from slag.

Table 17.2: Characteristics of steel production in Asturias

Aviles	
Area	830 ha
Capacity	2.5 million t/year
Source of raw materials	Iron ore by ship to ports at Gijon or Aviles. Coal by railway from port of El Musel.
Coke ovens	300 in batteries of 30. Production capacity 1.85 million t/year
Blast furnaces	4
Steel works	Fully automated with continuous casting machines, hot strip mills, tandem mills, temper mill, and finishing facilities.
Products	Range of steels, plate and sheet, bars, road track materials, cold rolled flat products, tinplate. By-products: Benzol, naphthalene, toluene, ammonium sulphate.
Gijon	
Area	600 ha
Capacity	2.2 million t/year
Source of raw materials	Iron ore by ship to Gijon, coal by rail and conveyer from El Musel.
Coke ovens	45 in 2 batteries. Production capacity of 1.05 million t/year.
Blast furnaces	2
Steel works	Continuous conventional casting and rolling mills.

Assessment

In both the Basque Country and Asturias the situation of the steelworks is typical of many other parts of Europe:

- the steel industry is facing over-capacity and will have close down some of its production in the near future;
- the steelworks and associated industries have caused soil, water and air pollution and are faced with the prospect of having to clean this up before new investment can be attracted to the area.

The steelworks at Gijon in Asturias is a modern, integrated plant built on a greenfield site with little waste deposition on or near the site. The by-products of steel production are sold; gas for thermal electricity production, slag for road making.

At Aviles, the older plant, by-products are similarly sold but there has been pollution of the estuary from steel making and other heavy industries.

There has been no reclamation of steelworks in Asturias, but should parts of the steelworks close there will be an opportunity for commencing reclamation at the demolition stage. This will allow the fate of demolition materials to be decided upon both on the basis of their saleability and their potential as a source of contamination if they are allowed to remain on site, and thus facilitate reclamation of the sites to an appropriate end-use.

REFERENCES

1 Ackman, T.E., Jones, J.R. (1991) Methods to identify and reduce potential surface stream water losses into abandoned underground mines, *Environ. Geol. Water Sci.*, 17, No.3, 227-232.

2 Ackman, T.E., Kleinmann, R.L.P. (1991) An in-line system for the treatment of mine water, *International Mine Waste Management News,* 1, No.3, 1-4.

3 Ambosetti, P., Capurso, R., Cavino, R., Gallo, G., Monicchia, R. and Salvatici, B. (1992) Storia, energia, ambiente, Un modello di centrale a Pietrafitta. Collaborazione deditoriale dell'ENEL. Perugia: Protagon editrice.

4 Annels, A.E. and Williams, K.P. (1988) An analysis of the mineral content and washing techniques for the Bargoed Tip. In *Land reclamation - a feasibility study* (Cardiff 1988), 27-60, Cardiff: University College of Wales.

5 Anon (1967) Report of the tribunal appointed to inquire into the disaster at Aberfan on October 21 1966. London: HMSO.

6 Anon (1980) Britain's tip heaps - an unlikely subject for conservation? *Earth Science Conservation,* No.18, 2-4.

7 Anon (1982) Proceedings of international symposium on mine spoil heaps and usage of derelict industrial land (Essen 1992). Essen: Kommunalverband Ruhrgebiet (in English).

8 Anon (1985) Guidelines for inventories of historic buildings and engineering and industrial structures, Washington, DC.: Historic American Building Survey/ Department of Interior, American National Park Service.

9 Anon (1990) Bodenwaschanlange - Steuerung und Uberwachung mit dem Monitor 77, *Chemie-Technik*, 19, No.4, 48-51.

10 Anon (1992) Hazardous waste landfills unearthed in Liguria, *Haznews,* No.52, 9. London: D. Coleman.

11 Anon (1992) Scottish steel works leaves massive clean-up bill, *The ENDS report,* No.215, 7.

12 Armishaw, R., Bardos, R.P., Dunn, R.M., Hill, J.M., Pearl, M., Rampling, T. and Wood, P.A. (1992) Review of innovative soil clean-up processes. Stevenage: Warren Spring Laboratory, UK Department of Trade and Industry.

13 Arz, P. (1988) The diaphragm wall chamber system. Inspection of the efficacy and long-term observations. In *Contaminated Soil '88* (Hamburg 1988), edited by K. Wolf, W.J. van den Brink and F.J. Colon, 613-615. Dordrecht: Kluwer.

14 Ash, H.J. (1983) The natural colonisation of derelict industrial land. Ph.D. Thesis, University of Liverpool.

15 Ayerst, J.M. (1978) The effect of compaction on the revegetation of spoil heaps, *Reclamation Review* 1, 27-30.

16 Ayres, A.J. (1985) Treatment of shallow underground fires. In *Failures in Earthworks, Conference Proceedings,* London: Institution of Civil Engineers.

17 Backes, C.A., Pulford, I.D., Duncan, H.J. (1986) Studies on oxidation of pyrite in colliery spoil I. The oxidation pathway and inhibition of the ferrous-ferric oxidation. *Reclamation and Revegetation Research,* 4, 279-291.

18 Bardos, P. (1992) The NATO/CCMS pilot study on demonstration of remedial actions/technologies for contaminated land and groundwater: Status December 1991. Stevenage, UK: Warren Spring Laboratory.

19 Barratt, P.A. (1990) Biosorption - a potential mechanism for the removal of radionuclides from nuclear effluent streams. In *USEPA Second Forum on Innovative Hazardous Waste Treatment Technologies: Domestic and International* (Philadelphia, PA. 1990), 20, Philadelphia, PA: USEPA.

20 Barratt, P.A. (1991) On-site bioremediation of contaminated land. In *Proceedings of Conference on Contaminated Land* (Birmingham, UK 1991). Northampton: Institute of Wastes Management.

21 Barratt, P.A. and Harold, P. (1991) *In situ* biological treatment of contaminated land - feasibility studies and treatment of a creosote contaminated site. In *Land Reclamation - an End to Dereliction?*, edited by M. C. R. Davies, 336-346. London: Elsevier.

22 Barry, D.L. (1985) Former iron and steelmaking plants. In *Contaminated Land, Reclamation and Treatment,* edited by M.A. Smith, 311-339, London: Plenum.

23 Barry, D.L. and Raybould, J.G. (1988) Landfill gas migration hazards. In *Mineworkings '88, Conference Proceedings,* (Edinburgh 1988), edited by M.C. Forde. Edinburgh: Engineering Technics Press.

24 Barton-Bridges, J.P. and Robertson, A.M. (1989) Design and reclamation of mine waste facilities to control acid mine drainage. In *Reclamation, A Global Perspective* (Calgary, Canada 1989), edited by D.G. Walker, C.B. Powter and M.W. Pole, 2, 717-728, Edmonton, Canada: Alberta Land Conservation and Reclamation Council.

25 BCAMD Task Force (1989) Draft acid rock drainage technical guide. Volume 1.

26 Beckett, M.J. and Cairney, T. (1992) Reclamation options. In *Contaminated Land, Problems and Solutions,* edited by T. Cairney, 68-83, Glasgow: Blackie.

27 Bedford, E. and Smith, A.J. (1988) Underground heating beneath Oakthorpe, Leicestershire, *Municipal Engineer*, 5 August, 167-181.

28 Bell S.E. (1978), Successful design for mining subsidence, In *Large Ground Movements and Structures, Conference Proceedings* (Cardiff 1977), London: Pentech.

29 van den Berg, R. (1992) Risk assessment of contaminated soil: Proposals for adjusted, toxicologically based Dutch soil clean-up criteria. In *Remediation of Contaminated Sites, Environmental Standards and Human Health Risk Assessment* (Perth 1992), Royal Australian Chemical Institute, W. A. Branch, Health and Safety and Environment Group.

30 Berkowitz, N. (1979) Gasification. In *An Introduction to Coal Technology*. Academic.

31 Bewley, R.J.F. and Theile, P. (1988) Microbiological reclamation of contaminated soil. In *Land Rec '88, Conference Proceedings* (Durham, UK 1988), edited by D. Williamson, 181-187,

32 Binns, W.O. and Crowther, R.E. (1983) Land reclamation for trees and woods. In *Reclamation '83 Conference Papers,* 23-28, Grays, Essex, UK: Industrial Seminars Limited.

33 Blaurock, H. (1982) Report on mine tips in the Ruhr. In *International Symposium on Mine Spoil Heaps* (Essen 1982), 13-18.

34 Blunt, S.M. (1991) Practical techniques for the revegetation of derelict land. M.Sc. Thesis, University of Liverpool.

35 Bohm, K. (1992) A thermal treatment for cleaning contaminated soil. In *Contaminated Land Treatment Technologies*, edited by J. F. Rees, 195-217. London: Elsevier.

36 Boraston, G.W. (1990) Revolving fluidized bed technology for the treatment of hazardous materials. Paper presented at *USEPA Second Forum on Innovative Hazardous Waste Treatment Technologies: Domestic and International* (Philadelphia, PA. 1990), Philadelphia, PA.: USEPA.

37 Bosman, D.J., Clayton, J.A., Maree, J.P., Adlem, C.J.L. (1990) Removal of sulphate from mine water with barium sulphide. In *Acid Mine Water in Pyritic Environments* (Lisboa, Portugal 1990), edited by P.J. Norton, 149-163, International Mine Water Association.

38 Bouroz, A. (1964) Les composants pétrographiques principaux des schistes houillers et leur signification sédimentologique. In *Vème Congrès Internacional de Stratigraphie et de Geologie du Carboniférе,* 289-301, Paris.

39 Bouska, V. (1981) The geochemistry of coal. Amsterdam: Elsevier,

40 Bradshaw, A.D. and Chadwick, M.J. (1980) The restoration of land. Oxford: Blackwell.

41 de Bruijn, P.J. and de Walle, F.B. (1988) Soil standards for soil protection and remedial action in the Netherlands. In *Contaminated Soil*

'88 (Hamburg 1988), edited by K. Wolf, W.J. van den Brink and F.J. Colon, 339-349, Dordrecht: Kluwer.

42 BSI (1963-1992) Methods for analysis of coal and coke. BSI 1016.

43 BSI (1988) Code of practice for the identification of potentially contaminated land and its investigation. Draft for Development. DD175.

44 Buck, F.A.M. and Smith C.A. (1990) Control of air emissions from soil venting systems. In *Physical/Chemical Processes. Innovative Hazardous Waste Treatment Technology Series, Volume 2*, edited by H. M. Freeman, 205-210, USA: Technomic.

45 Byers, W. (1990) Air stripping technology. In *Physical/Chemical Processes. Innovative Hazardous Waste Treatment Technology Series, Volume 2*, edited by H. M. Freeman, 19-31, USA: Technomic.

46 Cairney, T., Clucas, R. and Hobson, D. (1990) Evaluating subterranean fire risks on reclaimed sites. In *Treatment and Utilisation of Coal Mining Wastes, Conference Proceedings,* edited by A.K.M. Rainbow. Rotterdam: Elsevier.

47 Cañibano, J.G. and Leininger, D. (1987) The characteristics and use of coal wastes. In *Reclamation. Treatment and Utilization of Coal Mining Wastes* (Nottingham 1987), edited by A.K.M. Rainbow, 111-122, Amsterdam: Elsevier.

48 Carter, D. (1984) Glengarnock iron and steelworks, Ayrshire, *Landscape Design,* No.8, 52-54.

49 Caruccio, F.T. and Geidel, G. (1984) The prediction of acid mine drainage from coal strip mines. In *Proceedings of a Conference on Reclamation of Abandoned Acid Spoils* (Osage Beach 1984), Missouri: Department of Natural Resources Land Reclamation Commission.

50 Cass, R.M. (1983) Hard development potential and design techniques. Part 1: Land use. Site planning and reclamation principles. In *Reclamation of Former Iron and Steelworks Sites, Conference Proceedings* (Windermere 1983), edited by G.P. Doubleday, E1-E6.

51 Chadwick, M.J. (1974). Unterschied liche chemische Bestandteile im Abraum von Kohlenzechen im Hinblick auf vorzunehmende Bepflanzungen. In *Grüne Halden im Ruhrgebiet*, 17-21, Essen: Siedlungsverband Ruhrkohlenbesirk.

52 Chapman, D. (1992) Editor. Water quality assessments. London: Chapman and Hall.

53 Charles, J.A. (1993) Building on fill: Geotechnical aspects. Watford: Building Research Establishment.

54 CIRIA (In press) Engineering-based remedial measures. Volume VI. Physical containment and hydraulic measures. London: CIRIA.

55 Collins, D. (1975) The European Communities. The social policy of the first phase. Volume 1, The European Coal and Steel Community, 1951-1970. London: Martin Robertson.

56 Coppin, N.J. and Bradshaw, A.D. (1982) Quarry reclamation. Mining Journal Books.

57 Coppin, N.J. and Richards, I.G. (1990) The use of vegetation in civil engineering. London: Butterworths.

58 Costigan, P.A., Bradshaw, A.D. and Gemmell, R.P. (1981) The reclamation of acidic colliery spoil. I Acid production potential. *Journal of Applied Ecology,* 18, 865-878.

59 Council of Europe (1985) Architectural heritage reports and studies: Number 6. The Industrial heritage: what policies? In *Proceedings of the Colloquy* (Lyon 1985). Available from the Council of Europe Publications service BP 431 R6, F-67006 Strasbourg Cedex.

60 Council of Europe (1988) Architectural heritage reports and studies: Number 15. Mining engineering monuments as a cultural heritage. In *Proceedings of the Colloquy* (Bochum 1988). Available from the Council of Europe Publications Service BP 431 R6, F-67006 Strasbourg Cedex.

61 Couper, A. (1990) Dust monitoring overcomes a burning bing thought too hot to handle, Ramsey Bing Loanhead. In *Treatment and Utilisation of*

Coal Mining Wastes, Conference Proceedings, editor A.K.M. Rainbow, Rotterdam: Elsevier.

62 Couper, A. (1992) How an environmental charter is helping to cut pollution in a traditional coal mining area. In *European Environment Conference* (Nottingham 1992).

63 Couper, A. (1992) *Pers. comm.*

64 Crawford, C.B. and Burn, K.N. (1969) Building damage from expansive steel slag backfill, *Journal of the American Society of Civil Engineers, Soil Mechanics and Foundations Division,* 95, 1325-1334.

65 Crawford, C.B. and Burn, K.N. (1971) Building damage from expansive steel slag backfill, *Journal of the American Society of Civil Engineers, Soil Mechanics and Foundations Division*, 97, 1026-1029.

66 Dacey, P.W. and Colbourn, P. (1979) An assessment of methods for the determination of iron pyrites in coal mine spoil. *Reclamation Review,* 2, 113-121.

67 Dancer, W.S., Handley, J.F. and Bradshaw, A.D. (1977) Nitrogen accumulation in kaolin mining wastes in Cornwall, II. Forage legumes, *Plant and Soil*, 48, 303-314.

68 Davies, R.J. (1987) Weed competition and broad leaved tree establishment. In *Advances in Practical Arboriculture. Forestry Commission Bulletin.* 65, 91-98, London: HMSO.

69 Davison, J., Jones, S. (1990) Mine drainage bioremediation - the evolution of the technology from microbes to bio-carb. In *Proceedings of the 1990 Mining and Reclamation Conference and Exhibition* (Charleston, WV. 1990), edited by J.G. Skousen, J.C. Sencindiver and D.E. Samuel, 71-75, Morgantown, WV.: West Virginia University.

70 Delaine, J. (1990) Hazardous materials in demolition. In *Decommissioning and Demolition Conference Proceedings* (Manchester 1990), edited by I.L.Whyte, 92-100.

71 Department of the Environment (UK) (1987) Waste Management Paper Number 18: Asbestos wastes. London: HMSO.

72 Department of the Environment (UK) (1988) Treatment of disused mine openings. London: HMSO.

73 Derelict Land Reclamation Research Unit, University of York (1979) An assessment of the potential of derelict and industrial wasteland for the growth of energy crops. Report to the Energy Technology Support Unit fuels from biological materials programme; Project 3.

74 DIN (1984), Sludge and Sediments (Group S). Determination of leachability by water (S 4). *DIN 38 414*. Part 4. Berlin: DIN.

75 Doubleday G.P. (1971) Soil forming materials: their nature and assessment. In *Landscape reclamation Volume 1,* 70-83, Guildford: IPC Business Press.

76 Doubleday, G.P. (1972) Development and management of soils on pit heaps. In *Landscape reclamation Volume 2*, 25-35, Guildford: IPC Business Press.

77 Eakin, W.R.G. (1983) Hard development potential and design techniques, Part II - Some technical problems. In *Reclamation of Former Iron and Steelworks Sites, Conference Proceedings (Windermere 1983)*, edited by G.P. Doubleday, E7-E17.

78 Eakin, W.R.G. and Crowther, J. (1985) Geotechnical problems on land reclamation sites, *Municipal Engineer,* 2 Oct, 233-245.

79 Ebel, J.C., SACMI, Lens, France (1993), *Pers. Comm.*

80 Ellington, A., Burke, T. (1981) Europe: Environmental. The European Community's environmental policy. London: Ecobooks.

81 Elsworth, S. (1984) Acid Rain. London: Pluto Press.

82 Emery, J.J. (1974) A simple test procedure for evaluation the potential expansion of steel slag. In *Roads and Transportation Association of Canada, Proceedings of the Annual Conference* (Ottawa 1974), 90-103.

83 Emery, J.J. (1980) Assessment of ferrous slags for fill applications. In *Conference on Reclamation of Contaminated Land*, (Eastbourne 1979), F1/1-10, London: Society of Chemical Industry.

84 Emery, J.J. (1986) Slag utilisation in asphalt mixes. In *Roads and Transportation Association of Canada, Proceedings of the Annual Conference* (Ottawa 1986), C3-C19.

85 ENDS (1992) Dangerous substances in water, a practical guide. London: Environmental Data Services Limited.

86 Environmental Resources Limited (1987) Problems arising from the redevelopment of gas works and similar sites (Second Edition), prepared for the UK Department of the Environment. London: HMSO.

87 EPML (1992) Sites sidérurgiques Lorrains: Genèse d'un nouveau paysage. In *l'Etablissement Public de la Métropole Lorraine*. Lorrain: EPML.

88 European Commission (1990) Environmental policy in the European Community. Luxembourg: Office for Official Publications of the European Communities.

89 European Commission, DG XI (1992) European Community environment legislation. Volume 7: Water. Luxembourg: Office for Official Publications of the European Communities.

90 Evans J. (1987) Tree shelters. In *Advances in Practical Arboriculture. Forestry Commission Bulletin,* 65, 67-75. London: HMSO.

91 Ferguson. C.C. (1992) The statistical basis for spatial sampling of contaminated land, *Ground Engineering*, June 1992, 34-38.

92 Filion, M.P., Sirois, L.L., Ferguson, K. (1990) Acid mine drainage research in Canada, *CIM Bulletin*, 83, No.944, 33-40.

93 Fitzpatrick, V.F. (1987) *In situ* vitrification - a candidate process for *in situ* destruction of hazardous waste, 120-127. MD, USA: Hazardous Materials Control Research Institute, Monograph Series.

94 Fitzpatrick, V.F. (1988) *In situ* vitrification - an innovative melting technology for the remediation of contaminated soil. In *Contaminated Soil '88* (Hamburg 1988), edited by K. Wolf, W. J. van den Brink and F.J. Colon, 857-859, Dordrecht: Kluwer.

95 Fleming, G. (1991) Editor. Recycling Derelict Land. London: Thomas Telford.

96 Freeman, J.R., Sturm, J.W. and Smith, R.M. (1984) Acid base accounting of acid spoils. In *Proceedings of a Conference on Reclamation of Abandoned Acid Spoils* (Osage Beach 1984), Missouri: Department of Natural Resources Land Reclamation Commission.

97 van der Galien, W., Urlings, L.G.C.M. and van Osterom, W. P. (1988) Leaching of cyanides. In *Contaminated Soil '88* (Hamburg 1988), edited by K. Wolf, W. J. van den Brink and F. J. Colon, Dordrecht: Kluwer.

98 Gardiner, J. and Mance, G. (1984) United Kingdom Water Quality Standards arising from European Community Directives, Water Research Centre Technical Report TR 204. Stevenage: Water Research Centre.

99 Gemmell, R.P. (1983) Methods of establishing and monitoring vegetation on iron/steel and gasworks sites. In *Reclamation of Former Iron and Steelworks Sites* (Windermere 1983), edited by G.P. Doubleday.

100 Geoffrey Walton Practice (1991) Handbook on the design of tips and related structures, prepared for the Department of the Environment. London: HMSO.

101 Gibbons, J.J. and Soundararajan, R. (1988) The nature of chemical bonding between modified clay minerals and organic waste materials. *R. Am. Lab.*, 20, No. 7, 38-46.

102 Gieseler, G. (1988) Contaminated land in the EC. In *Contaminated Soil '88* (Hamburg 1988), edited by K. Wolf, W. J. van den Brink and F.J. Colon, 1555-1562, Dordrecht: Kluwer.

103 Glover, H.G. (1981) Water disposal from underground coal mining activities. Part 2 - Control. In *Mining and Water Pollution, The Institution of Water Engineers and Scientists, Scientific Section Symposium,* (Nottingham 1981).

104 Goodman, G.T. and Chadwick, M.J. (1978) (Editors) Environmental Management of Mineral Wastes, The Netherlands: Sithoff und Nordhoff.

105 Greenwood, R.D. and Moffat, J.D. (1982) Implementation techniques for more natural landscapes. In *An Ecological Approach to Urban Landscape Design*, edited by A.R. Urff and R. Tregay, Manchester: University of Manchester.

106 Grisham, G.R. and Sondermann, W. (1990) Extraction and washing of contaminated soils using high pressure jet grouting techniques. In *Superfund '90. Proceedings of the Eleventh Annual National Conference and Exhibition*, 745-747, Hazardous Materials Control Resources Institute.

107 Gunther, K. (1988) Surface coverings for tips and contaminated sites. In *Contaminated Soil '88* (Hamburg 1988), edited by K. Wolf, W. J. van den Brink and F.J. Colon, 561-577, Dordrecht: Kluwer.

108 Hackett, B. Editor (1977) Landscape reclamation practice. IPC Science and Technology Press.

109 Haigh, M. (1992) Degradation of reclaimed lands previously disturbed by coal mining in Wales: Causes and remedies, *Land Degradation and Rehabilitation,* 3, 169-180.

110 Hall. I.G. (1957) The ecology of disused pit heaps in England, *Journal of Ecology*, 45, 699-720.

111 Hall, D.W., Sandrin, J.A. and McBride, R.E. (1990) An overview of solvent extraction treatment technologies, *Environmental Progress,* 9, No.2, 98-105.

112 Harrison, W.H. (1987) Durability of concrete in acidic soils and waters, *Concrete*, February 1987, 18-24.

113 Hawkins, A.B. and Pinches, G.M. (1987) Cause and significance of heave at Llandough hospital, Cardiff. A case history of ground floor heave due to gypsum growth, *Quarterly journal of engineering geology*, 20, 41-57.

114 Healy, P.R., and Head, J.M. (1984) Construction over abandoned mine workings. London: CIRIA.

115 Hillmert, C. (1990) Evaluation of hazardous sites for the determination of priority and need of action. In *Contaminated Soil '90* (Karlsruhe 1990),

edited by F. Arendt, M. Hinsenveld and W.J. van den Brink, 527-534, Dordrecht: Kluwer.

116 Hoffmann-Kamensky, M. (1993) Polycyclic aromatic hydrocarbons (PAH) from atmospheric deposition in forest soils of the Ruhr area, Germany. In *Contaminated Soil '93* (Berlin 1993), edited by F. Arendt, G.J. Annokkée, R. Bosman and W.J. van den Brink, 173-174, Dordrecht: Kluwer.

117 Hojsholt, U. (1993) Priority setting in the administration of the Danish Act of Contaminated Sites. In *Contaminated Soil '93* (Berlin 1993), edited by F. Arendt, G.J. Annokkée, R. Bosman and W.J. van den Brink, 35-40. Dordrecht: Kluwer.

118 Hope, D. (1992) The Bromford Standard. In *Polluted and Marginal Land 92, Proceedings of Second International Conference on Construction on Polluted and Marginal Land* (London 1992), edited by M.C. Forde, 29-35, Edinburgh: Engineering Technics Press.

119 Hund, K., Schenk, B., Jacob, R. and Schulz-Berendt (1993) Chemical and microbiological determinations during on-site decontamination for a PAH-contaminated site. In *Contaminated Soil '93* (Berlin 1993), edited by F. Arendt, G.J. Annokkée, R. Bosman and W.J. van den Brink, 505-506, Dordrecht: Kluwer.

120 ICRCL (1986) Notes on the fire hazards of contaminated land. ICRCL Guidance Note 61/84 Second edition.

121 ICRCL (1987) Guidance on the assessment and redevelopment of contaminated land. ICRCL Guidance Note 59/83 Second edition.

122 ICRCL (1990) Notes on the restoration and aftercare of metalliferous mining sites for pasture and grazing. ICRCL Guidance Note 70/90. London: HMSO

123 IEA Coal Research (1992) Coal production prospects in the European Community. London: IEA.

124 IISI (1992) World steel in figures, 1992. Brussels: IISI.

125 IISI (1993) *Pers. comm.*

126 Institution of Civil Engineers (1977) Ground subsidence. London: National Coal Board.

127 Jecko, G. and Raguin, J. (1981) Simulation au laboratoire des interactions entre l'eau de pluie et le crassiers sidérurgiques, *Eau Ind.,* 57, 50, 53-62.

128 Johnson, M.S. (1984) The effects of gel-forming polyacrylamides on moisture storage in sandy soils, *Journal of Science, Food and Agriculture,* 35, 1196-1200.

129 Jones, D.L. (1991) Redevelopment of a former steelworks. In *Land Reclamation, an End to Dereliction? Conference Proceedings*, (Cardiff 1991), edited by M.C.R. Davies, 228-234, London: Elsevier.

130 Jones, K.C. (1988) Polynuclear aromatic hydrocarbons in the soil system: long-term changes, behaviour and current levels in the UK. In *Contaminated Soil '88* (Hamburg 1988), edited by K. Wolf, W.J. van den Brink and F.J. Colon, Dordrecht: Kluwer.

131 Jones, M.J. (1972) Editor. Geological, mining and metallurgical sampling. London: IMM.

132 Kaleri, C., Rogers, L., Spencer, C. and Weston, R. F. (1990) Critical fluid solvent extraction. In *USEPA Second Forum on Innovative Hazardous Waste Treatment Technologies: Domestic and International* (Philadelphia, PA. 1990), USA: USEPA.

133 Kempton G.T. (1992) The use of reinforcements to support road embankments over areas subjected to mining subsidence, In *Highways and Transportation,* 21-31, December.

134 Kennedy, C.D. (1988) Lanarkshire steelworks redevelopment. In *Land Rec '88 Conference Proceedings* (Durham 1988), edited by D.Williamson, 103-119.

135 Kent, M. (1982) Plant growth problems in colliery spoil reclamation - a review, *Applied Geography*, 2, 83-107.

136 Kleinmann, R.L.P. (1991) Acid mine drainage. An overview. In *Energy in the 90s, Proceedings of the ASCE Energy Division Speciality Conference*

on Energy (Pittsburgh, PA. 1991), edited by B.J. Hobbs, 281-286, New York: American Society of Civil Engineers.

137 Kleinmann, R.L.P., Erickson, P.M. (1987) Control of acid mine drainage: an overview of recent developments. In *Innovative Approaches to Mined Land Reclamation, Proceedings of the National Mined Land Reclamation Conference* (St Louis, MO. 1986), edited by C.L. Carlson and J.H. Swisher, 283-305, Carbondale, IL.: Southern Illinois University Press.

138 Kloke, A. (1988) Fundamentals for determining use-related, highest acceptable contaminant levels in inner city and urban soils. In *Contaminated Soil '88* (Hamburg 1988), edited by K. Wolf, W.J. van den Brink and F.J. Colon, 289-298, Dordrecht: Kluwer.

139 Kloke, A. and Eikmann, Th. (1991) Nutzungs- und schutzbezogene Orientierungsdaten für (Schad-)Stoffe in Böden, Mitteilungen des Verbandes Deutscher Landwirtshaftlicher Untersuchungs- und Forschungastalten (VDLUFA), Sonderdruck aus Heft 1/1991.

140 Kolb, W., Rohai, P. and Kiefhaber, P.K. (1993) Demolition concepts for former industrial sites: break-off strategy on the example of former foundry facility. In *Contaminated Soil '93* (Berlin 1993), edited by F. Arendt, G.J. Annokkée, R. Bosman and W.J. van den Brink, 135-136, Dordrecht: Kluwer.

141 Kollar, J. and Gunkel, P. (1983) Untersuchungen an LD-Schlacken, insbesonderere im Hinblick auf die Verwertung im Straßenbau, Technische forschung stahl, Kommission der Europaischen Gemeinschaften, Forschungsvertrag Nr. 7210.XA/105. EUR 8311 DE.

142 Kooper, W.F. and Timmer, J.L. (1988) Prediction of groundwater clean-ups in *Contaminated Soil '88* (Hamburg 1988), edited by K. Wolf, W.J. van den Brink and F.J. Colon. 1175-1185, Dordrecht: Kluwer.

143 Land Use Consultants (1992) Amenity reclamation of mineral workings. Report to the UK Department of Environment. London: HMSO.

144 Lawrence, R.W., Poling, G.W., Ritcey, G.M. and Marchant, P.B. (1989) Assessment of predictive methods for the determination of AMD potential in mine tailings and waste rock. In *Proceedings of the International*

Symposium on Tailings and Effluent Management, (Halifax, Canada 1989). edited by M.E. Chalkley, B.R. Conard, V.I. Lakshmanan and K.G. Wheeland.

145 Leach, B.A. and Goodyear, H.K. (1991) Building on derelict land. London: CIRIA.

146 Lee A.R. (1974) Blast furnace and steel slags. London: Edward Arnold.

147 Leininger, D., Leonhard, J., Erdmann, W. and Schieder T. (1987) Research on suitability of coal preparation refuse in civil engineering in the Federal Republic of Germany. In *Reclamation, treatment and utilisation of coal mining wastes* (Nottingham 1987), edited by A.K.M. Rainbow, 55-68, Amsterdam: Elsevier.

148 Levitt, E. (1981) Water disposal from underground coal mining activities. Part 1 - Sources. In *Mining and Water Pollution, The Institution of Water Engineers and Scientists, Scientific Section Symposium,* (Nottingham 1981).

149 Lister, L. (1960) Europe's Coal and Steel Community: an experiment in economic union. New York: Twentieth Century Fund.

150 van Luin, A.B. (1988) The treatment of polluted groundwater from the clean-up of contaminated soil. In *Contaminated Soil '88* (Hamburg 1988), edited by K. Wolf, W.J. van den Brink and F.J. Colon, 1167-1174, Dordrecht: Kluwer.

151 Luke, A. and Macpherson, T. (1983) Direct seeding: a potential aid to land reclamation in central Scotland, *Arboricultural Journal*, 7, 287-299.

152 Mackay, D.M. (1990) Characterisation of the distribution and behaviour of contaminants in the subsurface. In *Groundwater and Soil Contamination Remediation: Toward Compatible Science, Policy, and Public Perception*. (Report on a colloquium sponsored by the Water Science and Technology Board). Washington DC: National Academy Press.

153 MAFF (1973) Lime and liming, seventh edition. Bulletin 35. London: HMSO.

154 Malot, J. (1990) Vacuum extraction: effective clean-up of soils and groundwater. In *Physical/Chemical Processes. Innovative Hazardous*

Waste Treatment Technology Series, Volume 2, edited by H. M. Freeman, 33-43, USA: Technomic.

155 Martien, W.F.Y. and van Wachem, E.G. (1988) Soil covering systems as remedial action in contaminated housing areas in the Netherlands. In *Contaminated Soil '88* (Hamburg 1988), edited by K. Wolf, W. J. van den Brink and F.J. Colon, 597-599, Dordrecht: Kluwer.

156 Matyas, A.G. (1978) Utilization of steelmaking slag, *Iron and Steel Engineer*, 55, No.8, 29-30.

157 McCann, D.M., Culshaw, M.G., Bell, F.G. and Cripps, J.C. (1988) Reconnaissance methods for the location of abandoned mineshafts and adits. In *Mineworkings '88, Conference Proceedings,* edited by M.C. Forde (Edinburgh 1988). Edinburgh: Engineering Technics Press.

158 McKecknie Thomson, G. and Rodin, S. (1972) Colliery spoil tips after Aberfan. London: Institution of Civil Engineers.

159 McNeill, K.R. and Waring S. (1992) Vitrification of contaminated soils. In *Contaminated Land Treatment Technologies*, edited by J. F. Rees, 195-217, London: Elsevier.

160 McNicol Bruce, J. (1980) Cokemaking - a decade of progress. *Iron and Steel International,* August 1980, 197-208.

161 Meek, F.A. (1991) Evaluation of acid preventative techniques used in surface mining. In *Proceedings of the Symposium on Environmental Management for the 1990's* (Denver, CO. 1991), edited by D.J. Lootens, W.M. Greenslade and J.M. Barker, 167-172, Littleton, CO.: Society for Mining, Metallurgy and Exploration Ltd.

162 Metcalf, B. (1990) Establishing long-term vegetational cover on acidic mining waste tips by utilising consolidated sewage sludges. In *Acid Mine Water in Pyritic Environments* (Lisboa, Portugal 1990), edited by P.J. Norton, 255-265, International Mine Water Association.

163 Miller, R.N., Hinchee, R.E. and Vogel, C.C. (1991) A field-scale investigation of petroleum hydrocarbon biodegradation in the vadose zone enhanced by soil venting at Tyndall AFB, Florida. In *In Situ*

Bioreclamation, edited by R.E. Hinchee and R.F. Olfenbutter, 283-302. Butterworth-Heineman.

164 Ministerium für Umwelt Baden Württemberg (1988) Editor. Altlastenhandbuch, Teil I, Altlastenbewertung. Heft 18 der Reihe "Wasserwirtshaftsverwaltung": Stuttgart (2nd edition).

165 Moen, J.E.T. (1988) Soil protection in the Netherlands. In *Contaminated Soil '88* (Hamburg 1988), edited by K. Wolf, W. J. van den Brink and F.J. Colon, 1495-1503, Dordrecht: Kluwer.

166 Morgan, H. (1988) Setting trigger concentrations for contaminated land. In *Contaminated Soil '88* (Hamburg 1988), edited by K. Wolf, W. J. van den Brink and F.J. Colon, 327-337, Dordrecht: Kluwer.

167 Motz, H. (1985) Neuere Erkenntnisse bei der Verwertung von LD- und Hochofenschlacken. In *Eisenhüttenschlacken für den Straßenbau*, Referate der 4. Vortragsveranstaltung am 11. Dez 1985 in Düsseldorf. Forschungsgemeinschaft Eisenhüttenschlaken.

168 Müller, M., Forge, F., Liebeskind, M. and Schröder, H.F. (1993) Physical, chemical and biological methods of *in situ* treatment of soils contaminated with mineral oil. In *Contaminated Soil '93* (Berlin 1993), edited by F. Arendt, G. J. Annokkée, R. Bosman and W. J. van den Brink, 1419-1420, Dordrecht: Kluwer.

169 National Coal Board (1970) Spoil heaps and lagoons, National Coal Board Technical Handbook. Second draft. London: National Coal Board

170 National Coal Board (1971) NCB (Production) Codes and rules - tips. First draft. London: National Coal Board.

171 National Coal Board (1975) Subsidence engineers handbook. London: National Coal Board.

172 National Coal Board (1982) Treatment of disused mine shafts and adits. London: National Coal Board.

173 National Water Council (1981) Design and analysis of urban storm drainage, The Wallingford procedure. Volume 1: Principles, Methods and

Practice and Volume 4: The Modified Rational Method. London: National Water Council.

174 Neale, B.S. (1992) Safety and health requirements following European Community Directives. In *Decommissioning and Demolition, Conference Proceedings* (Manchester 1992), edited by I.L.Whyte, 1-8.

175 Nederlands Normalisatie-Instituut (1991) Soil: Investigation strategy for exploratory survey. Draft standard. UDC 628.516. NVN 5740.

176 Nixon P.J. (1978) Floor heave in buildings due to the use of pyritic shales as fill material, *Chemistry and industry*, 5, 160-164.

177 Nixon, P.J., Treadway K.W.J. and Harrison, W.H. (1980) Durability and protection of building materials in contaminated soils. In *Reclamation of Contaminated Land, Conference Proceedings* (Eastbourne 1979), 1-10, London: Society of Chemical Industry, UK.

178 Nunno, T.J., Hyman, J.A. and Pheiffer, T. (1989) European approaches to site remediation, *Hazardous Materials Control*, 2, No.5, 38-46.

179 OECD Road Research Group (1977). Use of waste materials and by-products in road construction. Paris: OECD.

180 O'Hara, J. and Williams, K.P. (1993) Behaviour of small water-only cyclones for recovery of the coal from colliery spoils. In *Proceedings of the Institute of Chemical Engineers Research Event, Volume 1* (Birmingham 1993) Volume 1, 296-298, Rugby, Institution of Chemical Engineers.

181 Onysko, S.J., Kleinmann, R.L.P., Erickson, P.M. (1984) Ferrous iron oxidation by *Thiobacillus ferrooxidans*: inhibition with benzoic acid, sorbic acid, and sodium lauryl sulfate, *Applied and Environmental Microbiology*, 48, No.1, 229-231.

182 Osborne, D.G. (1988) Coal preparation technology. London: Graham and Trotman.

183 Palmer, J.P. (1984) An investigation of the potential for the use of legumes on colliery spoil. PhD Thesis, University of York, UK.

184 Palmer, J.P. (1992) Decontamination of metalliferous mine spoil. Paper presented at *NATO/CCMS Conference on the Pilot Study on Downstream and Remedial Action Technologies for Contaminated Ground and Groundwater* (Budapest 1992).

185 Palmer, J.P. and Iverson, L.R. (1983) Factors affecting nitrogen fixation by white clover (*Trifolium repens*) on colliery spoil. *Journal of Applied Ecology*, 20, 287-301.

186 Palmer, J.P., Williams, P.J., Chadwick. M.J., Morgan. A.L. and Elias, C.O. (1986) Investigation into nitrogen sources and supply in reclaimed colliery spoil, *Plant and Soil*, 91, 181-184.

187 Palmer, M.E. (1978) Acidity and nutrient availability in colliery spoil. In *Environmental Management of Mineral Wastes*, edited by G.T. Goodman and M.J. Chadwick, 85-126, The Netherlands: Sithoff und Nordhoff.

188 Patel, Y.B., Shah, M.K. and Cheremisinoff, P. N. (1991) Methods of site remediation, *Pollution Engineering*, 22, No.12, 58-66.

189 Pearce, K.W. and Finnecy, E.E. (1983) The chemical nature of slags and cokeworks waste and resulting problems of land reuse. In *Reclamation of Former Iron and Steelworks Sites, Conference Proceedings* (Windermere 1983), edited by G.P. Doubleday, H1-H8.

190 Pennington, J.C. (1986) Feasibility of using mycorrhizal fungi for enhancement of plant establishment on dredged material disposal sites: a literature review. Misc. paper D-86-3, US Army Engineer Waterways Experiment Station, Vicksburg MA.

191 Petit, D. (1980) La vegetation des terrils du Nord de La France, ecologie, phytosociologie, dynamisme. Thesis, Docteur des Science Naturelles, L'Université Sciences et Techniques de Lille, Lille, France.

192 Petit D. (1982) Natural vegetation of the tips of North France - its connection with some spoil chemical parameters. In *International Symposium on Mine Spoil Heaps* (Essen 1982), 99-118, Essen: Kommunalverband Ruhrgebiet.

193 Pierard, R. (1982) Integration of mine tips into the urban landscape. In *International Symposium on Mine Spoil Heaps* (Essen 1982), 119-123.

194 Platt, J. and Hellewell, E.G. (1974) Coal tip recovery. In *Minerals and the Environment, Proceedings of the IMM International Symposium* (London 1974), London: Institution of Mining and Metallurgy.

195 Pulford, I.D. (1991) A review of methods to control acid generation in pyritic coal mine waste. In *Land Reclamation - an End to Dereliction?* (Cardiff 1991), edited by M.C.R. Davies, 269-278, London: Elsevier.

196 Putwain, P.D. and Evans, B.E. (1992) Experimental creation of naturalistic amenity woodland with fertiliser and herbicide management plus lupin companion plants. In *Aspects of Applied Biology Number 19: Vegetation management in forestry, amenity and conservation areas,* 179-186. Warwick, UK: Association of Applied Biologists.

197 Pye, K. and Miller, J.A. (1990) Chemical and biochemical weathering of pyritic mudrocks in a shale embankment, *Quarterly journal of engineering geology*, 25, 365-381.

198 Rainbow, A.K.M. (1989) Geotechnical properties of UK minestone. London: British Coal.

199 RETI (1993) *Pers. comm.*

200 RETI-CEE-BRGM (1992) Derelict land reclamation, soils decontamination. Final report, January 1992.

201 Richards, Moorehead and Laing Ltd (1991) Guidelines for the control of Japanese Knotweed, Cardiff: Welsh Development Agency.

202 Richards, Moorehead and Laing Ltd (1992) Constructed wetlands to ameliorate metal-rich mine waters. Review of existing literature. R&D Note 102 to National Rivers Authority.

203 Richards, Moorehead and Laing Ltd (1993) Model specification for the control of Japanese Knotweed, prepared for the Welsh Development Agency. Unpublished.

204 Richardson, G.V. and Barry, W.R. (1992) Potential sources of contamination during landfill site investigation. In *Polluted and Marginal Land '92, Proceedings of Second International Conference on Construction*

on Polluted and Marginal Land (London 1992), edited by M.C. Forde, 135-140, Edinburgh: Engineering Technics Press.

205 Rimmer, D.L. and Colbourn, P. (1978) Problems in the management of soils forming on colliery spoil. Newcastle, UK: University of Newcastle.

206 Rimmer, D.L. and King, J.A. (1988) Soil substitutes. In *Land Rec '88, Conference Papers* (Durham 1988), edited by D. Williamson, 169-174.

207 Ritcey, G.M. (1989) Tailings management - problems and solutions in the mining industry. Amsterdam: Elsevier.

208 Roberts Shultz, S., Oresik, W., Graham, D., Kiener, L., Levin, S., Trynowski, J. and Schultz, D. (1990) Superfund site sludge treatability studies. In *Proceedings of 1990 Speciality Conference* (Arlington, VA. 1990), 568-575.

209 Robinson Jones Partnership (1987) Working with nature. Cardiff: Welsh Development Agency.

210 Santoleri, J.J. (1990) Design and operation of thermal treatment systems for contaminated soils. In *Proceedings of the 1990 EPA/A&WMA International Symposium on Hazardous Waste Treatment: Treatment of contaminated soils* (Cincinnati, OH. 1990), 79-94, USA: USEPA.

211 Scarooson, D.W. and Flemming, E. (1990) Electrolytic treatment of waste pickling liquor. In *Innovative Hazardous Waste Treatment Technology Series, Volume 2, Physical and Chemical Processes,* edited by H.M. Freeman, 97-100, Technomic.

212 Schuldes, H. and Kübler, R. (1990) Öklogie und vergesellschaftung. Von *Solidago canadensis* und *gigantea, Reynoutria japonica* und *sachalinensis, Impatiens glandulifera, Helianthus tuberosa, Heracleum mantegazzianum.* Ihre Verbreitung in Baden - Württemberg sowie Notwendigkeit und Möglichkeiten ihrer Bekämpfung. In *Studie im Auftrag des Ministeriums für Umwelt* (Baden - Württemberg 1990), Baden -Württemberg: des Ministeriums für Umwelt.

213 Sibley, R.D. (1988) Soil gas geochemistry: a potential method for detecting old mine workings. In *Mineworking '88, Conference Proceedings*

(Edinburgh 1988), edited by M.C. Forde, Edinburgh: Engineering Technics Press.

214 Silver, M. (1989) Biology and chemistry of generation, prevention and abatement of acid mine drainage. In *Constructed Wetlands for Wastewater Treatment: Municipal, Industrial and Agricultural* (Chattanooga, TN. 1988), edited by D.A. Hammer, 753-760, Chelsea, MI.: Lewis.

215 Simmons, E. (1984) The effects of recultivation on the productivity of previously reclaimed colliery waste. Paper 25, *Symposium on the Reclamation, Treatment and Utilisation of Coal Mining Wastes* (Durham 1984).

216 Simmons, E. (1988) Cultivation and seeding techniques used in reclaiming land. In *Land Rec '88 Conference Proceedings* (Durham 1988), edited by D. Williamson, 189-194.

217 Singer, P.C., Stumm (1970) Acidic mine drainage: the rate determining step, *Science*, 167, 1121-1123.

218 Singh, G., Bhatnagar, M., Sinha, D.K. (1990) Environmental management of acid water problems in mining areas. In *Acid Mine Water in Pyritic Environments* (Lisboa, Portugal 1990), edited by P.J. Norton, 113-132, International Mine Water Association.

219 Skarzynska, K.M., Burda, H., Kozielska-Sroka, E. and Michalski, P. (1987) Laboratory and site investigations on weathering of coal mine wastes as a fill material in earth structures. In *Reclamation and utilization of coal mining wastes*, (Nottingham 1987), edited by A.K.M. Rainbow, 179-197, Amsterdam: Elsevier.

220 Skousen, J.G. (1991) Anoxic limestone drains for acid mine drainage treatment, *Green Lands*, 21, No.4, 30-35.

221 Skousen, J.G., Sencindiver, J.C., Smith, R.M. (1987) A review of procedures for surface mining and reclamation in areas with acid-producing materials. Morgantown, WV.: West Virginia University Energy Research Centre.

222 Sleeman, W. (1987) Colliery spoil in urban development. In *Reclamation, Treatment and Utilisation of Coal Mining Wastes*, (Nottingham 1987), edited by A.K.M. Rainbow, 163-178, Amsterdam: Elsevier.

223 Smith, M.A. (1985) Editor. Contaminated land: Report of the NATO/CCMS pilot study on contaminated land. New York: Plenum.

224 Smith, M.A. (1985) The effect of phenol upon concrete. *Magazine of Concrete Research,* 37, No.133, 234-237.

225 Smith, S.W. and Thomas, G.H. (1988) Problems of derelict iron and steel works complexes. In *Land Rec '88, Conference Proceedings* (Durham 1988), edited by D. Williamson, 97-102.

226 Sobek, A.A., Rastogi, V. and Benedetti, D.A. (1990) Prevention of water pollution problems in mining: the bactericide technology. In *Acid Mine Water in Pyritic Environments* (Lisboa, Portugal 1990), edited by P.J. Norton, 133-148, International Mine Water Association.

227 Soczó, E.R., Meeder, T.A. and Versluijs, C.W. (1992) Ten years of soil clean-up in the Netherlands. In *NATO/CCMS Pilot Study First International Conference, Evaluation of Demonstrated and Emerging Technologies for the Treatment and Clean-up of Contaminated Land and Groundwater (Phase II)* (Budapest 1992).

228 Sørensen, E., Bjerre, A.B. and Rasmussen, E. (1990) Soil recovery by net oxidation, *Environmental Technology,* 11, 429-436.

229 Soundararajan, R., Barth, E. F. and Gibbons, J.J. (1990) Using an organophilic clay to chemically stabilise waste containing organic compounds. *Hazardous Materials Control*, January-February 1990.

230 Staubitz, W.W., Surface, J.M., Steenhuis, T.S., Peverly, J.H., Lavine, M.J., Weeks, N.C., Sanford, W.E. and Kopka, R.J. (1989) Potential use of constructed wetlands to treat landfill leachate. In *Constructed Wetlands for Wastewater Treatment: Municipal, Industrial and Agricultural* (Chattanooga, TN. 1988), edited by D.A. Hammer, 735-742, Chelsea, MI.: Lewis.

231 Stevenson F.J. (1982) Organic forms of soil nitrogen. In *Nitrogen in Agricultural Soils* edited by F.J. Stevenson, 67-22, Madison, Wisconsin: American Society of Agronomy.

232 Strohmeier, G. (1992) Two routes to more ecological steelmaking, *Steel Times International*, November, 30-31.

233 Swanstrom, C. and Palmer, C. (1990) XTRAX - Transportable thermal separator for organic contaminated solids. Presented at *Second Forum on Innovative Hazardous Waste Treatment Technologies: Domestic and International (Philadelphia, PA. 1990)*. USA: USEPA.

234 Tandy, C.R.V. (1975) The landscape of industry. London: Leonard Hill Books.

235 Taylor, R.K. (1984) Composition and engineering properties of British colliery discards. London: National Coal Board.

236 Taylor, R.K. and Cripps, J.C. (1984) Mineralogical controls on volume change, In *Ground movements and their effects on structures,* edited by P.B. Attewell and R.K. Taylor, 268-302, Glasgow: Surrey University Press.

237 Thomas, G.H. (1983) Properties of iron and steel slags. In *Reclamation of Former Iron and Steelworks Sites, Conference Proceedings* (Windermere 1983), edited by G.P. Doubleday, C1-C8.

238 Thomas, G.H. (1992) *Pers. comm.*

239 Thyssen Umweltschutz (1992) *Pers. comm.*

240 Tomlinson, M.J., Driscoll, R., and Burland, J.B. (1978) Foundations for low-rise buildings. Watford: Buildings Research Establishment.

241 Tucker, I.B. (1988) Land reclamation in conjunction with tip washing. In *Land Rec '88, Conference Proceedings* (Durham 1988), edited by D. Williamson, 121-125.

242 Tucker, M.E. (1981) Sedimentary petrology, an introduction. Geoscience Texts, Volume 3. Oxford: Blackwell Scientific.

243 UNEP (1986) Guidelines for environmental management of iron and steel works, Industry and Environment Guidelines Series.

244 USEPA (1991) Innovative treatment technologies - Overview and guide to information sources. Report No. EPA/540/9-91/002, Section 4.

245 Veitch, P.J. (1988) A landscape view of reclamation. In *Land Rec '88, Conference Proceedings* (Durham 1988), edited by D. Williamson, 35-46.

246 Versluijs, C.W., Aalbers, T.G., Adema, D.M.M., Assink, J.W., van Gestel, C.A.M. and Anthonissen, I.H. (1988) Comparison of leaching behaviour and bioavailability of heavy metals in contaminated soils and soils cleaned up with several extractive and thermal methods. In *Contaminated Soil '88* (Hamburg 1988), edited by K. Wolf, W.J. van den Brink and F.J. Colon, 11-21, Dordrecht: Kluwer.

247 Visser, W., Olie, J.J., Bremmer, C. and v.d. Heuvel, M. (1993) The use of *in situ* measurement techniques for soil pollution problems. In *Contaminated Soil '93* (Berlin 1993), edited by F. Arendt, G.J. Annokkée, R. Bosman, and W.J. van den Brink, 631-639, Dordrecht: Kluwer.

248 VROM (1987) Environmental program of the Netherlands 1988-1991. The Hague: Staatsuitgeverij.

249 van de Vusse, A.C.E. and de Baas, H.J. (1993) Reducing inconveniences for residents during remedial actions. In *Contaminated Soil '93* (Berlin 1993), edited by F. Arendt, G.J. Annokkée, R. Bosman and W.J. van den Brink, 99-105. Dordrecht: Kluwer.

250 Waltham, A.C. (1989) Ground subsidence. Glasgow: Blackie.

251 Walton, G. and Cobb, A.E. (1984) Ground subsidence. In *Ground Movements and their Effects on Structures*, edited by P.B. Attewell and R.K. Taylor, Glasgow: Surrey University Press.

252 Wang, X., Yu, X. and Bartha, R. (1990) Effect of bioremediation on polycyclic aromatic hydrocarbon residues in soil, *Environmental Science and Technology*, 24, No.7, 1086-1089.

253 WDA (1991) *Pers. comm.*

254 Weale, A. (1992) The new politics of pollution. Manchester: Manchester University Press.

255 Weltman, A.J. and Head, J.M. (1983) Site investigation manual. London: CIRIA.

256 Wheatley, B.I. and Pooley, F.D. (1990) Production of zinc powder from arc and smelter flue dusts. In *Recycling of Metalliferous Materials*, 291-299, London: Institution of Mining and Metallurgy.

257 Whittaker, B.N., and Reddish, D.J. (1989) Subsidence - occurrence, prediction and control. Amsterdam: Elsevier.

258 Whyte, I.L. (1992) Demolition: a review for engineers. In *Decommissioning and Demolition, Conference Proceedings* (Manchester 1992), edited by I.L.Whyte, 132-141.

259 Wildeman, T.R. and Laudon, L.S. (1989) Use of wetlands for treatment of environmental problems of mining: non-coal-mining applications. In *Constructed Wetlands for Wastewater Treatment: Municipal, Industrial and Agricultural* (Chattanooga, TN. 1988), edited by D.A. Hammer, 221-231, Chelsea, MI.: Lewis.

260 Williams, K.P. (1992) Principles and application of physical particle separation for treatment of contaminated land. In *Contaminated land treatment technologies*, edited by J.F. Rees, 113-120, London: Elsevier.

261 Williams, P.J. (1975) Investigations into the nitrogen cycle in colliery spoil. In *The ecology of resource degradation and renewal*, edited by M.J. Chadwick and G.T. Goodman, 259-274, Oxford: Blackwell.

262 Wilson, P.J. and Wells, J.H. (1950) Coal, coke and coal chemicals. McGraw-Hill.

263 Wilson, R. and Ward, T. (1983) Demolition techniques and procedures. In *Reclamation of Former Iron and Steelworks Sites, Conference Proceedings (Windermere 1983)*, edited by G.P. Doubleday, D1-D13.

264 Winter, E.J. (1974) Water, soil and the plant. In *Science in Horticulture Series*. London: MacMillan.

265 Young, C.P. and Barber, C. (1980) Contaminated land - effects on water resources. In *Conference on Reclamation of Contaminated Land* (Eastbourne 1979), D5/1-11, London: Society of Chemical Industry.

266 Ziemkiewicz, P.F. (1990) Advances in the prediction and control of acid mine drainage. In *Proceedings of the 1990 Mining and Reclamation Conference and Exhibition* (Charleston, WV. 1990), edited by J.G. Skousen, J.C. Sencindiver and D.E. Samuel, 51-54, Morgantown, WV.: West Virginia University.

GLOSSARY

Acid mine drainage	Acidic water draining a mine or an area of land affected by mining, which is acidic as a result of the oxidation of pyrite (iron sulphide) to sulphuric acid.
Adit	A mine entrance tunnel, driven at a shallow angle, often in a hillside, and sometimes conveying drainage water from the mine.
Aerobic	Conditions under which oxygen is freely available.
Aggregate	Granular, inert materials which form a substantial park of concrete, asphalt *etc.* Examples include broken stone, gravel and sand.
Aggregates (soil)	Soil crumbs, composed of fine particles such as clays, held together by surface charge and/or soil polysaccharides and other organic molecules.
Air stripping	The process of enhancing the volatilisation of contaminants into the atmosphere from physical treatment of soils, groundwater and effluents.
Allowable bearing	The maximum allowable intensity of foundation pressure loading, taking into account bearing capacity, type of structure and limits on foundation settlement.
Alloy steels	Carbon steels to which various amounts of other metal elements, such as chromium and nickel, have been added.
Ammoniacal liquor	Spent scrubbing liquor from the coal carbonisation process, containing phenols and ammonia.
Anaerobic	Conditions under which gaseous and soluble oxygen is absent.
Angle of repose	Angle to the horizontal at which the sides of a heap of granular material will be at rest.

Anions	Negatively charged atoms or molecules. For example chloride (Cl^-), sulphate (SO_4^{2-}), carbonate (CO_3^{2-}), nitrate (NO_3^-).
Aquifer	Soil or rock that contains sufficient saturated permeable material to yield water in significant quantities.
Aromatic	Compounds related to benzene.
Artesian water	Hydrostatic pressure which occurs when water from pressure higher ground finds its way into permeable strata which have impermeable strata above and below. Artesian pressure may be sufficient to force water to the surface via a fissure or borehole.
Auger	A helically-shaped corkscrew-like tool designed to bore holes. The two types of earth auger are (a) the percussion type used in shell and auger boring, and (b) the helical screw type which may be power-driven or hand-operated.
Back stowing	Backfilling of underground workings with mine spoil.
Background concentration	The natural or normal concentration of a substance in the environment.
Batter	An artificial, uniform slope.
Bearing capacity	The bearing capacity of a soil is the foundation pressure which that soil is capable of supporting.
Bell pit	Method of mining using a bell-shaped pit comprising a vertical access shaft and enlarged workings at the base of the shaft (see Section 3.2.1).
Benthic organisms	Animals which live on the bed of river, lake or sea, at the solid-liquid interface.
Bentonite	A type of clay with a high affinity for water. Bentonite may be used to form an impermeable seal or barrier. Bentonite slurry may be used to support the sides of trench excavations, or as a lubricant for drilling.

Benzole — Crude benzene by-product of the coal carbonisation process.

Berm — A horizontal platform or ledge formed part way up a slope to improve stability or provide access.

Binder — Chemicals used to artificially stabilise contaminants within a soil or other contaminated matrix.

Bioaccumulation — The process by which certain substances are concentrated by living tissues so that they become present at higher concentrations in organisms further up a food chain.

Biodegradation — Microbially-mediated process by which enzymes sequentially degrade larger molecules to smaller ones, so altering both the chemistry and physical structure of a given substrate.

Bioengineering — The use of vegetation to perform an engineering function, such as erosion control or slope stabilisation.

Bioremediation — The use of microbial treatment methods to decontaminate soil and groundwater in a field situation.

Biosorption — The adsorption of ions to the surface of a microbial cell, or absorption into the cell.

Biotope — A small habitat within a large ecological community.

Biplanar slip — Slope failure involving shearing along two planes of differing orientation (see Figure 6.2(f)).

Blast furnace — A furnace used for smelting of iron ore to produce pig iron. Operates as a continuous process.

Bloomery — A furnace used for smelting of iron ore in a charcoal fire, now superseded by the blast furnace. Smelting was carried out as a batch process, and produced a mass of wrought iron known as a bloom.

Bored filter drain — Borehole drilled at a shallow angle to drain water from beneath an embankment (see Figure 6.4(c)).

Borehole	A cored hole, usually vertical, sunk to determine ground conditions or for the extraction of water.
Boring mill	Machine, usually water powered, used for boring large holes.
Break layer	A stratum of high permeability granular material within a containment layer, the purpose of which is to stop upward capillary movement of soluble contaminants.
Briquetting plant	Plant for producing blocks made from small pieces of coal.
By-product	Chemical residues arising as subsidiary products of a chemical industrial process *i.e.* products other than the primary products.
Cable percussion	Method of creating boreholes whereby an open-ended tube is driven into the ground by repeated lifting and dropping. Also known as 'shell and auger'.
Calorific value	The number of heat units obtained by the combustion of a unit mass of material.
Cap (shaft)	Structure placed preferably at rock level over a shaft to support the overburden above. Usually of reinforced concrete.
Carbon to nitrogen ratio	The ratio of carbon to nitrogen in a particular material. It is an indicator of the resistance of that material to decomposition (see Box 5.2).
Carbon steels	Steels whose properties are determined primarily by the percentage of carbon present. Carbon steels may, in addition to iron and carbon, contain small amounts of manganese, silicon, sulphur and phosphorus.
Carbonisation	Process by which coal is heated in oxygen-depleted conditions (ovens) to produce a carbonised product (coke), coal gas and various by-products.

Carcinogen — A substance which, when in contact with body tissues, can lead to the formation of a tumour or cancerous growth (carcinoma).

Casting pit — Area where the remelted iron or steel was cast into the finished product.

Cations — Positively charged atoms or molecules. All metallic elements form cations, as do certain molecules *e.g.* ammonia (cation: NH_4^+).

Chamaephytes — Woody or semi-woody perennial plants bearing their buds less than 250mm from the ground.

Clay — Fine-grained cohesive soil with particles less than 0.002mm in diameter.

Climax vegetation — The vegetation resulting from the complete process of ecological succession *i.e.* the successive change in vegetation composition after colonisation of a substrate.

Clinker — Incombustible residue from the burning of coal or coke consisting of fused ash.

Co-metabolite — Two or more substances that are metabolised by an organism when in association with one another.

Coal measures — Geological strata, consisting of beds of coal interstratified with shales, sandstones, limestones and conglomerates.

Coal preparation — Process by which coal is separated from non-coal materials present in the run-of-mine (see Run-of-mine).

Coal rank — Classification of coal according to carbon content.

Colliery spoil — The waste material produced during coal mining (see Section 5.3.1).

Competent rock — Sound rock strata capable of supporting shaft capping or other structures.

Composting	A process by which organic matter is aerated to allow a succession of microorganisms to degrade the organic residues to form compost; a partly degraded organic residue which may then be used as a soil conditioner.
Consolidation	A result of weathering and settlement whereby fine-grained particles fill the larger soil pore spaces, impeding drainage, aeration and root penetration.
Consolidation	Ground settlement resulting from consolidation due settlement to the weight of the overlying material and the dissipation of excess porewater.
Contaminated land	Land which contains substances (contaminants) which, when present in sufficient quantities or concentrations, are likely to cause harm, directly or indirectly; for example to humans, the environment or building materials.
Creep	Slow continuous downwards and outwards movement of soil materials on a slope.
Crest (of a slope)	Top edge of a slope.
Crown hole	Surface cavity resulting from the underground collapse of mine workings.
Culvert	A covered channel or large pipe for carrying a watercourse below ground.
Cupola	A furnace used to remelt pig-iron in the production of cast iron.
Derelict land	Land so damaged by industry and other development that it is incapable of beneficial use without treatment.
Desk study	The simple collection and collation of existing factual information.
Detection limit	The lowest concentration of a compound which can be analysed by a given analytical technique.

Differential settlement	The difference in settlement between any two points on a foundation or loaded area of soil.
Dip	The slope angle of geological strata relative to the horizontal.
Discard	The unwanted fraction from a mineral processing operation *e.g.* the material, consisting mainly of shales, remaining after recovery of coal from colliery spoil by washing operations.
Drainage blanket	Layer of free-draining material placed beneath an embankment (see Figure 6.5(a)).
Drawdown	A lowering of the water level in a lagoon or reservoir.
Drift	Mine tunnel driven at a shallow angle into a coal or mineral seam from a level ground surface.
Drift mining	Method of mining using inclined tunnels driven within the seam.
Dynamic compaction	Method of compaction of soil involving the repeated dropping of a heavy weight.
Earthworks	All on-site operations that relate to excavation, transport and deposition of soil or rock.
Ecotype	A genetic grouping within a species having specific characteristics or adaptations in particular ecological conditions.
Electrical conductivity	The ratio of electrical current flowing per unit cross sectional area of conductor to the applied electrical field. The electrical conductivity of soils and waters is an indicator the salinity of the soil or water.
Embodied energy	The energy required to produce a material.
End tipping	Placing of material by tipping from lorries or dump trucks.

Engine house	Building used to accommodate a steam engine or other power source.
Enzyme	A biologically produced protein catalyst that accelerates the conversion of one compound (or compounds) to another (or others).
Ex situ treatment	Decontamination treatment process applied to excavated soil or extracted groundwater in a batch system.
Exothermic	Accompanied by the evolution of heat.
Factor of safety (engineering)	Factor of safety for slope stability is the ratio of the resisting forces divided by the disturbing forces (see Box 6.3).
Finery	A type of forge used for decarbonisation of the pigs of iron to produce wrought iron.
Flame ionisation	An instrument which measures volatile organic detector (FID) compounds by burning them in a hydrogen flame, the electrical conductivity of which is measured by inserting two high voltage electrodes into the flame.
Flowslide	A water-borne movement of material, mainly in suspension, containing a proportion of water in excess of that required to fill the voids in the loose material.
Flue dusts	The particulate matter removed from the gas produced by a furnace.
Fluidised bed	A bed of solid particles which has the properties of a fluid, created by passing a fluid (*e.g.* a gas) upward through the bed at sufficient velocity to separate the particles from one another so they are supported by the fluid.
Fly ash	Ash with a fine particle size produced from pulverised coal burned in power stations, or other organic residues in incinerations.

Foundation	1. The soil or rock upon which a building or other structure rests. 2. The structure of brick, stone, concrete, wood or steel which transfers the building load to the ground.
Foundry	A place where metals are remelted prior to further processing.
Framework directive	An EC Directive which sets out a framework for legislation on a particular subject, and which has effect via daughter directives setting out particular requirements.
Froth flotation	A process for separation of mineral particles which uses reagents which adhere preferentially to particles containing the desired mineral (*e.g.* coal).
Galvanising	Coating of steel or iron with zinc. The zinc protects the iron or steel from corrosion.
Gas chromatography (GC)	An analytical method which separates mixtures of compounds by partitioning the compounds between a stationary (liquid) and a mobile (gas) phase.
Gasification	Process by which coal is heated in oxygen-depleted conditions to produce coal gas for use as a fuel.
Gasworks	Coal carbonisation plant whose primary product was gas, produced from heating coal.
Geophysical techniques	Non-invasive techniques for the investigation of the nature of materials below the surface of the ground. They rely on differences in the density, electrical resistance, magnetic properties, electrical resistance, electrical conductivity *etc.*, between different materials.
Geophytes	Plants which bear their perennating buds below the ground surface *e.g.* bulbs or corms.
Geosynthetic	Synthetic polymer extruded in sheet form.

Geotextile	Synthetic or natural permeable fabric used in conjunction with soil and vegetation, principally for erosion control, filtration, soil reinforcement and drainage.
Greenbelt	An area of open, undeveloped land around a town or city, designated for preservation.
Grinding mill	Enclosed area with water-powered grindstones for sharpening tools.
Ground investigation	Investigation of the nature of materials below the surface of a site.
Groundwater	Sub-surface water contained within the saturated zone.
Grouting	Treatment of unstable ground by injecting a slurry of cement/pulverised fuel ash into the voids.
Growth factor	A compound which, when present in low concentrations, stimulates growth in an organism, and, when absent, can retard growth.
Gulley (geomophology)	A drainage channel formed by running water.
Hardstanding	Surface ground cover with hard wearing properties and little or no permeability to water *e.g.* tarmacadam, concrete.
Haul road	Temporary road constructed to enable materials to be moved more easily about a site.
Headframe	Structure above a shaft or adit supporting the winding gear. Usually made of steel, but sometimes of concrete or brick.
Heapstead	A large building at the mouth of a shaft or adit housing the head frame and coal- or ore-handling equipment.
Heave	Upward displacement of the ground that may arise as a result of excavating, surcharging, installing displacement piles or by chemical action.

Heavy metal	A term used to describe a range of dense, toxic metals which may be present in soluble or particulate form.
Hemicryptophyte	Plants which bear their perennating buds at the surface.
Herringbone drains	Drains laid in a herringbone pattern (see Figure 6.5(b)).
Horse gin circle	Winding equipment powered by a horse walking around a circular area.
Hot-spot	Area of higher concentration of contamination than present in the majority of the ground within the site under investigation.
Humus	Amorphous black or dark-brown material which results from the decomposition of organic matter in soils. A mixture of macromolecules based on benzene, carboxylic and phenolic acids.
Hydrocyclone	A small cyclone extractor for removing suspended matter from a flowing liquid by means of the centripetal forces set up when the liquid is made to flow through a tight conical vortex. Used to separate solids present in the liquid into coarse- and fine-grained fractions.
Hydrophobic	The property of a substance which makes it repel water and therefore resist solubilisation.
In situ treatment	Decontamination or other treatment process applied within undisturbed soil or groundwater to alleviate a contamination or pollution problem.
Indigenous	Referring to organisms which have developed naturally in a given area, and not as a consequence of artificial introduction.
Isopachytes	Lines connecting points of equal depth of cut or fill in earthworks; usually computer-generated. Isopachyte drawings indicate the location and depth of cut and fill areas, with the zero isopachyte indicating the boundary between cut and fill.

Jig	A device for concentrating ore according to the specific gravity of its constituent minerals.
Kiln	Stone built furnace, usually set against a hillside, used for removing impurities from the ore or for making lime (CaO).
Lagoon	Settling pond for material deposited from solution or suspension.
Landfill	The site of waste disposed by landfilling *i.e.* controlled tipping into ground repositories such as disused quarries, pits *etc.*
Leaching	Action of a solvent, usually water, on solutes in a permeable medium, such that solutes are released and lost from the medium.
Leaching tests	Tests where a solid material is brought into contact with a liquid. Carried out to determine the solubility in the liquid of substances present in the solid.
Legume	A plant of the Fabaceae (Leguminosae) family, characterised by the ability to fix atmospheric nitrogen.
Liquefaction	Change of a saturated granular soil into a fluid mass as a result of shock loading.
Longwall mining	A method of mining involving total extraction of a coal seam along an advancing coal face usually several hundred metres in length.
Magnetite	A type of iron ore which has magnetic properties.
Microclimate	The climatic conditions over a small area, differing from the general climatic conditions of a region by virtue of local characteristics such as vegetation cover, exposure or shelter *etc.*
Mineralisation	The conversion of organic compounds in the soil to simpler 'mineral' forms *e.g.* ammonium. Bacteria and fungi are usually involved in mineralisation.

Mining ventilation	Ventilation of mine workings to ensure an adequate supply of oxygen and prevent a build-up of mine gases.
Movement joint	Flexible joint within a building structure designed to accommodate limited movement.
Mud run	Movement of solids carried along by a flow of water.
Mutagenic	The ability of a chemical to cause a genetic mutation in a living organism.
Mycorrhiza	Symbiotic association of a soil fungus with the roots of a plant.
Naturalistic	Vegetation which is established and managed to vegetation resemble natural vegetation such as woodland or species-rich grassland.
Neutralisation	Adjustment of the pH of a medium by the addition of acid or alkali to produce a pH near 7.
Neutralising capacity	The extent to which a material neutralises acidity. Usually expressed as calcium carbonate or calcium oxide equivalents.
Nitrification	The oxidation and conversion, by certain bacteria, of ammonium to nitrate.
Nitrogen fixation	The conversion of free nitrogen gas into plant-available compounds by symbiotic bacteria or fungi living within nodules on the roots of plants, particularly legumes, or by free-living blue-green bacteria.
Nitrogen capital	The reserve of nitrogen in the soil.
Nurse species	Plant species which are included in a sown or planted mixture to shelter the slower growing components, or to provide some quick protective growth. They usually only survive a short time before being superseded by long-term components.

Opencast mining	Mining of shallow mineral deposits from surface excavations after removal of rock and other overburden.
Overburden	Material overlying a mineable deposit, but excluding soil horizons, which is removed separately.
Ozone	Powerful oxidising agent comprising three oxygen atoms, produced by the action of ultraviolet radiation or electrical discharge on gaseous oxygen.
Percentile	The value below which occurs a specified percentage of the observations in an ordered set of observations.
Pernicious weeds	Undesirable perennial plants which interfere with the use of land, are difficult to control and are able to spread rapidly unless controlled. Subject to national legislation in some countries.
pH	A measure of relative acidity and alkalinity. Low pH values (<7) indicate acidic conditions. High pH values (>7) indicate alkaline conditions.
Phaneophytes	Plants which bear their perennating buds freely in the air at heights above 250mm. They are mostly woody plants, trees and shrubs.
Phosphate fixation	The formation of virtually insoluble compounds of phosphate, which prevents the uptake of phosphorus by plants.
Photogrammetry	That branch of land surveying in which plans and maps are based on and prepared from photographs, mainly taken from the air.
Photomontage	The combination of photography and perspective artwork, used to give a realistic impression of the appearance of a proposed development.
Phytotoxic	Referring to chemicals that, when present in elevated concentrations, exhibit toxic effects on plants.

Pickling	Cleaning of a metallic surface by immersion in acid.
Piezometer	An open or closed tube or pipe with a porous element, sealed into the ground and used to record porewater pressures or static groundwater level.
Pig bed	Level floor adjoining a blast furnace where the molten iron or steel was poured into channels and cast in sand to form pigs.
Pig iron	Metallic iron produced by a blast furnace which contains 3-4% carbon and other impurities such as silicon, phosphorus, sulphur and manganese.
Piling	Columns of reinforced concrete, steel or sometimes timber, driven or bored into the ground to transmit foundation loads to suitable bearing strata.
Pioneer	Plant species which are able to colonise and grow on unimproved substrates or in exposed harsh conditions. Sometimes used to improve conditions for other species.
Piping	Formation of underground water channels due to internal water erosion.
Pitch	Distillation residue of coal tar. A solid or semi-solid black resinous substrate containing high molecular weight organic compounds.
Polluter pays	Legal principle which vests liability for decontamination costs with those responsible for a given pollution incident.
Pollution	The introduction into the environment of substances or energy liable to cause harm to human health, to living resources and ecological systems, damage to structures and amenity or infringement of legitimate uses of the environment.
Ponding	Pooling of water in hollows on a surface or due to a lack of capacity within a drainage system.

Pore pressure	Internal water pressure within a saturated soil.
Portal	Entrance structure to a mine tunnel.
Preloading	Consolidation of fill material by temporary surcharging, with the aim of increasing the bearing capacity of the ground before construction takes place.
Pump and treat	Decontamination methods which extract contaminated groundwater and treat the water in a batch or continuous decontamination process.
Pyrite	Iron pyrites, FeS_2, present in many colliery spoils, coal and some iron ores.
Radial drains	Drains laid in a radial pattern.
Raft foundation	Reinforced concrete foundation extending beneath the whole of a building to spread the loading uniformly into the underlying strata, and limit the effects of mining subsidence or differential settlement.
Refractory linings	The interior lining of a furnace which is in contact with hot gases, molten metals and slags and must therefore be made of materials which are capable of resisting high temperatures, abrasion and chemical action *i.e.* refractories.
Regrading (engineering)	The movement of surface soil and mineral deposits to change the shape of the land surface.
Regression (of plants)	The death or decline of vegetation caused by progressively worsening substrate conditions *e.g.* acidity or toxicity.
Reinforced strip or pad (engineering)	Reinforced concrete foundations beneath load-bearing foundation walls or columns to spread the loading uniformly into the underlying strata and limit the effects of mining subsidence or differential settlement.
Relief well	Borehole drilled at the foot of an embankment to relieve artesian water pressure or high pore pressure (see Figure 6.4(b)).

Retaining wall	Wall constructed to provide lateral support to the ground or to resist pressure from a mass of stored material.
Return period	The period of time between likely occurrences of a storm of a given intensity.
Risk	The chance or possibility of adverse consequences occurring. Risk combines the concepts of hazard *i.e.* harm, and probability.
Roasting and calcining area	Large level area where the ore and fuel were mixed and burnt to drive off impurities and improve the quality or grade of the ore.
Rockhead	Interface between subsoil and underlying bed rock.
Rolling mill	Enclosed area where iron or steel was rolled into its final shape.
Room and pillar mining	Method of mining in which unworked pillars were left in place to support the overburden (see Section 3.2.2).
Rotational slip	Slope failure involving a curved slip surface (see Figures 6.2(a) to (d)).
Run-of-mine	The material brought to the surface from underground prior to separation into coal and discard.
Seam (geology)	A discrete layer of coal or other minerals.
Semis	Semi-finished or intermediate steel products such as ingots, slabs, blooms and billets.
Settlement	Downwards movement of a structure due to subsidence of the ground beneath.
Settling pond	Lagoon used to allow sediment to settle out prior to discharge of water into a watercourse.

Shaft — A vertical means of entry into the ground, usually associated with mining.

Shale — A consolidated clay rock which possesses closely-spaced well-defined laminae. Present in coal measures it can form the main constituent of mine discard.

Shear strength — The level of stress in a soil at which shear failure occurs.

Sheet piling — Closely set piles, usually interlocking steel sections, driven into the ground to provide an earth retaining structure.

Silt trap — Chamber within a drainage system used to collect grit and sediment.

Site assessment — The process by which all relevant information concerning a site is compiled and evaluated prior to reclamation.

Site audit — Assessment of a site for environmental damage, liability and opportunities, prior to a reclamation and/or remediation scheme.

Site investigation — Site assessment activities carried out on site.

Slag — The solid waste material from a metallurgical furnace. Slag is generally a rock-like material formed by fusion of impurities present in the metal with a flux, such as limestone.

Slumping — Localised slope failure leading to bulging at the toe of the heap.

Slurry — Fluid mixture of fine-grained material and water *e.g.* cement slurry.

Smelting — The reduction of iron ore to metallic iron using heat.

Soakaway — A pit or chamber used to allow surface drainage water to percolate into the surrounding ground.

Soil (engineering)	Naturally occurring loose or soft deposits and similar unnatural deposits such as discard.
Soil mechanics	The application of engineering principles to soil.
Soil structure	The physical arrangement of soil particles and soil aggregates, water and air spaces, which maintains drainage paths, water storage and root penetrability.
Soil washing	Physical separation techniques using soil-water slurries to separate contaminants from a soil in a soil washing plant.
Solid solution	The substitution of atoms within a crystal lattice without alteration of the lattice structure.
Solidification	Treatment process by which chemical amendment of a soil or slurry will produce a solid matrix which is highly resistant to leaching and erosion.
Specific gravity (SG)	The ratio of the mass of a given volume of a material to the mass of an equal volume of water at a temperature of 4°C. The term relative density is also used.
Spent oxide	Chemically complex residue resulting from the use of iron oxide to purify coal gas. Contains high concentrations of sulphur compounds and cyanide.
Spoil heaps	Heaps of colliery or other waste material produced in mining.
Spontaneous ignition	Ignition due to self-heating within spoil material.
Stabilisation	Referring to a chemically-contaminated matrix undergoing chemical treatment to reduce the potential for the movement and leaching of contaminants.
Staging	Horizontal platform formed within a mine shaft, usually of steel or timber.

Stand pipe	An open vertical pipe connected to a borehole or pipeline such that the head of water in the pipe or borehole cannot exceed the length of the stand pipe. Used for monitoring groundwater levels and sampling water and gas.
Steam coal	This type of coal found usually within the bituminous to anthracite range, with a carbon:hydrogen ratio of approximately 20, is specifically suitable for steam generation in power stations. It is also sometimes used in smelting and coke making.
Stratification	Deposition of material in distinct layers.
Subsidence	Downwards movement of the ground *e.g.* due to the effects of collapsed mineworkings.
Substrate (plant)	The medium in which plants are growing or are required to grow.
Succession	Natural sequence or evolution of plant communities, each stage dependent on the preceding one and on environmental and management factors.
Support pillar	Pillar of unworked coal or mineral left in place to support the roof of the workings. Large support pillars may be used to avoid disruption beneath sensitive buildings or at shaft locations.
Surcharge loading	Imposed loading applied to the ground surface.
Surface pan	The hard, flat surface formed on soils by weathering and rainfall.
Surface slip	Slope failure due to surface slippage of granular material (see Figure 6.2(e)).
Tailings	The fine-grained particulate matter present in the liquid waste from colliery spoil washing or coal preparation.
Teratogenic	Causing abnormal development of an animal foetus.

Terracing	Formation of a series of horizontal platforms or berms on a slope.
Texture (soil)	The particle size distribution of the substrate, assessed as the ratio of particles in specific size classes, which governs some physical properties of the substrate.
Therophyte	Annual plants which pass the unfavourable season as an embryo or seed.
Tip (mining)	A deposit of discard.
Tippler	Equipment for gripping and rotating coal or ore trucks to discharge the contents.
Toe (of a slope)	Bottom of a slope.
Tolerance (plant)	The ability of an organism to maintain active growth in the presence of elevated concentrations of toxic chemicals which would adversely affect the growth of other organisms.
Topsoil	Upper humic part of a soil profile.
Tramway	Railway (other than standard gauge) used in a mine or works for transporting materials in wagons.
Transpiration	The process by which water is moved from plant roots to the leaves and released by evaporation.
Trial pit	Excavation carried out from the ground surface to determine the nature of the ground or the existence of underground structures and services.
Underpinning	Method of stabilising existing structures by the provision of new foundations placed beneath the existing foundations and taken down to suitable bearing strata using mass or reinforced concrete.
Utility services	Gas, water, electricity and telephone services.

Vacuum extraction	Treatment process for contaminated soil and groundwater whereby volatile contaminants are extracted from the ground via vacuum pumping.
Vadose zone	The unsaturated part of a soil profile above the water table.
Verification testing	Testing carried out after reclamation or decontamination works to evaluate and demonstrate their effectiveness.
Vitrification	Process by which intense heat and subsequent cooling converts inorganic mineral particles and associated chemicals into a fused glass-like matrix.
Void ratio (engineering)	Ratio of the volume of voids to the volume of soil particles.
Walkover survey	An initial visual inspection of a site.
Washery discard	Waste material from coal washing operations.
Water table	The standing water level within the ground below which all pores and fissures are saturated.
Water table, perched	Water table maintained above the general standing water level in the ground below, usually by an impervious stratum.
Winding gear	Engine and equipment used for operating a cage or skip in a mine shaft.
Windthrow	Uprooting of trees, by strong wind.

ABBREVIATIONS

AC	Alternating Current
ALD	Anoxic Limestone Drain
AMD	Acid Mine Drainage
ANC	Acid Neutralising Capacity
BCAMD	British Columbia Acid Mine Drainage (Task Force) (Canada)
BOD	Biological Oxygen Demand
BOD_5	Biological Oxygen Demand (5 day test)
BOS	Basic Oxygen Steel
BRE	Building Research Establishment (UK)
BRGM	Bureau de Recherche Géologique et Minière (France)
BTEX	Benzene, Toluene, Ethyl benzene, Xylene
CCMS	Committee for Challenges to Modern Society
CEE	Communauté Economique Européenne
CIRIA	Construction Industry Research and Information Association (UK)
COD	Chemical Oxygen Demand
CV	Calorific Value
DIN	Deutsche Industrie-Norm
EC	European Community or European Commission
ECSC	European Coal and Steel Community
EEB	European Environmental Bureau
EEC	European Economic Community
ENDS	Environmental Data Services Ltd (UK)
Epm	Ecart Probable Moyen
EPML	Etablissement Publique de la Métropole de Lorraine
FF	French Francs
FID	Flame Ionisation Detector
GC	Gas Chromatography
GL	Guide Level
HMSO	Her Majesty's Stationery Office (UK)
HPLC	High Performance Liquid Chromatography
ICRCL	Inter-Departmental Committee on the Redevelopment of Contaminated Land (UK)
IEA	International Energy Agency

IISI	International Iron and Steel Institute
ILS	In-Line System
LD	Linzer - Dusenverfahren (steel process)
MAC	Maximum Admissible Concentration
MAFF	Ministry of Agriculture, Fisheries and Food (UK)
MTR	Maximum Tolerable Risk
NATO	North Atlantic Treaty Organisation
NCB	National Coal Board (UK)
NPK	Nitrogen, Phosphorus and Potassium
OECD	Organisation for Economic Cooperation and Development
PAH	Polyaromatic Hydrocarbon
PCB	Polychlorinated Biphenol
PFA	Pulverised Fuel Ash
pH	Unit of acidity or alkalinity on a logarithmic scale of 1-14
PTFE	Polytetrafluoroethylene
RETI	Régions Européennes de Technologie Industrielle
SDA	Scottish Development Agency
SEC	Substances Extractable with Chloroform
SG	Specific Gravity
SLS	Sodium Lauryl Sulphate
spp.	Species (plural)
TNO	Organisation for Applied Scientific Research (The Netherlands)
UNEP	United Nations Environmental Programme
USEPA	United States Environmental Protection Agency
v/v	Volume for volume
VDLUFA	Verbandes Deutscher Landwirtschaftlicher Untersuchungs- und Forschungsanstalten (Germany - Association of German Agricultural Investigation and Research Institutes)
VFA	Volatile Fatty Acid
VROM	Ministry of Housing, Physical Planning and the Environment (The Netherlands)
WDA	Welsh Development Agency
WRC	Water Research Centre (UK)
w/w	Weight for weight

INDEX